Teacher's Book

Karen Morrison
Lisa Greenstein

Great Clarendon Street, Oxford, OX2 6DP, United Kingdom

Oxford University Press is a department of the University of Oxford.

It furthers the University's objective of excellence in research, scholarship, and education by publishing worldwide. Oxford is a registered trade mark of Oxford University Press in the UK and in certain other countries.

British Library Cataloguing in Publication Data

Data available

ISBN: 978-1-382-01018-4

1 3 5 7 9 10 8 6 4 2

Paper used in the production of this book is a natural, recyclable product made from wood grown in sustainable forests. The manufacturing process conforms to the environmental regulations of the country of origin.

Printed in Great Britain by CPI

Acknowledgements

The publisher and authors would like to thank the following for permission to use photographs and other copyright material:

Cover: Matthieu Nivesse. **Photos: p9:** ann_isme/Shutterstock; **p20:** Oxford University Press.

Artwork by Daniel Limon, John Haslam, Q2A Media, Integra Software Services, Pantek Media, and Oxford University Press.

Every effort has been made to contact copyright holders of material reproduced in this book. Any omissions will be rectified in subsequent printings if notice is given to the publisher.

Cover activities

The following activities are based on the Level 5 Pupil book and Workbook cover image. You can use these stimulus questions according to the children's learning to date.

Number

Continue the sequence: Can the children see any number patterns? Ask them to record the sequence of flowers on each plant (3, 6, 9, 12), and branches on the rounded trees (1, 2, 3, 4). Can they identify the rules and continue the patterns?

Spotting doubles and halves: Can the children see any double or half relationships? Encourage them to look at the groups of fish and numbers of flowers. Can they record the double or half relationships as calculations?

Geometry

Shapes around us: How many different 2D and 3D shapes can the children see? Can they identify 2D faces on the 3D shapes? Encourage children to describe the properties of the shapes. Are the 2D shapes regular or irregular? What kind of triangles are the sails? Can they identify right, acute or obtuse angles in the picture? What angle do the flying birds make? How many parallel, perpendicular, horizontal and vertical lines can the children find? Are there any symmetrical objects? Encourage the children to compare the heights and widths of objects, such as the tree trunks.

Contents

How Nelson Maths Works

LEVEL	PUPIL BOOK	WORKBOOKS	TEACHING SUPPORT	DIGITAL CONTENT ON OXFORD OWL
STARTER LEVEL				For all levels:
1				• Digital versions of the Pupil Books, Workbooks and Teacher's Books
2				• Assessment support
3				• Parent notes
4				• Curriculum mapping and planning guides
5				• Vocabulary support
6				

Access the digital content for this course online at
www.oxfordowl.co.uk

How to use Nelson Maths

Nelson Maths is a comprehensive maths programme for children aged 4–11. The course covers the following five strands: Number, Measure, Geometry, Statistics and Algebra to ensure full coverage of your chosen curriculum. The course is also full of opportunities to develop children's problem-solving skills through carefully designed questions and activities.

Nelson Maths is made up of seven levels (one for each year group) and Level 5 contains 18 units, which should ideally be taught in order. The units have been arranged in a careful progression, designed to support children to build their skills and knowledge and make connections across the different strands. Your children may progress through some units more quickly than others, but it is recommended that you spend approximately two weeks on each unit.

Children each have a **Pupil Book** and write-in **Workbook**. The main lesson activities are included in the **Pupil Book**, and the **Workbook** includes extra practice and consolidation activities. The **Teacher's Book** contains detailed teaching support for each unit, which will help you to introduce new concepts, revise prior learning, deepen children's understanding of a topic and encourage mathematical conversations and discussion.

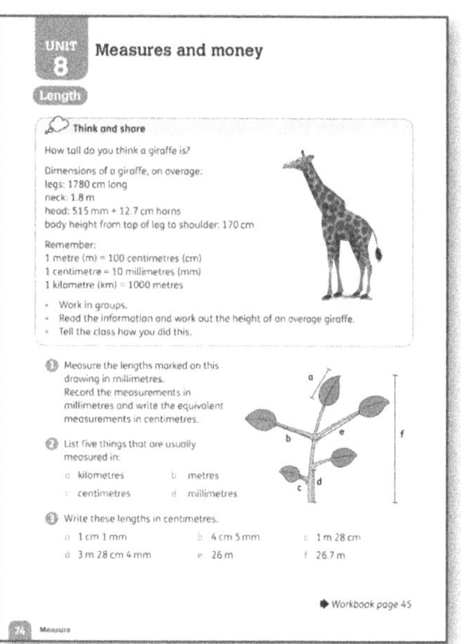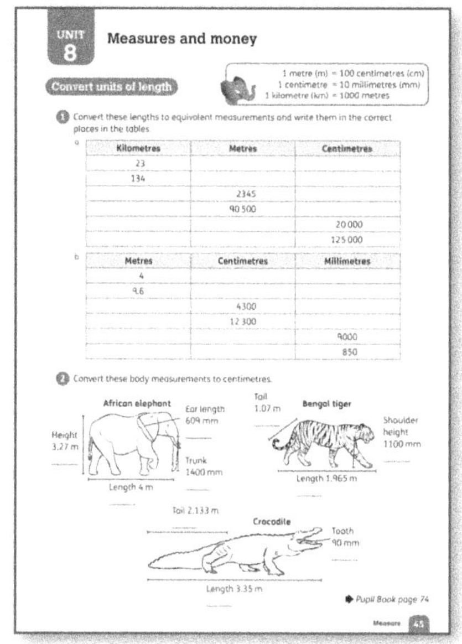

Each unit in the **Pupil Book** is supported by a corresponding unit in the **Workbook**.

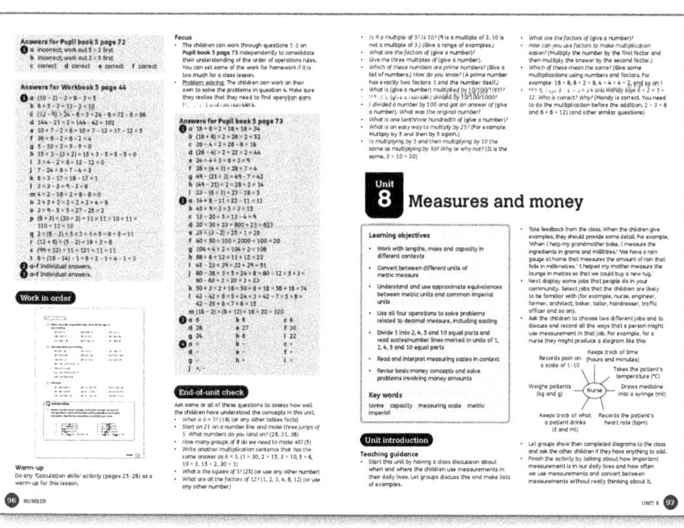

The **Teacher's Book** provides detailed teaching guidance for each unit and answers to the activities.

Read the relevant unit of the Teacher's Book before you begin teaching it. It's important to familiarise yourself with the learning objectives, key vocabulary and resources you need before each lesson. The following pages will give you a good understanding of how to use the Pupil Book, Workbook and Teacher's Book together.

In addition to the printed materials, you can also find lots of supplementary digital resources online on Oxford Owl. Please see page 9 for more information.

Using the Pupil Books

The **Pupil Book** units are designed to be worked through in order, following the discussion prompts and activities suggested in the **Teacher's Book**, and using the linked **Workbook** pages to help children to consolidate their understanding through independent practice. The following features of the **Pupil Books** are designed to help you get the most out of your lessons.

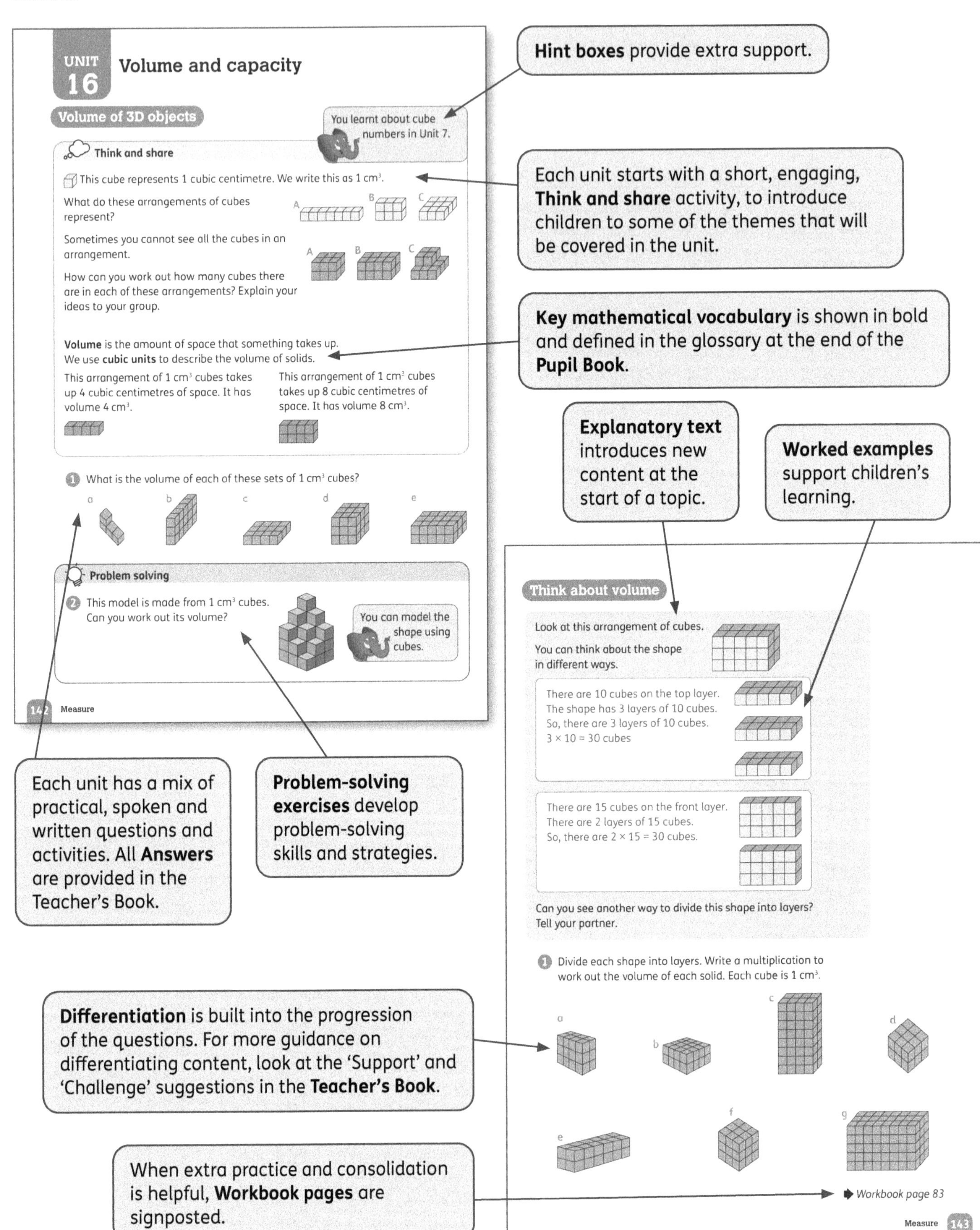

Hint boxes provide extra support.

Each unit starts with a short, engaging, **Think and share** activity, to introduce children to some of the themes that will be covered in the unit.

Key mathematical vocabulary is shown in bold and defined in the glossary at the end of the **Pupil Book.**

Explanatory text introduces new content at the start of a topic.

Worked examples support children's learning.

Each unit has a mix of practical, spoken and written questions and activities. All **Answers** are provided in the Teacher's Book.

Problem-solving exercises develop problem-solving skills and strategies.

Differentiation is built into the progression of the questions. For more guidance on differentiating content, look at the 'Support' and 'Challenge' suggestions in the **Teacher's Book.**

When extra practice and consolidation is helpful, **Workbook pages** are signposted.

The first unit in each Level is called **Think maths**. It encourages a growth mindset and builds resilience by teaching children that mistakes are a positive part of their learning journey. It prepares children to think mathematically and make connections across maths.

Think maths

Think about how you learn

THINK MATHS prepares you to think mathematically, value mistakes, and learn maths with a growth mindset.

Think and share

Sharla made this **pattern** using paper squares that measure 4 cm by 4 cm. She has cut some of the squares. She has two squares left over.

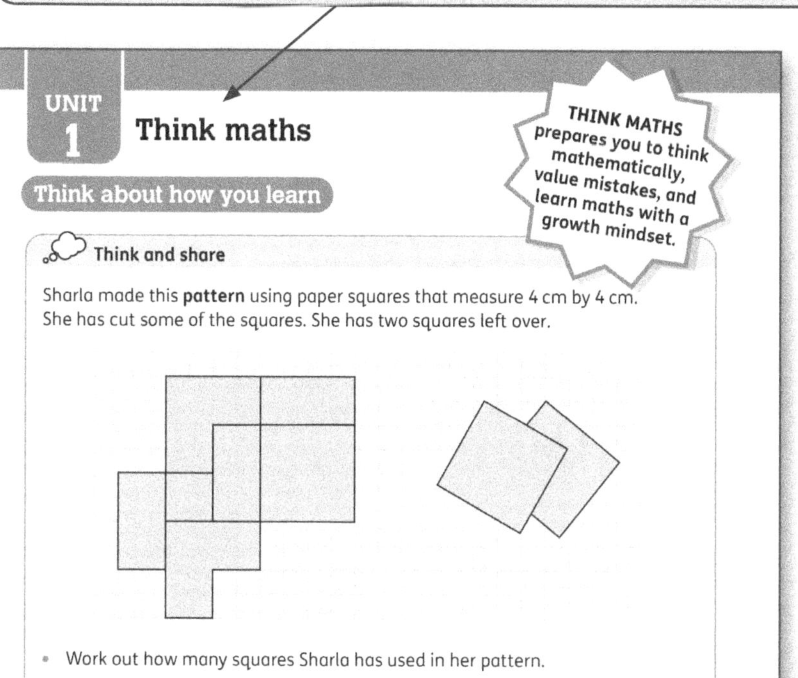

- Work out how many squares Sharla has used in her pattern.
- How did you work this out? Tell your group.
- Did you talk to yourself while you were working it out?
- How can talking to yourself while you work help you?
- How can talking to others while you work help you?
- Can you make the pattern into a square with the two squares le

1. Copy this shape. How many ways can you find to shade $\frac{1}{8}$ of the shape?

   Work on your own to start with and then share your ideas with your group.

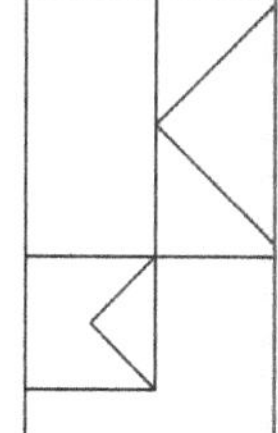

2. Think about the problems on this page and the discussions you had. Tell your group two interesting things that you heard or learnt.

Each book has three **Mixed practice** reviews – one per 'term'. These formative assessment questions help you to assess children's understanding. You may choose to revisit some concepts after the practice.

Mixed practice 3

1. A farmer has 21 rows of peach trees. There are 35 trees in each row. How many trees does she have altogether?

2. There are 19 houses in my road. There are 15 paving stones in front of each house. How many paving stones are in the road?

3. $83 \times 20 = 1660$. Show how you can use this fact to work out these calculations without doing any multiplication or division.

   a 21×83 b 19×83 c $1660 \div 20$ d $1660 \div 83$

4. An elevator at Burj Khalifa, the tallest building in the world, can carry 25 passengers at a time. The elevator went up to the top of the building 42 times in one day. What is the maximum number of passengers it could have carried on that day?

5. Jess wants to divide 800 by 25.

   This is what she writes in her book:

   $800 \div 25$
   $800 \div 100 = 8$
   $8 \times 4 = 32$

   a Is her answer correct?

   b Show how she could set out her working more clearly.

6. Look at this line graph carefully.

   a What does the red dashed line across the graph show?

   b What is the highest temperature measured in this period?

   c What was the coldest temperature measured on Tuesday?

   d Between which two times did the temperature rise the fastest? How do you know this?

   e Estimate the temperature at 3 p.m.

   f At what times was the temperature −2.5 °C?

   g What is the temperature 4 degrees lower than −3 °C?

Mixed practice **Answers** are provided in the **Teacher's Book**.

Using the Workbooks

The **Workbooks** provide essential practice to help children consolidate their learning. They provide new questions and activities linked to the concepts introduced in the **Pupil Books**. They are designed for independent use, making them ideal for homework or additional classwork. We recommend that children complete each page in the **Workbook** after they have completed the corresponding lesson in the **Pupil Book**. When extra practice and consolidation is helpful, **Workbook** pages are signposted at the end of the Pupil Book lesson.

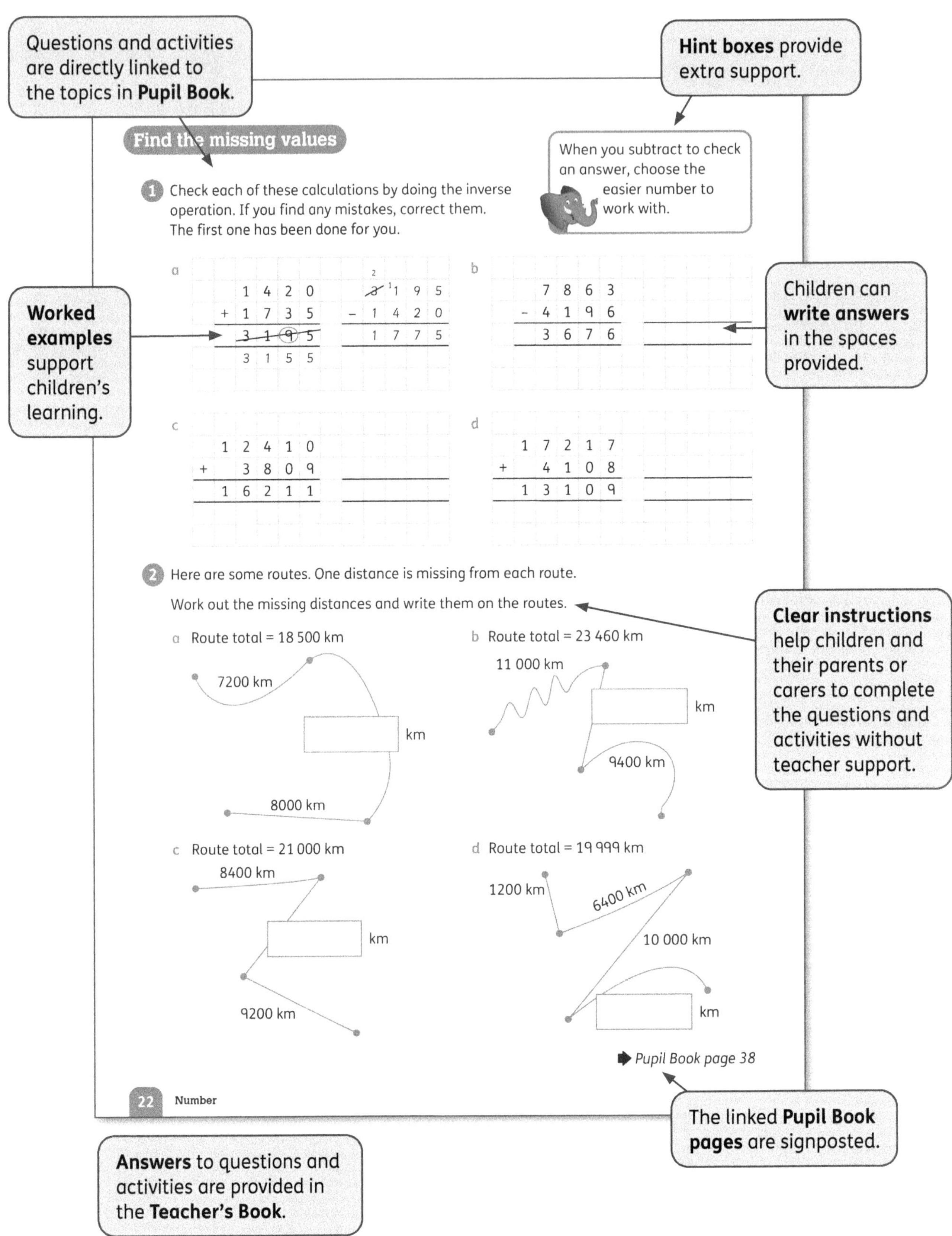

The linked **Pupil Book** pages are signposted.

Answers to questions and activities are provided in the **Teacher's Book**.

Using the Digital Content

This edition of *Nelson Maths* is supported by additional digital resources available online on **Oxford Owl**. These include:

- Digital versions of the Pupil Books, Workbooks and Teacher's Books to support planning and for front of class display
- A set of printable assessments with a progress tracking tool and mark scheme
- Guidance for you to share with children's parents or carers about their child's learning, including support for homework and ideas for incorporating maths into everyday life
- Vocabulary support
- Curriculum correlation charts
- Planning support

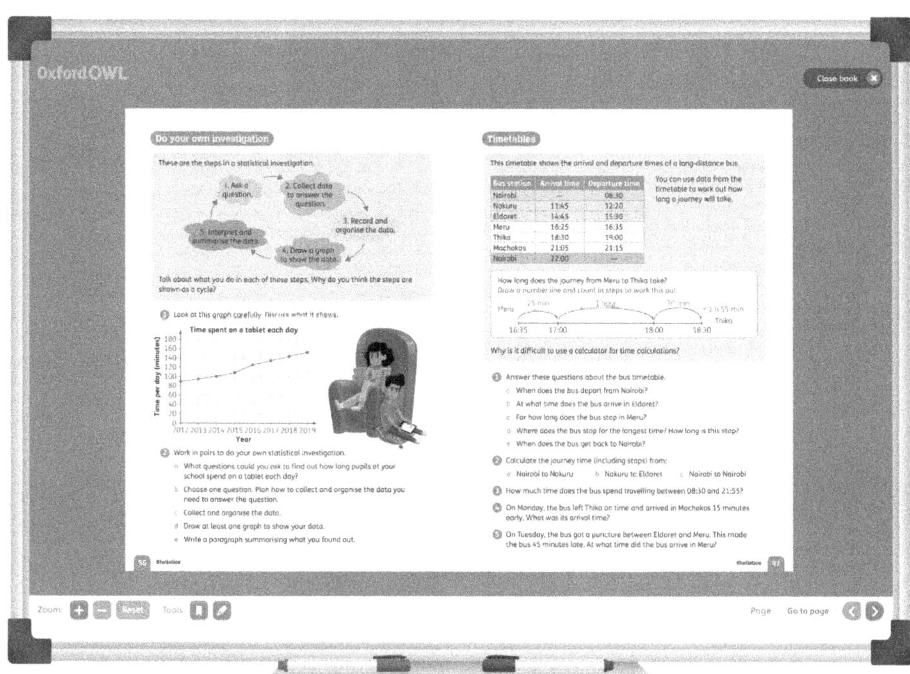

Using the Teacher's Book

The **Teacher's Book** provides step-by-step lesson notes for each unit in the **Pupil Book** and **Workbook**, including learning objectives, answers to the questions activities and suggestions for additional challenge and support.

Before you start teaching the units, take some time to familiarise yourself with these useful sections at the start of the **Teacher's Book**. These four sections are designed to help you get the most out of the *Nelson Maths* teaching materials:

1. **Introduction** – find out more about the mathematical strands, skills and fundamental principles that underpin *Nelson Maths*. Get ideas and support for teaching maths in an inclusive way that acknowledges individual differences and human diversity.

2. **Teaching approach** – guidance for teaching *Nelson Maths* effectively, including:
- how to develop robust learners with a growth mindset and an understanding of the importance of making mistakes
- the importance of the do-talk-record model that underpins the lessons
- using engaging teaching methods, from investigations and number talks, to exploring maths in real life contexts.

3. **An environment for exploration and play** – practical support on organising your classroom space; approaching whole-class, group and individual activities; and using manipulatives and games to deepen children's learning.

4. **Activity bank of warm-ups and mental maths** – a bank of engaging mental maths, support, and consolidation activities to use with your class.

Teacher's Book: using the unit pages

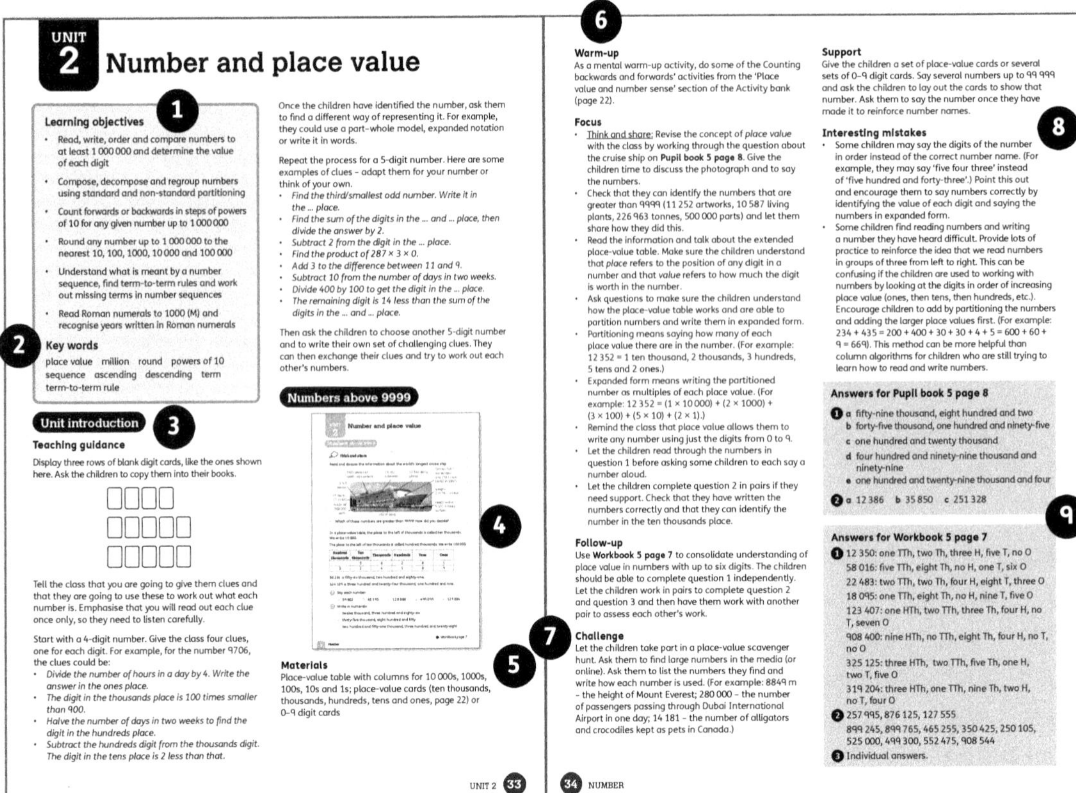

UNIT 2 — Number and place value

(1) Learning objectives

- Read, write, order and compare numbers to at least 1 000 000 and determine the value of each digit
- Compose, decompose and regroup numbers using standard and non-standard partitioning
- Count forwards or backwards in steps of powers of 10 for any given number up to 1 000 000
- Round any number up to 1 000 000 to the nearest 10, 100, 1000, 10 000 and 100 000
- Understand what is meant by a number sequence, find term-to-term rules and work out missing terms in number sequences
- Read Roman numerals to 1000 (M) and recognise years written in Roman numerals

(2) Key words

place value million round powers of 10 sequence ascending descending term term-to-term rule

(3) Unit introduction

Teaching guidance

Display three rows of blank digit cards, like the ones shown here. Ask the children to copy them into their books.

Tell the class that you are going to give them clues and that they are going to use these to work out what each number is. Emphasise that you will read out each clue once only, so they need to listen carefully.

Start with a 4-digit number. Give the class four clues, one for each digit. For example, for the number 9706, the clues could be:
- *Divide the number of hours in a day by 4. Write the answer in the ones place.*
- *The digit in the thousands place is 100 times smaller than 900.*
- *Halve the number of days in two weeks to find the digit in the hundreds place.*
- *Subtract the hundreds digit from the thousands digit. The digit in the tens place is 2 less than that.*

Once the children have identified the number, ask them to find a different way of representing it. For example, they could use a part–whole model, expanded notation or write it in words.

Repeat the process for a 5-digit number. Here are some examples of clues – adapt them for your number or think of your own.
- *Find the third/smallest odd number. Write it in the … place.*
- *Find the sum of the digits in the … and … place, then divide the answer by 2.*
- *Subtract 2 from the digit in the … place.*
- *Find the product of 287 × 3 × 0.*
- *Add 3 to the difference between 11 and 9.*
- *Subtract 10 from the number of days in two weeks.*
- *Divide 400 by 100 to get the digit in the … place.*
- *The remaining digit is 14 less than the sum of the digits in the … and … place.*

Then ask the children to choose another 5-digit number and to write their own set of challenging clues. They can then exchange their clues and try to work out each other's numbers.

Numbers above 9999

(5) Materials

Place-value table with columns for 10 000s, 1000s, 100s, 10s and 1s; place-value cards (ten thousands, thousands, hundreds, tens and ones, page 22) or 0–9 digit cards

UNIT 2 — 33

(6) Warm-up

As a mental warm-up activity, do some of the Counting backwards and forwards' activities from the 'Place value and number sense' section of the Activity bank (page 22).

Focus
- <u>Think and share:</u> Revise the concept of *place value* with the class by working through the question about the cruise ship on **Pupil book 5 page 8**. Give the children time to discuss the photograph and to say the numbers.
- Check that they can identify the numbers that are greater than 9999 (11 252 artworks, 10 587 living plants, 226 963 tonnes, 500 000 parts) and let them share how they did this.
- Read the information and talk about the extended place-value table. Make sure the children understand that *place* refers to the position of any digit in a number and that *value* refers to how much the digit is worth in the number.
- Ask questions to make sure the children understand how the place-value table works and are able to partition numbers and write them in expanded form.
- Partitioning means saying how many of each place value there are in the number. (For example: 12 352 = 1 ten thousand, 2 thousands, 3 hundreds, 5 tens and 2 ones.)
- Expanded form means writing the partitioned number as multiples of each place value. (For example: 12 352 = (1 × 10 000) + (2 × 1000) + (3 × 100) + (5 × 10) + (2 × 1).)
- Remind the class that place value allows them to write any number using just the digits from 0 to 9.
- Let the children read through the numbers in question 1 before asking some children to each say a number aloud.
- Let the children complete question 2 in pairs if they need support. Check that they have written the numbers correctly and that they can identify the number in the ten thousands place.

Follow-up

Use **Workbook 5 page 7** to consolidate understanding of place value in numbers with up to six digits. The children should be able to complete question 1 independently. Let the children work in pairs to complete question 2 and question 3 and then have them work with another pair to assess each other's work.

(7) Challenge

Let the children take part in a place-value scavenger hunt. Ask them to find large numbers in the media (or online). Ask them to list the numbers they find and write how each number is used. (For example: 8849 m – the height of Mount Everest; 280 000 – the number of passengers passing through Dubai International Airport in one day; 14 181 – the number of alligators and crocodiles kept as pets in Canada.)

Support

Give the children a set of place-value cards or several sets of 0–9 digit cards. Say several numbers up to 99 999 and ask the children to lay out the cards to show that number. Ask them to say the number once they have made it to reinforce number names.

(8) Interesting mistakes
- Some children may say the digits of the number in order instead of the correct number name. (For example, they may say 'five four three' instead of 'five hundred and forty-three'.) Point this out and encourage them to say numbers correctly by identifying the value of each digit and saying the numbers in expanded form.
- Some children find reading numbers and writing a number they have heard difficult. Provide lots of practice to reinforce the idea that we read numbers in groups of three from left to right. This can be confusing if the children are used to working with numbers by looking at the digits in order of increasing place value (ones, then tens, then hundreds, etc.). Encourage children to add by partitioning the numbers and adding the larger place values first. (For example: 234 + 435 = 200 + 400 + 30 + 30 + 4 + 5 = 600 + 60 + 9 = 669). This method can be more helpful than column algorithms for children who are still trying to learn how to read and write numbers.

(9) Answers for Pupil book 5 page 8

1. a fifty-nine thousand, eight hundred and two
   b forty-five thousand, one hundred and ninety-five
   c one hundred and twenty thousand
   d four hundred and ninety-nine thousand and ninety-nine
   e one hundred and twenty-nine thousand and four

2. a 12 386 b 35 850 c 251 328

Answers for Workbook 5 page 7

1. 12 350: one TTh, two Th, three H, five T, no O
   58 016: five TTh, eight Th, no H, one T, six O
   22 483: two TTh, two Th, four H, eight T, three O
   18 095: one TTh, eight Th, no H, nine T, five O
   123 407: one HTh, two TTh, three Th, four H, no T, seven O
   908 400: nine HTh, no TTh, eight Th, four H, no T, no O
   325 125: three HTh, two TTh, five Th, one H, two T, five O
   319 204: three HTh, one TTh, nine Th, two H, no T, four O
2. 257 995, 876 125, 127 555
   899 245, 899 765, 465 255, 350 425, 250 105, 525 000, 499 300, 552 475, 908 544
3. Individual answers.

34 — NUMBER

(10) End-of-unit check

Ask some or all of these questions to assess how well the children have understood the concepts in this unit.
- *What are the ten digits we use for representing all numbers?* (0, 1, 2, 3, 4, 5, 6, 7, 8, 9)
- *What makes it possible for us to represent large numbers, using only the 10 digits?* (Place value: the position of a digit determines its value.)
- *Put the correct sign, < or >, between these numbers.* (Give the children a pair of numbers.)
- *How do you order a set of numbers? Which part of each number do you look at to help you? Why?* (Look at the highest value place value first. If these are the same, look at the next highest place value and so on.)
- *What is the value of each digit in this number?* (Give the children some 5-, 6- and 7-digit numbers.)
- *How do you say this number?* (Write down some numbers in numerals for the children to say.)
- *Draw a part–whole model to show 1 234 509.* (part–whole model with whole 1 234 509 and parts 1 000 000, 200 000, 30 000, 4000, 500 and 9)
- *Write this number in expanded notation.* (Give the children some numbers.)
- *What is wrong with this expanded notation: 3 445 098 = 3 000 000 + 400 000 + 50 000 + 90 + 8?* (40 000 is missing and 50 000 should be 5000)
- *What rule was used to make this sequence?* (Provide some different number sequences.)
- *Is there any other way of carrying on this sequence? How?*
- *What comes next in this sequence? Why?*
- *Write these numbers in order of increasing size: 230 954, 230 949, 230 994, 230 998, 230 940.* (230 940, 230 949, 230 954, 230 994, 230 998)
- *What is 10 more/10 less/100 more/100 less/1000 more/1000 less than (a given number)?*
- *What is 2375 rounded to the nearest 10?* (2380)
- *What is 423 544 rounded to the nearest 1000?* (424 000)
- *Write these numbers using Roman numerals: 100, 250, 400, 512, 900, 930.* (C, CCL, CD, DXII, CM, CMXXX)
- *What number is represented by each Roman numeral?* (Give some Roman numerals with values between 1 and 1000.)
- *Find three Roman numbers that are the same front to back and back to front.* (For example: XXX, CXC, MCM.)
- *Naresh says that DLX and DXL are the same number because Roman numerals don't use place value. What mistake has he made?* (He has not noticed that the order of the letters is different. A smaller value written before a larger value means that the smaller value is subtracted from the larger value. DLX = 560 and DXL = 540)

UNIT 3 — Properties of shapes

Learning objectives

- Revise the names and properties of polygons
- Use side lengths and angle sizes to classify polygons as regular or irregular
- Identify 3D shapes including cubes and cuboids from 2D representations (nets)
- Identify, describe and sketch 3D shapes in different orientations
- Know that angles are measured in degrees: estimate and compare acute, obtuse and reflex angles
- Compare angles, estimate and measure angles in degrees and draw angles of a given size
- Know that angles at a point total one whole turn (360°) and that angles at a point on a straight line total half a turn (180°), and use these facts to calculate missing angles
- Identify other multiples of 90°
- Recognise and describe spatial patterns and solve related problems
- Recognise the use of symbols to represent unknown quantities in calculations

Key words

polygon vertex, vertices trapezium heptagon nonagon net angle acute angle obtuse angle reflex angle scalene triangle isosceles triangle equilateral triangle square number

Unit introduction

Materials

Large sheet of paper or card; pencils; crayons and marker pens

UNIT 3 — 45

---

(1) Learning objectives are listed at the beginning of each unit.

(2) Key words match the important words highlighted in bold in the **Pupil Book**.

(3) Unit introduction activities introduce the topics in the unit, engage children and find out what they already know.

(4) The **linked Pupil Book and Workbook pages** are signposted.

(5) Materials children need for the lesson are listed.

(6) Most lessons are broken down into three stages:
- **Warm-up** – introduce the topic and unlock prior learning
- **Focus** – develop a firm understanding of concepts and skills by working through the **Pupil Book**
- **Follow-up** – recap the key learning points and extend the learning with further activity suggestions in the **Workbook**.

(7) Where relevant, **Challenge** and **Support** sections give guidance on deepening the learning or making it more accessible, so that every child is catered for.

(8) It is important for children to see mistakes as part of their learning journey. The **Interesting mistakes** section highlights common errors and misconceptions that children may have, along with what to look for and advice on how to support children.

(9) Answers for **Pupil Book** and **Workbook** questions are provided to make marking easy.

(10) An **End-of-unit check** allows you to assess how well children have understood the concepts in each unit, so you can monitor children's understanding throughout the course. A set of formative assessment questions or a closing activity are provided with answers.

Level 5 Scope and sequence

Unit	Unit title	Strand	Learning objectives	Key words
1	Think maths	Think maths	• Establish a classroom environment conducive to thinking and working mathematically • Establish key messages of growth mindset mathematics • Consolidate planning and problem-solving skills	
2	Number and place value	Number	• Read, write, order and compare numbers to at least 1 000 000 and determine the value of each digit • Compose, decompose and regroup numbers using standard and non-standard partitioning • Count forwards or backwards in steps of powers of 10 for any given number up to 1 000 000 • Round any number up to 1 000 000 to the nearest 10, 100, 1000, 10 000 and 100 000 • Understand what is meant by a number sequence, find term-to-term rules and work out missing terms in number sequences • Read Roman numerals to 1000 (M) and recognise years written in Roman numerals	place value, million, round, powers of 10, sequence, ascending, descending, term, term-to-term rule
3	Properties of shapes	Geometry	• Revise the names and properties of polygons • Use side lengths and angle sizes to classify polygons as regular or irregular • Identify 3D shapes including cubes and cuboids from 2D representations (nets) • Identify, describe and sketch 3D shapes in different orientations • Know that angles are measured in degrees: estimate and compare acute, obtuse and reflex angles • Compare angles, estimate and measure angles in degrees and draw angles of a given size • Know that angles at a point total one whole turn (360°) and that angles at a point on a straight line total half a turn (180°), and use these facts to calculate missing angles • Identify other multiples of 90° • Recognise and describe spatial patterns and solve related problems • Recognise the use of symbols to represent unknown quantities in calculations	polygon, vertex, vertices, trapezium, heptagon, nonagon, net, angle, acute angle, obtuse angle, reflex angle, scalene triangle, isosceles triangle, equilateral triangle, square number
4	Addition and subtraction	Number	• Add and subtract mentally with increasingly larger numbers • Use appropriate laws of arithmetic to simplify calculations • Add and subtract numbers with more than four digits, including using formal written methods • Use rounding to check answers to calculations and determine, in the context of a problem, levels of accuracy • Solve addition and subtraction multi-step problems in context, deciding which operations and methods to use and why	inverse operation, multiple, area, estimate, approximate
5	Decimals and percentages	Number	• Extend the place-value table to include thousandths, and relate these to tenths and hundredths • Understand and explain the value of any digit in a decimal with up to three decimal places • Compose, decompose and regroup numbers with up to three decimal places using standard and non-standard partitioning • Round decimals (two decimal places) to the nearest whole number and to one decimal place • Read, write, order and compare numbers with up to three decimal places • Recognise the per cent symbol (%) and understand that per cent relates to 'number of parts per hundred' • Write a percentage as a fraction with a denominator of 100 and as a decimal • Use waffle diagrams to show percentages of a whole (100 parts) and answer questions about the proportions shown • Read and write decimal numbers as fractions • Solve problems using decimals, fractions and equivalent percentages	thousandth, percentage, whole, decimal equivalent fraction, denominator, ascending, descending

Unit	Unit title	Strand	Learning objectives	Key words
6	Time	Measure	• Understand that there are time zones and compare times in different zones • Solve problems involving converting between units of time • Understand time intervals of less than one second • Find time intervals in seconds, minutes and hours that bridge 60 • Recognise that time intervals can be expressed as decimals or using mixed units	
7	Multiplication and division 1	Number	• Build increasing fluency in multiplication table facts and related division facts • Multiply and divide numbers mentally drawing upon known facts • Identify multiples and factors, including finding all factor pairs of a number, common factors of two numbers and lowest common multiples • Express numbers as a product of two factors • Know and use the vocabulary of prime numbers, prime factors and composite numbers • Work out whether a number up to 100 is prime and recall prime numbers to 19 • Multiply and divide whole numbers by 10, 100 and 1000 • Recognise and use square numbers and cube numbers and use notation for squared and cubed • Solve problems involving multiplication and division including using their knowledge of factors and multiples, squares and cubes • Understand that the four operations follow a particular order • Use laws of arithmetic to simplify calculations	multiple, common multiples, lowest common multiple (LCM), square numbers, squared, cube numbers, factor, common factor, highest common factor (HCF), composite number, prime number
8	Measures and money	Measure	• Work with lengths, mass and capacity in different contexts • Convert between different units of metric measure • Understand and use approximate equivalences between metric units and common imperial units • Use all four operations to solve problems related to decimal measure, including scaling • Divide 1 into 2, 4, 5 and 10 equal parts and read scales/number lines marked in units of 1, 2, 4, 5 and 10 equal parts • Read and interpret measuring scales in context • Revise basic money concepts and solve problems involving money amounts	tonne, capacity, measuring scale, metric, imperial
9	Perimeter and area	Measure	• Measure and calculate the perimeter of composite rectilinear shapes in centimetres and metres • Calculate and compare the area of rectangles using standard units of measure • Use the properties of rectangles to deduce related facts and find missing lengths and angles • Estimate the area of irregular shapes • Understand that shapes with the same perimeter can have different areas and vice versa	perimeter, area, square unit, formula, composite shape
10	Statistics	Statistics	• Plan and carry out a statistical investigation to answer a set of related statistical questions • Record, organise and represent data using appropriate tables, diagrams and graphs • Solve comparison, sum and difference problems using information presented in a line graph • Find the mode, median and range of a set of data • Complete, read and interpret information in tables, including timetables	frequency table, bar chart, bar-line chart, dot plot, Carroll diagram, Venn diagram, line graph, axis, axes, mode, median, range

Unit	Unit title	Strand	Learning objectives	Key words
11	Fractions	Number	• Compare and order fractions whose denominators are all multiples of the same number • Identify, name and write equivalent fractions of a given fraction, represented visually, including tenths and hundredths • Find equivalent fractions and understand that they have the same value and the same position on a number line • Recognise mixed numbers and improper fractions and convert from one form to the other and write mathematical statements > 1 as a mixed number • Understand that proper fractions can act as operators and find non-unit fractions of quantities • Add and subtract fractions with the same denominators and denominators that are multiples of each other • Multiply fractions by a whole number • Solve problems involving multiplication, including scaling by simple fractions	numerator, denominator, improper fraction, mixed number, equivalent fractions
12	Position, direction and movement	Geometry	• Compare the relative position of coordinates, with or without a grid • Use knowledge of 2D shapes and coordinates to plot points to form lines and shapes in the first quadrant • Identify, describe and represent the position of a shape following a reflection or translation, using the appropriate language, and know that the shape has not changed • Reflect 2D shapes in both horizontal and vertical mirror lines to create patterns on square grids • Use knowledge of reflective symmetry to identify and complete symmetrical patterns	coordinates, reflection, mirror line, translation, symmetrical, coordinate grid
13	Multiplication and division 2	Number	• Use known facts and laws of arithmetic to simplify multiplication and division calculations • Estimate and multiply numbers with up to 4 digits by 1-digit or 2-digit whole numbers, including using long multiplication for the 2-digit multipliers • Estimate and divide numbers with up to 4 digits by a 1-digit number using the formal method of short division • Solve problems involving addition, subtraction, multiplication and division and a combination of these, including interpreting remainders appropriately for the context • Solve problems involving multiplication and division, including problems involving simple rates	place value, inverse, short division, remainder, rate
14	Work with negative numbers	Number	• Interpret negative numbers in context • Count backwards and forwards with positive numbers, including through zero • Add and subtract integers, including where one integer is negative	positive numbers, negative numbers, minus sign, thermometer, degrees Celsius
15	Calculate with decimals	Number	• Estimate, add and subtract numbers with the same number of decimal places • Apply place-value knowledge to known additive and multiplicative number facts (scaling facts by one tenth or one hundredth) • Estimate and multiply decimals by a 1-digit whole number • Solve problems involving decimals with up to three decimal places	decimal, whole, tenths, hundredths, place value, square unit
16	Volume and capacity	Measure	• Estimate volume and capacity using appropriate standard units • Use 1 cm^3 blocks to estimate and work out the volume of different 3D shapes • Use all four operations to solve problems involving volume and capacity	volume, cubic unit
17	Ratio and proportion	Number	• Understand that a proportion compares a part to a whole and that a ratio compares part to part of two or more quantities	ratio, proportion
18	Probability	Statistics	• Understand that a proportion compares a part to a whole and that a ratio compares part to part of two or more quantities • Position events on a simple probability scale • Understand that some outcomes are more likely than others and that some outcomes are certain, while others are impossible • Carry out simple probability experiments and use tables, graphs and other diagrams to represent and interpret the results	probability, chance, certain, likely, equally likely, unlikely, impossible, probability scale, probability experiment, possible outcomes

Section 1 Introduction

This Teacher's book is designed to support the component parts of *Nelson Maths* Level 5.

Sections 1–4 of this Teacher's book (pages 14 to 29) provide background and reference information, tips on classroom organisation, and a bank of activities and games to use for teaching practically. Section 5 provides lesson notes for the Pupil book and Workbook unit-by-unit and page-by-page.

Strands and skills

In Levels 1 to 6, the content is divided into strands:
- Number – numbers and the number system, and calculating
- Measure – money, length, mass and capacity, and time
- Geometry – shape, position and movement
- Statistics – organising, categorising and representing data
- Algebra – use of formulae, variables and equations

In this course, problem solving is not treated as a separate strand but is integrated into the materials at all levels.

Fundamental principles

This series makes the following assumptions about the teaching of mathematics:
- Children need concrete (practical, hands-on) experiences in order to acquire sound mathematical understanding. Like adults, children learn best when they investigate and make discoveries for themselves.
- Children refine their understanding and develop conceptual structures by talking about their own thinking and what they have done.
- Individual children develop at different rates. While some will find certain elements of mathematics difficult, others will understand them quickly.
- Children learn in a variety of ways, so mathematics teaching should provide a rich and wide variety of experiences. Children need plenty of opportunities to apply what they have learnt and to relate their mathematics work to other areas of the curriculum and their daily lives.
- Children will become more mathematically able if they are allowed to develop reliable personal ways of working. The formal recording used by mathematicians comes with time and experience, and we cannot expect young children to understand it or use it themselves before they are ready to do so. The conventions (formal methods) should be taught only once children are confident in their own knowledge, concepts and skills.

- Children learn mathematics most effectively when they enjoy what they are doing and when they can see the relevance of what they are learning.

This course reflects current thinking about the most effective ways of teaching and learning mathematics at primary level. It recognises the professionalism of the teacher, and acknowledges that teachers are the best judges of which experience(s) are most appropriate at different stages for the children in their classes. For this reason, this course does not impose a rigid, inflexible structure. Instead, it provides a wide variety of practical activities and games linked to clearly defined purposes and objectives. The teacher selects according to the needs of their classes, groups and individuals.

Individual differences and inclusivity

Everyone learns at their own pace, and in different ways. We also need to recognise that each child enters the classroom with their own experience, their own identity, and their own hopes, fears and curiosities.

This course recognises individual differences and aims to give children the chance to explore the world of mathematics and solve problems in their own way. To achieve this, we believe that each child needs to feel that maths is for them too. Therefore, examples and illustrations need to include a wide range of children of different genders and ethnic, cultural and linguistic backgrounds and should not promote stereotypes.

- When giving children additional maths problems to solve:
  - Use examples of girls doing things that are traditionally considered 'for boys'.
  - Use examples of carers (including dads as the primary carers) as well as working mothers and female leaders.
- Be aware of children's differences and acknowledge different bodies: not all children have ten fingers and ten toes, and some may not be able to walk or run as easily as others.
- Avoid comparisons of height or weight that some children may experience as body-shaming.
- Celebrate slower workers in your classroom as much as faster workers. Praise reasoning and focused working rather than speed. Never focus on speed-testing or 'who can find the answer the fastest'. Inclusivity means acknowledging that our slower, deeper thinkers are some of our best mathematical minds too. Maths is not about speed. It is about reasoning. Please avoid any games that use countdown clocks or timers which turn mental maths into a race or speed test.

- Praise different ways of working. Some children need to move and fidget and play in order to solve problems. Include those who think visually or rely on practical manipulatives and drawings to solve their problems.
- Struggle and confusion are key parts of the process of maths learning. Never dismiss a child's struggle as evidence that they cannot do maths or that they are not mathematically minded.

- When you are planning classroom activities, bear in mind the abilities and needs of your group and change or adapt activities as necessary. For example, it may be more appropriate to count sets of ten counters, or use ten frames than to use fingers for counting. When activities expect children to identify, sort or match colours, adapt these activities for children who have colour-blindness, by focusing on pattern rather than colour.

Section 2 Teaching approach

A growth mindset

In recent years, scientists have made important discoveries about how the brain learns and grows. A key discovery is the notion of neuroplasticity or brain plasticity. Brain plasticity is the brain's ability to change connections and neural pathways. Put simply, this means that the brain can learn and grow throughout a person's life.

These discoveries have led many educators to talk about 'growth mindset' versus 'fixed mindset'. People with a fixed mindset believe their abilities are determined and fixed, for example by genetics or by factors outside their control. A fixed mindset can lead to fears of making mistakes. Children with fixed mindsets may believe that they are either 'good at' or 'bad at' maths, and may easily get frustrated when they find problems challenging. We need to help children to understand that our maths ability changes all the time as we learn and grow. By getting children to understand and believe that growth, development and improvement are always possible, we can foster a 'growth mindset', in which children are not afraid of tackling difficult problems.

Getting rid of maths myths
Every child has the potential to learn maths, irrespective of their existing level of mathematical ability. The ability to learn maths is also not related to other traits such as gender, race or ethnicity. There is no such thing as a 'maths person'. It is extremely important that teachers understand that all children are able to learn to think mathematically. Maths is not a special subject and does not require any special natural 'talent'. Some children may grasp concepts more quickly than others – just like in other subjects – but giving them the time to develop their understanding, allows all children to achieve highly in mathematics.

The importance of mistakes
Scientists have also studied the brains of people attempting to solve problems. They found something that might seem surprising – that more neural connections and pathways form when we make mistakes than when we get things right! Although getting correct answers is usually praised by teachers, it is key that we also start understanding the importance of mistakes. Mistakes offer a valuable opportunity to explore why and how a concept makes sense. You can explore mistakes in a productive way with your class by:
- frequently reminding the children that making mistakes shows we are learning something new
- ensuring that the children know that mistakes (and questions) are welcome in your classroom
- discussing mistakes with interest and curiosity
- instead of always asking a question that invites a correct answer, sometimes phrasing questions to ask the children to identify the *incorrect* answers and then explain why they don't work.

You can read more about growth mindset in books by Carol Dweck and Jo Boaler, or search for these online.

The do–talk–record model

The learning framework for this course can be summarised as: *do–talk–record*. In other words, the starting point is always to explore and investigate concepts. Talking, discussing and explaining follows. Finally, written work can record and represent the concepts that the child has already explored practically and in discussion.

Doing
The first step in developing and understanding concepts is to let children:
- handle and manipulate apparatus
- play games
- investigate patterns and rules using real objects
- model situations and problems using real objects.

Children need time to play and explore in this way before they are expected to communicate about their work.

Talking
Children can make sense of what they have been doing by discussing:
- what they have done
- why they have done it
- what they have found out.

This allows them to generalise concepts and ideas from their particular experiences. The teacher's role is to create situations for discussion and to ask open-ended but directed questions. Most of the activities in this Teacher's book will help you to facilitate discussion and will encourage the children to listen to each other and experiment with different ways of thinking about and solving problems.

Recording

By Level 5, children are likely to have refined their skills and knowledge to some degree and they will have developed strategies that they find easy and useful for solving problems. They may still need to use informal and very personal methods (jottings) of recording steps in a process, or keeping track of what they have done. Jottings are an important step in moving towards non-standard methods of calculation (such as diagrams and jumps on a number line) that give the children a foundation for more concise standard written methods of recording.

It is very important that you allow, and in fact encourage, children to make use of jottings as they work. Here are some possible ways of doing this in the classroom:

- Do jottings of your own as you work out solutions. For example, if you are demonstrating how to calculate 144×5 you might jot the following on the board to show how you are thinking:

  1440

  720

- Talk through the jottings as you make them. For example, *144 times 10 is 1440, half of that is 720.* This modelling process helps children to see that jottings are important and useful.
- Make space for jottings in the children's exercise books. You can reinforce the importance of jottings as a means of showing your working by encouraging the children to jot as they work. If you only allow jotting on scrap paper, children may think it is not as important or valuable as their 'real' work in their book.
- Limit the use of prepared sheets with boxes for answers and no space for jotting down steps.
- Do activities where jotting is the point of the activity, for example, ask children to represent $\frac{3}{4}$ visually in as many ways as possible, or ask them to work out problems where they will need to jot down interim steps to keep track of the process: for example, *How many ways can you find of making one dollar using any combination of 50 cent and 10 cent coins?*
- Ask children to share their jottings and compare them to show that there are different methods of working. This can help the children to see that some strategies are more efficient than others and, in turn, refine their own thinking. In the 'make a dollar' task above you may find that some children draw coin combinations, others list them and those who are more confident may make a table and work more systematically. All of these methods may provide the correct answers, but obviously some will take longer than others.

In the early stages of using apparatus in a new way, recording may take the form of drawings or words and drawings. Some children will gradually find this time-consuming and will simplify their recording independently. Others may need your suggestions and encouragement. As a teacher, you will need to work out carefully when a child is ready to use a standard mathematical symbol or format, so that recording is based on full comprehension.

Although in Level 5 you will teach children some standard written methods for operations on larger numbers, it remains crucial that you do not force children into formal and standard methods of recording calculations before they have fully grasped the process and are confident in the methods.

Exploring and investigating

Traditionally, primary mathematics has almost always tended towards short, directed tasks that result in 'right' or 'wrong' answers. The activities in this course provide a balance between short, fairly self-contained activities and open-ended investigations that may sometimes require the children to collect information, make their own observations, ask questions or complete activities at home. Most of the activities are designed to develop children's awareness of the range of mathematical possibilities open to them when tackling a mathematical task.

As much as possible, allow the children to take control, make decisions and explore the many avenues that can arise from a simple starting point. Always encourage children to ask *What if … ?* and *Why?* when investigating. These questions may lead to new challenges, fresh understanding and the development of new skills.

Many investigations have no final 'answer' or easily accessible generalisation for the children. Some have a simple pattern or rule that may be discovered and explained. However, many children will want to know why certain patterns repeat, and offer explanations about the rules that govern them. This is the first step towards generalisation, and you should encourage this by asking questions, for example: *Why is the same number added each time?* or *Can you guess what will happen next?*

The value in investigations is in children following them as far as they can, and in the new skills that they acquire on the way. For some children, the early, often concrete, experimentation is enough to give them confidence, and increase their enjoyment of using already acquired skills.

Use investigations to introduce new concepts. For example, you can introduce number patterns by presenting number chains (sequences), and letting the children work out what comes next or what is missing. Introduce geometric patterns through explorations of colour arrangements using pattern blocks, tiles or geoboards. The children can explore the relationship between area and perimeter, and between volume and the dimensions of cuboids.

As the children develop an investigative approach, help them to work systematically, and remind them of the question they are working to answer. Making new observations and discoveries along the way is encouraged, but ensure that these do not eventually distract from the question they are working on. Ask questions such as: *What does this show us? What do you notice? What do you wonder? How can you use this to work out the answer? What other information do you need? How can you find it?* and so on.

Number talks

Number talks are open-ended, meaning they present a question that has multiple possible answers, and multiple possible strategies for solving.

Children can use agreed hand signals as shown below, instead of calling out or putting their hands up. However, these may not be culturally appropriate in all countries, and you may want to come up with alternative signals that are suitable for your class.

- **Step 1: Present the problem and give time for independent thinking**
  First present the problem, then ask the children to think to themselves quietly how they could solve it. They can use hand signals (one finger raised means 'I have a way to solve it') to show when they have an idea.

I have an idea, and one way to solve it.

Note that when children have one finger raised with an idea, you can encourage them to think of more ways to solve it, and to indicate when they have more possible ways of doing it. They can add more fingers as shown.

I have an idea and two ways to solve it.

- **Step 2: Share in pairs**
  In pairs, children share their suggestions for how to work it out. Note that explaining how they solved a problem is as important as (or more important than) the final answer. They can use a hand signal to indicate they are finished, such as a fist to the chest.

I have finished.

- **Step 3: Children share their strategies**
  Call on individual children to share their strategies with the class. Discuss each strategy as the child demonstrates for the class. Continue using hand signals (thumb and little finger out for 'I agree';

holding out both hands in front, palms down, and moving the hands in a criss-crossing motion for 'I disagree'; one more finger raised for 'I have another way to do it') etc. so that children can engage with the discussion and take turns to share their own working.

I agree

- **Step 4: Discussion**
  The discussion will take place as other children show that they agree or disagree, and offer alternative strategies.

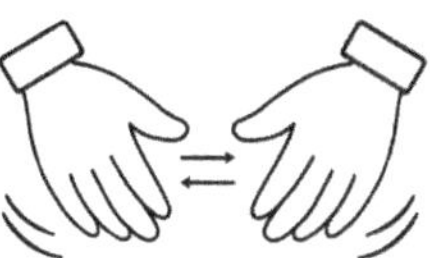

I respectfully disagree

Principles of the number talk strategy

The number talk strategy aims to get all children to engage actively and creatively with mathematics. To achieve this engagement, it is important to remember the 'big ideas' behind the number talk:

- *Value the contributions of all children.* Throughout the discussion help them to articulate their thoughts and reach a clearer understanding of how they are working things out. The emphasis is on the process and understanding, not on reaching the right answer fast.
- Children often answer in a roundabout way, with a very general idea or 'sense' of what they want to say. Use questions to *clarify, refine and explore* the meaning behind each response.
- Encourage children to explore *conceptual explanations*, rather than procedures.
- Gently *explore interesting mistakes* because these provide opportunities for deeper understanding. We can often better understand how something does work by reaching a clearer understanding of how it doesn't work.
- Praise children for effort, for making sense of a problem and for *persistence in the face of difficulty*, rather than for efficiency. We like to tell children that if it is difficult, that tells us we are in the place of learning. If it is easy, that means we've already learnt it and it's time to find more challenge.
- Create a *safe community space* where children feel they can share their ideas without criticism or judgement.
- Encourage *sharing ideas*, rather than competition. Children should understand that learning comes from connecting with other people's ideas and building on them.

Finding out more about number talks

It is important to remember that a number talk is a format or technique, not a specific lesson or set content.

- View number talks in action, shared by teachers, on the Internet.
- Find posters and strategies for number talks through an online image search.
- Books on number talks include: *Number Talks: Whole Number Computation, Grades K-5: A Multimedia Professional Learning Resource* by Sherry Parrish. *Classroom-Ready Number Talks for Kindergarten, First and Second Grade Teachers* by Nancy Hughes.

Numberless word problems

Many teachers fear problem solving because they present a problem, and then watch as children try to guess which operation to use, or which numbers to arrange into a number sentence. Numberless word problems offer a useful strategy for encouraging children to think and understand questions before looking for the 'maths answers'.

Here is an example. Say you want children to practise their addition. Instead of giving a problem in traditional form, we start with a numberless form of the same problem, and work very gradually towards presenting the numbers.

> *There are some children in the playground. Some more children come out to play. Now there are many children in the playground.*

The numberless word problem does two things:
- It removes the numbers.
- It removes the question.

This forces children to slow down and think about the problem before they can attempt to solve it.

To present a numberless word problem, follow these steps:
- **Step 1:** Present the word problem you want to use, but remove the numbers and question. So, reword it using words such as *some, more, joined, went away, took away, fewer* and so on. Ask the children to tell you what is happening in the story.
- **Step 2:** Ask children to think about what the question could be in this story. You will need to give plenty of thinking time, and invite the children to be 'question detectives', guessing from the clues what we are going to ask. They may have a variety of suggestions, such as: *How many were in the playground to start with? How many children came out? How many were there altogether?* Once they have identified the possible questions, you can reveal the question, but still without numbers.

> *There are some children in the playground. Some more children come out to play. Now there are many children in the playground.*
> *How many children came out to play?*

- **Step 3:** Have the children identify what information they need in order to solve the problem. First, ask the children if they can answer the question. They cannot – because they don't have the information they need. Let them suggest what they need to find out in order to solve it. Give them the first piece they need, by crossing out *some* and putting a number in:

> *There are 127 children in the playground. Some more children come out to play. Now there are many children in the playground.*
> *How many children came out to play?*

Again, ask if they can answer the question. When they say no, let them work out what other information they need to solve it. Let them work out what else they need to know. Finally, cross out *many* and change it to a number (e.g. 151).
- **Step 4:** As with number talks, give the children time to solve the problem using their own strategies. Follow the steps of a number talk as you let them share and discuss their strategies.

Mathematics in real life

Some children may struggle to understand the relevance of mathematics in their everyday lives. This course places great emphasis on making children aware of the relevance of mathematics in real life.

In this Teacher's book, you will find ideas for using the child's own environment as a stimulus for mathematical activities. The Pupil book and Workbook frequently require the children to look at the mathematics in the classroom, the playground and their own homes. Each set of activities and problems requires new skills and fresh understanding. Many questions are open-ended or have no exact solution, and the children are asked to make predictions, generalisations and estimates, and to evaluate their own answers. Encourage these skills in all areas of the curriculum. Children use their understanding of mathematics at home and at school, in situations such as sorting toys or other items, telling the time, looking for and making patterns, helping to prepare food, and playing board and other games.

In school

In school, there are many opportunities for you to teach mathematics through familiar situations, so that the children understand why it is useful and appreciate the order and sense that mathematics gives to life. For example, the children can identify the date each day, as well as the time at various points throughout the lesson. Registration, dinner money, timetables, sorting and putting away equipment will provide a range of relevant experience in data work, measures, geometry as well as number.

In play

Children of all ages should have opportunities to play both in and out of school. This offers them the freedom to explore new situations, to make discoveries for themselves and to be creative. Unfamiliar mathematics equipment should be introduced through play, with the children exploring the functions and possibilities of the materials. A good example of this is to experiment with pairs of compasses by drawing patterns and pictures before using them as mathematical instruments.

At home

Part of the teacher's role is to involve parents and guardians in the children's learning. Parents can do more than supervise their children's homework. Many activities can involve the parent in the child's learning and provoke mathematical discussion and language at home.

Encourage parents to extend their children's mathematical understanding through playing board and card games, and by encouraging them to help with normal home activities such as cooking, gardening, cleaning and organising the home, drawing up plans and measuring when redecorating, and estimating how many or how much when shopping.

Many children will voluntarily help and encourage younger siblings in games and getting organised. Family visits and holidays give children opportunities to see environments different from their own, and to experience time and distance. They are also likely to be budgeting pocket money, saving for special things and predicting how long it will take to afford treats.

Children may play with computer games that require a variety of mathematical skills. They might see and use a wide range of other electronic equipment at home, too, which also demands mathematical skills. Many children will be responsible for their own timekeeping and have a degree of responsibility for others.

Some homes will not actively encourage girls to use construction kits, computers or calculators, and some parents will not be confident of their own mathematical skills or understanding. As a teacher, you can help a great deal by making explicit the mathematical content of everyday experiences and activities.

You can find additional parent notes for each level on Oxford Owl. See page 9 for more details.

Section 3 An environment for exploration and reasoning

Organising the space

All teachers have their own preferences about how best to organise the available space. However, here are some useful guidelines for any classroom.

Storage

Always store equipment so that the children have easy access to it and you can check it periodically. Label all items clearly and encourage the children to make their own decisions about what they need. From the very beginning, insist that the children pack up and return equipment to the correct storage containers or shelves.

A mathematics centre

Create a mathematics centre in your classroom. This does not have to be where the equipment is stored, but it will be a bright and attractive part of the classroom with displays of children's work and other mathematical stimuli. The centre is a place for children to go at odd moments in the day, to be challenged with mathematics-related stories, questions and activities.

Questions and activities should be provided by both teachers and children for interactive problem solving, for example: *The answer is 3. What was the question?*, inviting children to draw or write out their suggestions.

A number pattern or sequence, on a series of cards organised by the children, may be 'secretly' altered by the teacher, and children have to discover what has changed, and put it right.

Recycled resources

Many household items can be repurposed to make mathematical equipment. Invite the children and their families to collect some of these materials at home in order to keep your classroom well resourced:

- bottle caps
- egg boxes
- clean glass jars
- buttons
- plastic caps from juice or milk cartons
- plastic yogurt pots
- beads
- natural objects such as shells, seed pods, stones for the children to arrange in size order.

Store the materials neatly in plastic containers or jars.

Whole class teaching

Often you will work with your whole class, especially at the beginning and end of a lesson. The course offers plenty of ideas for this kind of approach. The children may be seated together on chairs or on the floor, arranged in a circle for a discussion, or clustered around you for a demonstration. If you are calling on individuals to show things to the group, make sure there is enough space at the front for them to come up, and make sure everyone can see and hear the volunteer child's answer.

Set up some classroom rules at the beginning of the year and go over these at the beginning of each term or even each week if necessary. Classroom rules might include listening when others are speaking and waiting turns. Many teachers find it useful to use signs for *agree*, *disagree* and *I know the answer* rather than having the children raise their hands or shout out. The traditional approach of raising hands can be intimidating for quieter children or those who take a little longer to think about their answers. Hand signals allow teachers to give more time to those in the group who need it. The illustrations on page 17 show some commonly used hand signals.

Group work

You can group the children in similar or mixed-ability groups, to suit the purpose of the work. This gives the children opportunities to collaborate and to discuss their work with each other and with you. It allows peer teaching to take place and for the work to be matched to their needs. It also allows you to work simultaneously with a number of children and this minimises the need for repeated explanations to individuals. Group teaching is an effective form of classroom organisation for both teacher and children.

Working individually or in pairs

At times it may be appropriate for the children to work as individuals or in pairs, to provide extra help to children who need it, or to stimulate and challenge children who have grasped a concept quickly. Working individually gives the children the opportunity to concentrate on their own thinking, to develop this through investigations and problem solving, and to experiment with materials. Children working in pairs have the opportunity to develop collaborative skills, to play games together and to share ideas in an investigation.

Manipulatives give children an opportunity to experience the use of mathematics through concrete materials and objects. There are suggestions for different kinds of manipulatives in the unit-by-unit teaching guidance section of this book.

Counting apparatus
Interlocking cubes

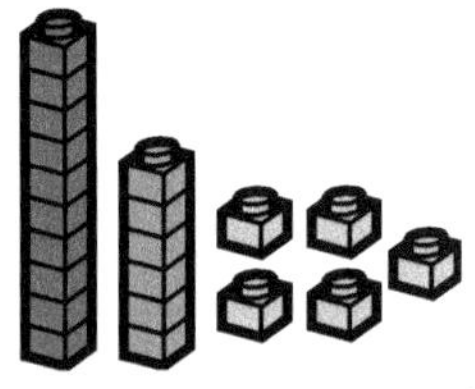

Interlocking or 'snap' cubes

Interlocking cubes are an invaluable resource for learning to count, especially as the children learn their number bonds up to ten, and then begin to work with tens and ones as they work with numbers up to 100.

Numicon (number frames)

Numicon

Numicon is another powerful and versatile concrete resource that supports children to visualise and work with numbers. Each hole in a shape represents a number.

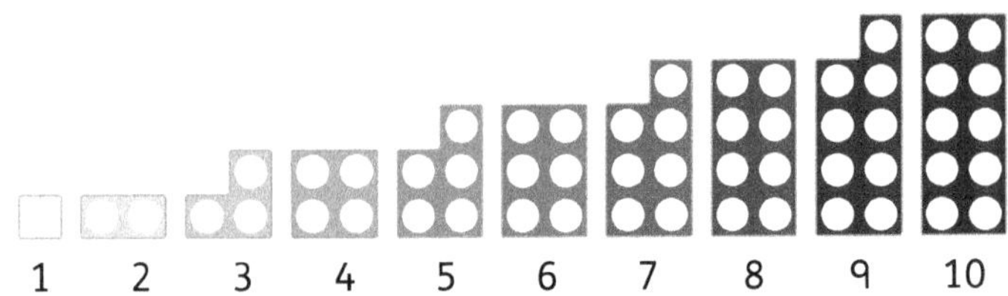

The children can use the Numicon shapes to count, represent, compare and calculate using fractions, decimals and ratios.

Base-ten blocks

Base-ten blocks are a concrete resource made up of blocks of varying sizes. Each block can represent a different value. For whole numbers, they represent:

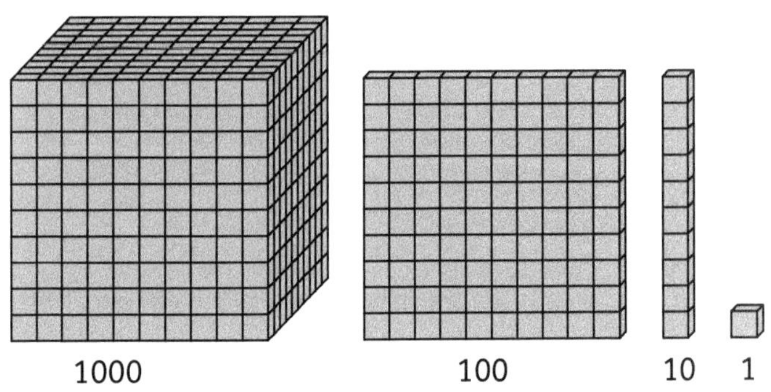

They can also be used to introduce decimals, representing, ones, tenths, hundredths and thousandths respectively.

Base-ten blocks help children to represent numbers, compare and calculate. They are particularly effective when used to support partitioning for example, into tens and ones.

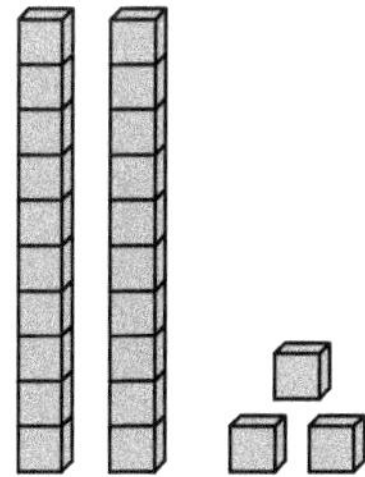

Representing 23 as two tens and three ones using two 10-rods and three ones cubes

Using base-ten blocks and place-value charts helps children see how numbers change when multiplied and divided by powers of ten. Subtracting using base-ten blocks is a great, 'hands-on' way to introduce and support exchanging.

Bar models

Drawing bar models is excellent problem-solving strategy that children can use in both primary and secondary school. A problem that may otherwise seem very challenging can become more accessible when you 'draw' it.

Children should be encouraged to both interpret bar models and draw their own when solving word problems.

There are two main types of models: the part-part-whole model and the comparison model. Which you choose to use will depend on the problem you are trying to solve.

For example, Tyra has 32 and she give 13 to Esme. How many marbles does she have left?

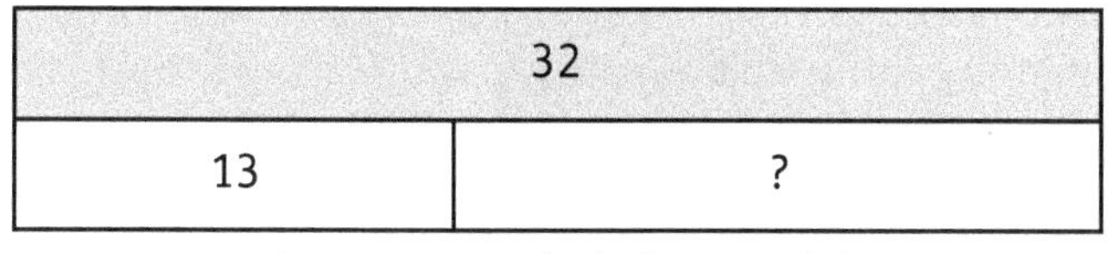

A part-part-whole bar model

Lewis completed 142 laps of the running track in one week. Joshua completed 52 laps in the same week. How many fewer laps did Joshua complete than Lewis?

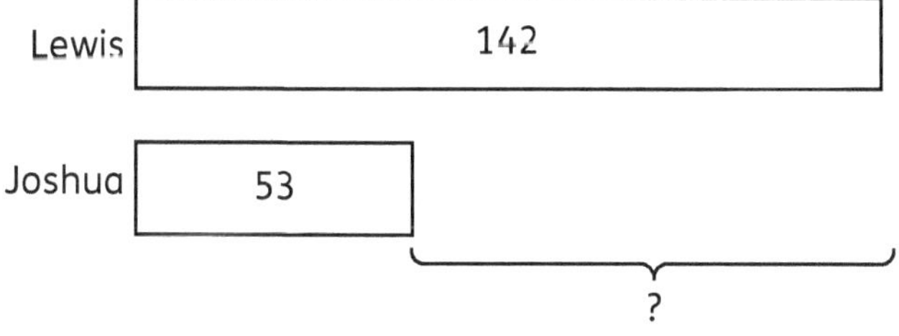

A comparison bar model

Place-value cards

Place-value cards have an 'arrow' or point on the right-hand side. Children can organise the cards horizontally or vertically to represent numbers in expanded notation. They can overlap cards and line up the arrows to form multi-digit numbers.

If any children have not previously worked with place-value cards, you will need to teach them how to use them. Begin by pointing out the arrows on the cards. Explain that these arrows always go on top of each other when you are making a number, for example:

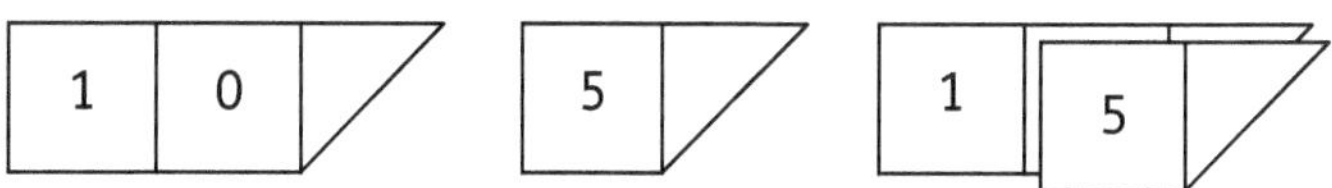

Place-value cards are an important teaching and learning resource and it would be useful to have a set available for each child. If possible, laminate the cards to make them more durable. (If you are making a set for each child, you may like to send the cards home for parents or carers to cut out.)

Place-value cards for decimals have an 'arrow' or point on the left-hand side, and a decimal point. For example:

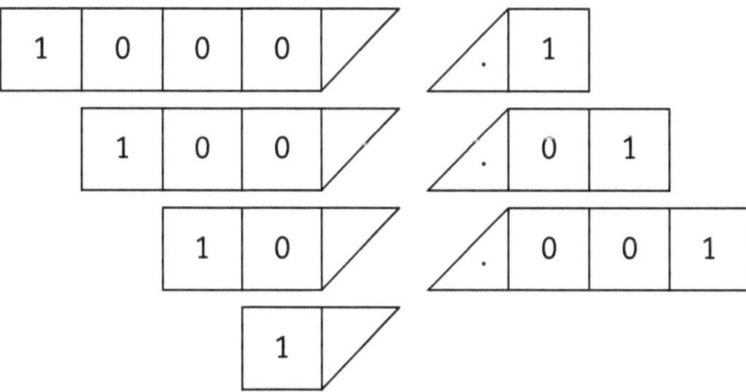

You can combine cards to make a decimal value:

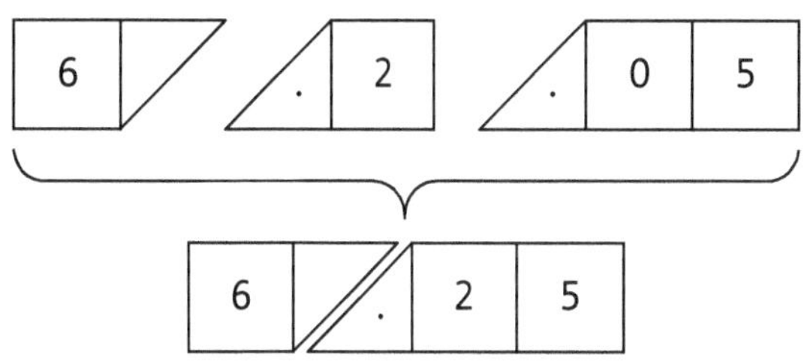

You can find printable place-value cards online.

Place-value tables

Children can use place-value tables to build numbers using counters or blocks. If possible, make place-value tables on card and laminate them to make them more durable.

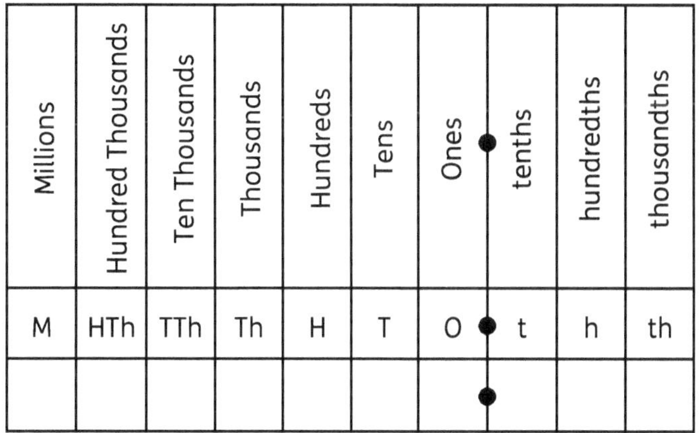

Millions	Hundred Thousands	Ten Thousands	Thousands	Hundreds	Tens	Ones	tenths	hundredths	thousandths
M	HTh	TTh	Th	H	T	O	t	h	th

Section 4 Activity bank of warm-ups and mental maths

This activity bank includes a range of mental maths activities, as well as some suggestions for support and consolidation activities. Most of the ideas are for number work and operations, although there are also some suggestions for the other strands. You can use this as a resource for activity ideas for your class. The unit-by-unit section in the next chapter refers back to specific activities in this section as unit introductions, suggestions for extra support and consolidation.

Try to include ten minutes of mental maths activities each day. This section includes many examples that you can use as they are or adapt to suit your own classroom. We have tried to provide a range of different types of activities (factual recall, games, grids, tables, problem solving and puzzles) to show some of the ways in which you can approach the mental maths part of the lesson. However, this is not a definitive list and some activities will appeal more to some classes and teachers than others.

> Please avoid any games or activities that use countdown clocks or timers to turn mental maths into a race or speed test. Speed testing can be damaging to children's mathematical development and cause maths anxiety.

Place value and number sense

Counting backwards and forwards

Do lots of activities in which the children have to count in given steps. Vary these according to what you are doing in class and the number range that the children are working in. For example:

- *Count in ones from 9999 to 10 020*
- *Count in hundreds from 25 100 to 44 900*
- *Count in tens from 1150 to 1300*
- *Count back in tens from 15 000 to 14 500*

Adding and subtracting tens and hundreds

Ask questions based on counting backwards and forwards using a number line marked from 0 to 10 000 in intervals of 500 (0, 500, 1000, 1500, etc.). Some possible questions are:

- *What is 100 more than 6000? (6100)*
- *What is 100 less than 1500? (1400)*
- *What is 100 less than 10 000? (9900)*
- *What is 100 more than 3500? (3600)*
- *What is 100 more than 4900? (5000)*

Write a number from words

Listen carefully: Read out some numbers and ask children to write them in digits. For example:

- *twenty two thousand, four hundred and thirty-five* (22 435)
- *forty one thousand, nine hundred and three* (41 903)
- *ten thousand and ninety-nine* (10 099)
- *sixty three thousand and seven* (63 007)
- *five thousand, seven hundred.* (5700)

Write the value of the given digit: Read out some numbers and ask children to write down the value of one particular digit. For example, what is the value of 3 in each of these numbers?

43 516 (three thousand)	*23 678* (three thousand)
1366 (three hundred)	*44 213* (three ones)
1493 (three ones)	*32 908* (thirty thousand)

(Make sure that the numbers you use only have one digit with the value you are looking for!)

Find the number

Children work in pairs. Ask each child to write down a number without showing their partner (you can select the range, for example, any number less than 5000). The children then try to guess what number their partner has written down by asking questions that can only be answered by yes or no. For example: *Is it odd? Does it have three digits? Are there more ones than tens?* And so on. Once the children have asked 10 questions, they should try to guess the number.

If they can't guess it, their partner can give them a clue and allow them to ask two more questions before guessing again. For example: *My number is between 300 and 350. It has no tens.*

5- and 6-digit numbers

Say the number: Write any 5- or 6-digit number on the board. For example: 302 645

- Ask children to say the number aloud.
- Ask different children to say how many hundred thousands, ten thousands, tens, ones etc. there are.
- Point to a digit and ask children to say its value.
- Ask children to reverse the digits and say the number.
- Ask children to make **five** different numbers using digits from the given number in any order.
- Ask them to exchange these and say each other's numbers aloud.
- Repeat the place-value questions using numbers the children have made.

Digit cards: Write a selection of 5- and/or 6-digit numbers on cards. If you make numbers with similar digits, it makes the activity a little more demanding. Display the number cards randomly on the board. For example:

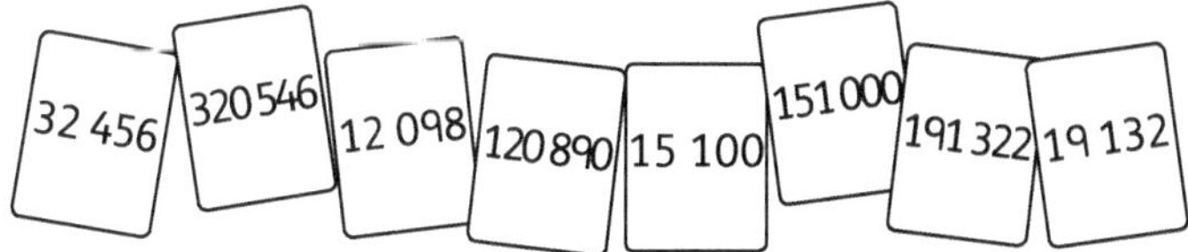

- Say numbers aloud in words and ask children to identify the number you said, for example: thirty-two thousand, four hundred and fifty-six.
- Ask the children to reverse the digits and say the numbers aloud. (If you are going to do this, be clear about what they should do with a 0 in the ones place or do not use any numbers with 0 in the ones place.)

Count forwards and backwards in tens, hundreds and thousands:
Children work in pairs. Give them instructions such as:
- *Count forwards in thousands from 40 000 to 50 000.*
- *Count forwards in tens from 15 890 to 16 020.*
- *Count backwards in hundreds from 6300 to 5200.*

Make the smallest or largest possible number: Give groups of children a set of up to six mixed digit cards. Ask the groups to make:
- the smallest possible number
- the largest possible number
- the smallest possible odd/even number
- the largest possible odd/even number.

Number sentences
Compare 3-digit numbers: Ask each child to write down a 3-digit number. Write some 3-digit numbers on the board. Ask the children to make number sentences using your numbers and the number they have written down using the <, > or = signs.

Find the difference: Ask the children to use mental strategies and jottings to find the difference between the numbers in their number sentences. Spend some time talking about the strategies they suggest.

Compare 5- and 6-digit numbers: Ask each child to write down a 5- or 6-digit number. Choose one child to come to the front of the class and display their number.
- Ask, *Whose numbers are greater than this?* Ask the children to hold up their numbers.
- Repeat for 'less than'.
- Choose one child at random. Ask them to come up and stand to the left or right of the other child (depending on whether their number is less or greater) and display their number. Choose other children, to come up and position themselves relative to the numbers already on display.

Directed numbers
Write a list of positive and negative temperatures on the board. For example:

12 °C   −3 °C   0 °C   −5 °C   −4 °C   9 °C   5 °C   −7 °C
−1 °C   1 °C   −9 °C

- Draw a blank number line on the board to represent a thermometer with only the start and finish marked.
- Ask the children to find the lowest temperature. Write this on the left-hand side of the scale. Repeat for the highest temperature, writing it on the right-hand side of the scale.
- Children volunteer to come up to the board. Ask them to choose one of the temperatures and then position it on the scale as accurately as they can. Once all the numbers have been placed, discuss whether there are any inaccurate placements. Let the children decide and suggest how to move the numbers if necessary.

This activity can be adapted to work with decimals, fractions, mixed numbers and whole numbers.

Making numbers
How many different numbers can you make? Give the children some possible digits for each place value and ask them to work out how many numbers are possible with a given number of digits. For example:
- *How many 3-digit numbers can you make if*
  - *the hundreds place can have: 2, 3, 4 or 5 (400)*
  - *the tens place can have: 1, 2, 3, 4, 5 or 6 (540)*
  - *the ones place can have: 0, 1, 2, 3? (360)*

Make these 6-digit numbers: Prepare a set of 40 cards with the digits 0–9 repeated four times. Shuffle these and deal six cards at random to small groups of children. Ask the children to use all the cards to make:
- the greatest possible number
- the smallest possible number
- the number as close as possible to 999 999
- a number less than 500 000.

You can vary the task by changing the instructions and changing the number of digits. (For example, *Make the largest possible 3-digit number.*)

Spot the rounding error: Write several 4-digit numbers on the board. Round each to the nearest 10, 100 or 1000 (choose one place value to round to per activity). Make sure some of the rounded values are incorrect.
For example: (rounding to the nearest 100)

2345 → 2300 (correct)
2662 → 2660 (incorrect: 2700)
4129 → 4200 (incorrect: 4100)
3888 → 4000 (incorrect: 3900)
3999 → 4000 (correct)

Ask the children to find the incorrectly rounded numbers and to correct them.

Repeat this using different numbers and rounding to different place values.

Fractions, decimals and percentages
Give each group of children a set of number cards with the digits from 0–9. Ask them to select two cards at random and write them as the numerator of a fraction with a denominator of 100. Repeat this five or six times.

They then complete a table like this for each fraction they make. This example uses digit cards 6 and 0.

Fraction	Decimal	Percentage
$\frac{60}{100}$	0.6	60%

Rounding and estimating

Rounding mentally
Draw a grid like this one on the board. If you are going to reinforce rounding to the nearest ten, make sure that the numbers all have a value other than 0 in the ones place.

456	1275	499	109
3245	6501	1295	1082
3509	8024	8019	876
103	562	901	1052

- Ask children to copy the grid, and rewrite the numbers after rounding them to a given place value (for example, the nearest 10, the nearest 100).
- Alternatively, you can work through the grid, pointing at each number in turn and ask individual children to round it to the nearest ten.
- Repeat this for different place values.
- If possible, you could display a large a photograph of a number of items (for example, a tray of beans, stitches in a knitted jumper or bees on a hive) and ask the children to estimate how many there are. Encourage them to explain how they find their answers.

Find the closest total
Write a set of six numbers on the board. For example, a set of 3- and 4-digit numbers such as these:

593   3199   5804   7083   908   427

Ask the class to estimate which two numbers should be added to get the total closest to a given number. For example, *Which two should you add to get the total closest to 10 000?* (3199 + 7083 = 10 282)

Find the closest difference
You can adapt the previous activity for subtraction.

Make the target number
Children work in pairs. Ask them to write down any 5 digits. (You can vary the task by stipulating that digits can or cannot be repeated.)

- Write a target 5-digit number on the board, for example 40 000. Ask the children to arrange their digits to make the number closest in value to this target number.
- Children can compare their numbers and decide whose is closest.
- Repeat using different target numbers.

Wipe out calculator game
Children work in pairs.

- One child enters a 4-, 5- or 6-digit number on a calculator. The other child should try to wipe out the digit in a given place value without changing any other digits. For example, if the number is 345 234, and you ask the children to wipe out the hundreds digit, the child needs to subtract 200.
- Repeat, with the other child choosing the number.

Round decimals
Use number lines and/or measuring tapes/scales to practise rounding decimals and measurements to the nearest whole number. Display a number line, for example:

20   30

Then ask questions such as:

- *Is 23.6 closer to 23 or 24?* (24)
- *Is 28.5 closer to 28 or 29?* (It's exactly half way between them but would round up to 29)
- *Which whole number is closer to 29.25 – 29 or 30?* (29)
- *Pete's little brother weighs 22 kilograms to the nearest kilogram. Is it possible for his brother's actual mass to be: 22.3 kg?* (yes) *22.19 kg?* (yes) *22.9 kg?* (no) *22.54 kg?* (no) *23.01 kg?* (no)

The questions can be adapted to different values and different measurements.

Make decimal fractions
Ask two children to each pick a digit-card (1–9).

- Ask the children to combine these digits to make two decimal fractions. For example, if they pick 2 and 7 they can make 2.7 and 7.2.
- Ask the children to round each decimal fraction to the nearest whole number.

Mental problem solving

Continue the sequence
Write the number sequence 3, 6, 12 … on the board.

- Ask the children to silently decide what the next two numbers in the sequence should be.
- Then write 3, 6, 12, 24, 48 and 3, 6, 12, 21, 33 on the board.
- Ask the children to say which solution they think is correct and why.
- If they don't say that both are possible, then tell them that they are and ask them to work out how the sequences have been continued (doubling for sequence 1 and adding consecutive multiples of 3 for the second one).
- Give the children a few more similar examples to complete. Ask them to find two ways of completing each one. For example:

  30, 60, 120 … (30, 60, 120, 240, 480, … : doubling); 30, 60, 120, 210, 330, … : adding consecutive multiples of 30)

4, 8, 16 … (4, 8, 16, 32, 64, … : doubling; 4, 8, 16, 28, 44, … : adding consecutive multiples of 4)

5, 10, 20 … (5, 10, 20, 40, 80, … : doubling; 5, 10, 20, 35, 55, … : adding consecutive multiples of 5)

Periods of time

Write a time period on the board, for example, 16 days.

- Ask the children to estimate and write down how many weeks this is.
- Repeat this for different units. For example:

4 weeks – how many days? (28 days)
8 months – approximately how many weeks? (32 weeks)

Mental addition and subtraction

Draw a series of grids like this on the board to reinforce and practise addition and subtraction.

47	19	23
10	15	44
50	27	31

Highlight a start and an end number. Ask the children to work out what operation is required to get from the starting value to the end value. You can jump vertically, horizontally or diagonally and move one or two places. For example:

- *from 47 one down to 10* (subtract 37)
- *from 47 across diagonally to 31* (subtract 16)

You can use this as a game in which the children work in pairs to move on a grid and answer each other's questions. They can use a calculator to check the answers. You can make this activity more difficult by increasing the size of the grid and extending the number range.

Problem solving

Display this variation of the traditional rhyme 'As I was going to St Ives'

As I was going to St John's
I met a man with seven sons
Every son had seven sacks
Every sack had seven cats
Every cat had seven kittens
Kittens, cats, sacks, sons
How many were going to St John's?

Ask the children to work out the total number of kittens, sacks, cats and sons. Ask them to explain how they worked it out. (7 sons, 49 sacks, 343 cats, 2401 kittens; multiply each number by 7) Change the activity by changing the numbers. For example, *I met a man with twenty sons …*

Number combination games

These should involve more than one operation.

For example:
- *Use each digit 1, 3, 4, 6 and 9 once only to make a calculation that gives 3280.*

To find digits that work, do a calculation yourself. 91 × 36 + 4 = 3280. Then write them in size order for the children.

Letter–number codes

Give each letter of the alphabet a numerical value. This can be prices, such as A = $1, B = $2, C = $3 and so on up to Z = $26. Ask:
- *What is the value of your first name in this code?*
- *What is the value of your family name?*
- *Can you find a word that is worth $50?*
- *Can you find a word that is worth more than $75?*
- *Can you make a word worth at least $200?*

Logic puzzles

These can require mental calculations. For example:
- *I have a 7-litre and a 5-litre container. How can I use these to accurately measure 4 litres of water?* (Fill the 7, pour it into the 5 leaving 2. Empty out the 5. Pour the 2 litres from the 7 into the 5. Fill the 7. Pour from the 7 to fill the 5. This will use 3 litres and leave 4 litres in the 7-litre container.)
- *How can you measure 1 litre with a 3-litre and a 5-litre container?* (Fill the 3 and pour the water into the 5. Fill the 3 again and pour the water into the 5 again. This will use 2 litres and leave 1 litre in the 3-litre container.)

Make a dollar

Ask the children to work out how many ways there are to make a dollar using any combination of 1c, 2c, 5c, 10c, 25c and 50c coins. This activity encourages children to work systematically and to use jottings to keep track of their thinking. The most successful children will start either with the largest value coins or the smallest, like this:

50 + 50
50 + 25 + 25
50 + 25 + 10 + 10 + 5
50 + 25 + 10 + 5 + 5 + 5

You can make this problem easier by limiting the coin combinations.

Mathematical vocabulary

Give the children word problems to be solved mentally. For example:
- *Find the number that can be increased by 3 to make 21.* (18)
- *What is the product of 4 and a number 5 greater than 4?* (36)
- *What is the square of the difference between 16 and 7?* (81)
- *What number do you get if you halve the product of 9.8 and 100?* (490)
- *What is the sum of 345 and double 90?* (525)

Calculation skills

Multiplying on a grid

Draw an empty 2 × 6 grid on the board and ask the children to copy it.

Fill in the numbers from 1–12 on the grid on the board in random order. For example:

6	1	10	11	8	4
2	5	9	7	3	12

Give the children a multiplier (for example × 7, which suits work for converting weeks to days). Ask them to fill in the result of multiplying each number on your grid by 7.

Mental addition and subtraction
The operations on these grids are given across and down. Children can work in any order to fill in the missing values.

+ 7 ↓	+ 9 →				
	15	24			
	22	31			

− 7 ↓	− 5 →				
	59	54			
	52	47			

- For subtraction, make sure that you start with a large enough value to avoid negative numbers at too early a stage.
- You could prepare a variety of these grids on laminated cards.
- You can adapt this task to include decimal values.

Doubling
Give one child a number, for example 7, and ask them to double it.
- Go round the class and see how far you can get with doubling the number.
- Change the pattern by adding or subtracting a given number to get a different starting value. For example, once the children get to 448, say something like: *Next student, subtract 8.*

Halving
Repeat the previous task, but ask the children to halve a given number. Remember that halving is more difficult than doubling, as the children will get to fractional values as soon as they have to halve an odd number. When this happens, you can change the pattern by getting them to add a value such as $9\frac{1}{2}$.

Target number games
Give the class a target number. Either:
- ask the children to suggest different ways of getting to this number using whichever operations they like, or
- ask everyone to find ten different ways of getting to the number.

For example, if the target number is 86, the children could make the target by $80 + 6$, $90 - 4$, 8.6×10, $\frac{860}{10}$ and so on.

To make the game more challenging you can give instructions for operations. For example, *Make the target by*
- *adding three numbers*
- *doubling a number and then subtracting.*

Decimal times-tables
Give the children the first five multiplications to establish a pattern involving decimals, for example:

    1 × 2.5 (2.5)

    2 × 2.5 (5)

    3 × 2.5 (7.5)

    4 × 2.5 (10)

Ask them to write the answers and to continue the pattern.

Find the sum
Ask the children to find the sum of all the numbers from 1 to 10, 1 to 25, 1 to 50 and so on, using their own strategies. (1 to 10: 55; 1 to 15: 325; 1 to 50: 1275)

Fill in the gaps
Provide incomplete multiplication tables such as these and ask the children to find the missing values. This requires them to think strategically and to use inverse operations to find the missing values.

×			4
			12
			8
6	36	30	

×		8	
20	1800		320
		336	
		392	

- You can generate many of these tables yourself, but remember that you need at least one horizontal and one vertical multiplier and at least four values for the children to find the missing values.
- You can extend these to a higher number range as the children become more competent at multiplication.

Odd one out
Provide a sheet of operations that form equivalent pairs with one odd one out. These could be equivalent fractions, decimal fractions paired with mixed numbers, or percentages of amounts.

- Ask the children to find the odd one out and to explain how they decide. For example:

$\frac{3}{10}$ of 100	$\frac{9}{10}$ of 20	10% of 90
80% of 80	30% of 250	25% of 256
$\frac{25}{100}$ of 72	$\frac{1}{2}$ of 150	$\frac{1}{2}$ of 60

(10% of 90 = 9; it is the only answer that does not make a matching pair.)

Same total

Make a set of cards with the digits 1 to 7 written on them. Display these on the board. Draw a grid like this one on the board.

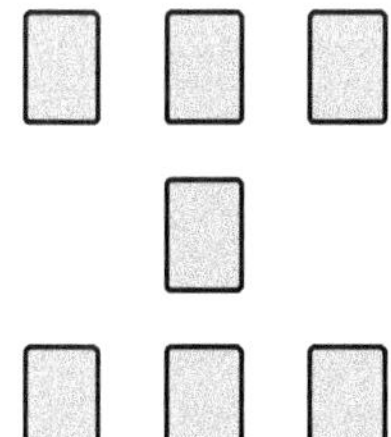

Ask the children to find ways of organising the digits so that the totals along any line (horizontal, vertical and diagonal) are the same. Encourage them to explain how they thought this out. (A possible answer is top row: 5, 1, 6; middle row: 4; bottom row: 2, 7, 3.)

Arrange the digits

Make a set of cards with the digits 1 to 6 written on them. Draw a grid like this on the board.

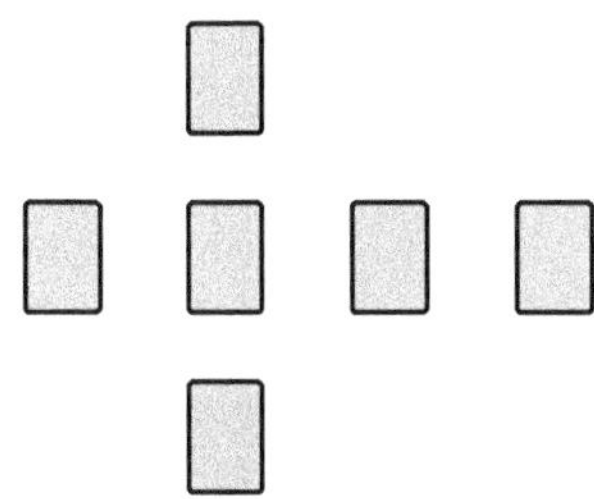

Ask the children to find ways of arranging the digits so the sum of the vertical digits (the column) is equal to the sum of the horizontal ones (the row). (A possible answer is vertical digits: 2, 5, 6; horizontal digits: 3, 5, 4, 1)

You can make this more difficult by changing the instructions. For example:

- *Arrange the digits so the sum of the horizontal row is 2 greater than the sum of the vertical column. (A possible answer is vertical column: 5, 1, 4; horizontal row: 2, 1, 3, 6)*
- *Arrange the digits so the sum of the row is double the sum of the column. (A possible answer is vertical column: 4, 3, 1; horizontal row: 2, 3, 5, 6)*

Spinners

Give each pair of children number cube a spinner with single-digit numbers on it. They take turns to spin the spinner number cube and multiply the value they spin by 19, 21 or 25. They should spin 5 numbers each and record their answers.

You could ask the children to:

- *Add up the products for the 5 calculations.*
- *Work out the difference between the highest and lowest scores.*

Minus the money

The children work in pairs. Each student starts with $10. They take turns to spin a spinner and multiply the result by 10 (or any number of your choice). They must subtract that number of cents from their total. They continue to play until one player has no money left.

Magic squares

Prepare a set of magic squares that each have one incorrect value. Ask the children to find the incorrect number and explain how they decided that it was incorrect. They should also suggest what the correct value is.

28	35	30
33	31	29
32	26	34

71	73	69
66	68	70
67	72	65

93	92	97
98	94	89
91	96	95

The highlighted number is the incorrect one, this is for your information only. (The correct numbers are 27, 64 and 90.)

Equal sums

Show the class an arrangement of blank digit cards. Ask them to:

- *Arrange the digits from 1 to 6 so that the sum along each line of 3 cards is equal. (A possible answer is top row: 3; middle row: 5, 4; bottom row: 1, 6, 2)*

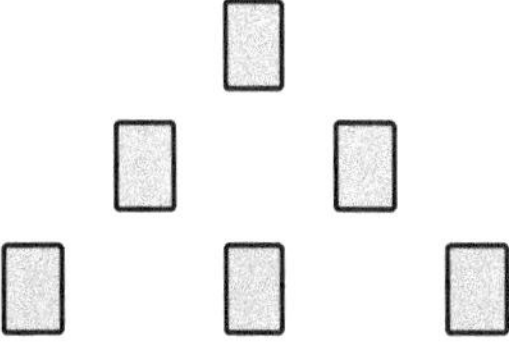

- *Arrange the digits 1 to 9 so that the sum of each line is equal.*

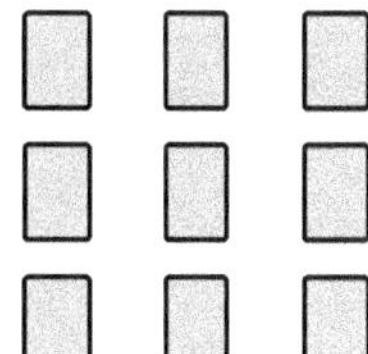

(A possible answer is top row: 6, 1, 8; middle row: 7, 5, 3; bottom row: 2, 9, 4)

- *Arrange the digits 1 to 7 in these circles so that the sum on each line is equal.*

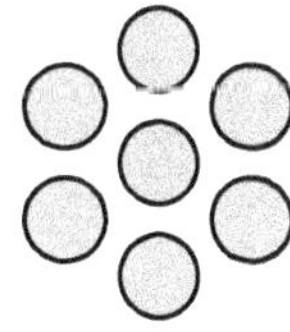

(Possible answer: 1 in the middle; clockwise from top: 5, 3, 2, 4, 6, 7)

- *Use each digit from 1 to 7 once only to make a calculation with the answer 100. (A possible answer is $5 \times 6 \times 4 - 7 \times 3 + 2 - 1$)*

Multiples, factors and divisibility rules

Write a starting number on the board, for example, 64.

- Ask the children to jot down as many multiplications as possible to make this number.
- Repeat for other starting numbers including some prime numbers or decimals.

Multiplication grid

Point to an empty cell and invite individual children give the product. For example:

×	1	2	3	4
3				
6				
8				
9				

Division grid

Use a multiplication grid with the products filled in but some of the numbers missing from the header row and the header column. Point to a number in the grid and ask children to fill in the missing number.

×	1		3	
	3	6	9	12
6	6	12	18	24
	8	16	24	32
9	9	18	27	36

Three or four in a row

This is a variety of noughts and crosses that uses the equivalence between fractions, decimals and percentages. The children play in pairs. Give each pair a prepared grid with suitable families of fractions on it. For example:

$\frac{1}{10}$	20%	$\frac{1}{2}$	$\frac{45}{100}$
$\frac{6}{100}$	30%	$\frac{4}{100}$	16%
$\frac{9}{100}$	$\frac{8}{10}$	$\frac{3}{5}$	$\frac{4}{10}$
45%	80%	$\frac{55}{100}$	$\frac{28}{100}$

- The children take turns to claim cells by offering an equivalent fraction (or decimal or percentage).
- The children can place counters on the cells they claim. The aim is to make a row of three (or four) while preventing their partner from making a row or three (or four).

Calendars and time

Calendar patterns

The dates each month on a calendar offer a number of sequences. In a row, the dates count up in ones. The columns offer numbers with a difference of 7 and on diagonals, the difference is either 6 or 8. The sum of 3 consecutive numbers will be 3 times the first number + 3 times the difference between them.

- Provide a blank calendar for a month with some dates filled in and ask the children work out the missing dates (and only those). For example, fill in the first and then let the children work out the dates going diagonally.

Finding times earlier and later than a given time

Prepare a table like this one for display in the classroom:

$\frac{1}{2}$ hour earlier	Time	15 minutes later	$\frac{3}{4}$ hour later	24-hour time	Time 4 hours ahead
	3.15 p.m.				
	2.30 a.m.				
	Half past five in the afternoon				
	10 to 6 in the morning				

Point to different cells in the table and invite the children to give the missing values.

Geometry and measures

Making towers

Use a variety of solid objects (a few boxes, a can and a ball) to build a small tower in the classroom. Give the children questions to answer.
For example:
- *How many faces are there altogether in this structure? How did you work this out?*
- *How many corners (vertices) are there? How did you work this out?*

Angles

Angles on a straight line: Prepare a set of angles.

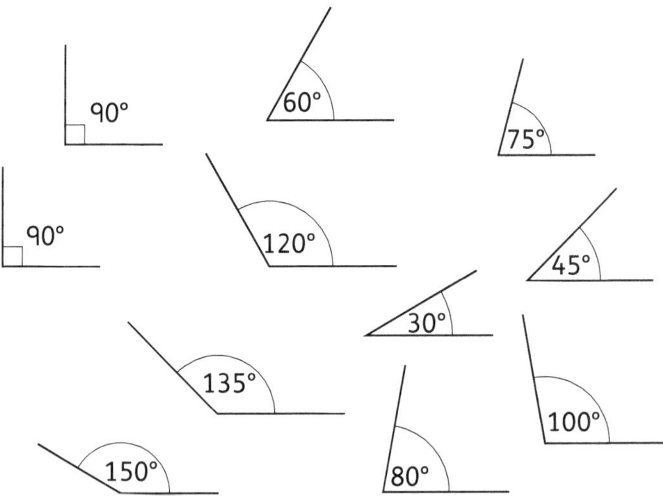

- Ask the children which angles can be put together to fit onto a straight line. 90° and 90°; 120° and 60°; 150° and 30°; 100° and 80°; 135° and 45°)
- To make it more difficult, include combinations such as 124° + 56° or 38° + 142°.

Missing angle game: Give children one angle on a straight line and ask them to find the size of the other one to make a total of 180°. For example:

90 + ☐ (90)

50 + ☐ (130)

34 + ☐ (146)

Types of triangles

Draw a variety of different types of triangles on the board (acute- and obtuse-angled scalene, right-angled scalene, acute- and obtuse-angled isosceles, equilateral) and give their perimeters in different units.

- Identify the type of triangle and ask the children to work out possible lengths for the sides. They will need to apply the properties of the different triangles.

Units of measurement

Multiplying and dividing by 10: Give the children a variety of tables like this to complete.

	›× 10 → ÷ 10
2.3	
8	
	14
	40

Multiplying and dividing by 100 and 1000: Give the children a variety of different tables like the one above, that involve multiplying and dividing by 100 and 1000.

Converting metric units: The previous two activities can be adapted to give the children practice at converting between metric units of measurement.

Learning objectives

- Establish a classroom environment conducive to thinking and working mathematically
- Establish key messages of growth mindset mathematics
- Consolidate planning and problem-solving skills

Key words

pattern

Teaching guidance

This short introductory unit aims to establish important attitudes in your class. It prepares both children and teachers for the work ahead. You should be able to complete this unit in two or three lessons.

As teachers of maths, it is important for us to examine our own ideas about who can do maths and about how mathematical ability and learning develop. It is also important to make sure the children understand that their own beliefs affect how they learn (or don't learn). From the earliest levels, this course aims to find real, experiential ways for the children to understand the following ideas:

- Asking questions is the best way to learn.
- Making mistakes is a key part of brain growth and learning.
- Everyone's brain has the potential to learn maths. There is no such thing as a special 'maths brain'.

If we simply tell the children these things, they will not believe us. We need to provide evidence of the ways in which the brain grows and learns. We also need to show them that maths can include everyone.

Think about how you learn

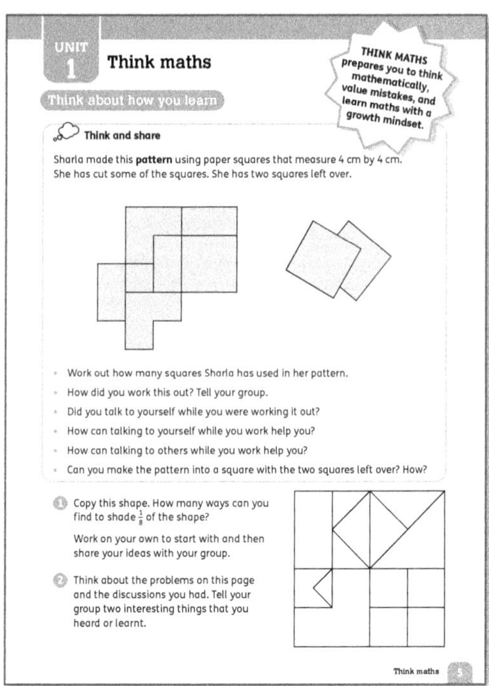

Materials

Squared paper; rulers; pencils; coloured pencils; scissors

Focus

- <u>Think and share:</u> Place the children in pairs. Turn to **Pupil book 5 page 5** and ask the children to consider the problem about *pattern*. Allow them to physically model the problem using squared paper if they need to.
- Take feedback from the class to check answers and to discuss how they worked them out.

(e.g. Sharla used 7 squares and cut and overlapped them to make the pattern.

e.g. To work this out, I used paper squares to make the diagram for myself; I counted the numbers of sections in the diagram.

e.g. I talked to myself to help me to organise my thoughts.

e.g. I talked to others to share ideas and methods.

To make the square, use one whole 4 cm by 4 cm square, and cut the other into two 2 cm by 2 cm squares and a 2 cm by 8 cm rectangle.)

- Encourage the children to share some of their self-talk. Some of the self-talk might be negative, which will give you a good opportunity to discuss how limiting negative self-talk can be.

- Explain that when we try to solve problems, we use our brains. The harder we think about the problem, the more connections we use and make in our brains, which is how we learn.
- Give each child a sheet of squared paper to complete question 1. Copying the design will provide good practice in using a ruler and pencil. The children also have to make decisions about how to draw the shape accurately. Once they have copied the design, let them spend some time on their own using different colours to shade sections that are $\frac{1}{8}$ of the shape.
- Check that everyone has done some shading. Then put the children into groups to talk about how they worked to solve the problem.
- Ask the children to complete question 2. Each child considers what they heard and learnt and shares two interesting things with their group. Take feedback from the class about what the children found interesting.

Answers for Pupil book 5 page 5

1 Each type of shading shows $\frac{1}{8}$ of the shape.

2 e.g. Talking about maths can help you do maths; there can be more than one solution to a problem.

Talk to yourself

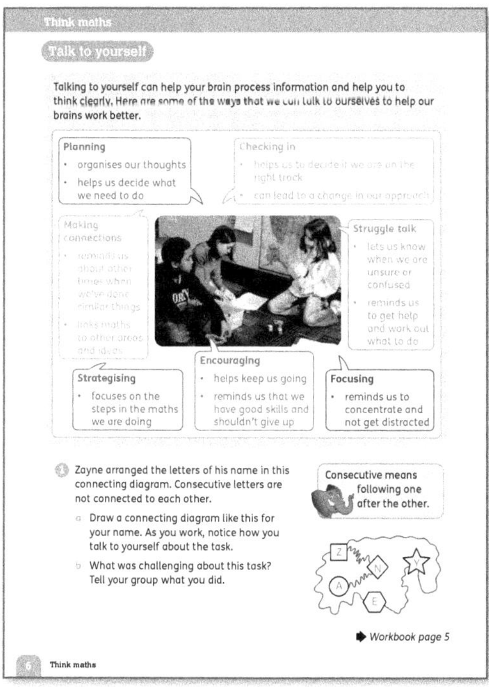

Materials

Paper; coloured pens

Warm-up

Remind the class that our brains are like powerful computers and that they process millions of pieces of information as we go about our daily lives. Our brains are also elastic – they can grow and become more powerful depending on our own behaviour. (The children have done several activities on brain growth since Level 1, so you could ask them to recall some of these.)

Focus

- Turn to **Pupil book 5 page 6** and ask the children to read through the different ways in which we talk to ourselves. Ask them to give examples for each category. For example, for 'Strategising', they might say: 'I think drawing a number line will help me order the numbers. A bar model might be useful here. The word "sum" tells me that I should add.'
- Instruct the children to look at Zayne's connecting diagram in question 1. Point out that the letter Z is connected to every other letter except A. Discuss the task with the class. Emphasise that the children must pay attention to their self-talk as they work. Give them time to draw their own connecting diagrams.
- In their groups, the children talk about the task and how they overcame challenges.
- Take class feedback. You may like to display the connecting diagrams in the classroom.

Follow-up

Use **Workbook 5 page 5** as a journal activity. Tell the children that they should choose the activity they found most challenging and write examples in the table of their self-talk during the activity. Explain that this is a personal account and that they won't need to share it. When they have done this, they can highlight any self-talk messages that they think are negative and write more positive versions of these statements to help them with their maths work.

Interesting mistakes

Children with short names may not know what to do. Children with two-letter names will have to draw two discrete circles with no connections. Children with three-letter names can only connect the first and last letters. Then they could draw a connecting diagram for a city or country name of their choice.

Answers for Pupil book 5 page 6

1 a Children's connecting diagrams for their own names.

   b e.g trying to not connect consecutive letters.

How we talk about maths matters

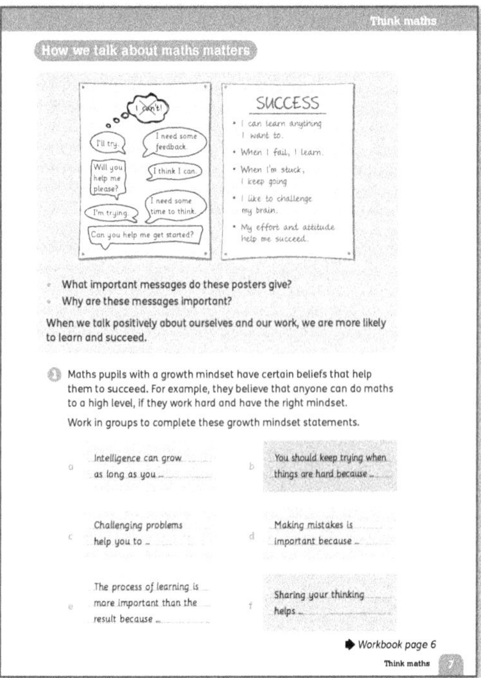

Focus

- Turn to **Pupil book 5 page 7** and ask the children to read the posters. If you have similar posters on display in your classroom, include them in this activity. Take feedback about the messages on the posters and why these are important for helping us to learn successfully. Remind the children that if they *believe* they can do something, there is a much better chance that they *will* be able to do it.
- Remind the class that there are two different kinds of mindsets – fixed and growth.
- People with a fixed mindset believe you are born clever or not clever and that you cannot change that.
- People with a growth mindset know that you can become more intelligent and better at things. These people are more successful at learning new things because they believe that it is possible and that mistakes are just part of the process of learning.
- Place the children in groups and let them decide how to complete the statements in question 1 using a growth mindset. Let different groups share the completed statements.

Follow-up

Let the children work on their own or in pairs to complete **Workbook 5 page 6**. When the children have written their positive messages, ask them to leave their books open on their desks. The children can then walk around the class and read the other children's positive messages.

Interesting mistakes

Adults and children carry implicit beliefs and stereotypes about mathematics. These assumptions may come from families or friends, from earlier experiences of the subject, from teachers and from popular culture and the media. Here are a few myths and stereotypes:

- 'Maths is a difficult subject.'
- 'Only some people are truly good at maths.'
- 'If I get things wrong, it means I'm useless at maths.'
- 'Some people have a maths brain and others are better at creative/language subjects.'
- 'Maths is a boys' subject.'
- 'If I ask a question, everyone will know I don't understand. They might think I'm stupid.'
- 'Smart kids get everything right.'

None of these myths are true. Encourage the children to question them by focusing on the process of learning and the importance of asking questions and making mistakes.

Answers for Pupil book 5 page 7

1 Individual answers, for example:

a Intelligence can grow as long as you work hard.

b You should keep trying when things are hard because with perseverance you will succeed.

c Challenging problems help you to grow your knowledge about maths.

d Making mistakes is important because your brain grows as a result.

e The process of learning is more important than the result because it helps your brain make connections.

f Sharing your thinking helps you to see other ways of solving a problem.

End-of-unit check

Give the children a table of fixed mindset statements and ask them to replace each one with a more positive, growth mindset statement. Here is one set of fixed mindset statements you could start with. You can find many more online.

Instead of this …	I can say this …
I can't do any better.	I can improve my understanding.
I just don't understand this.	I can solve simple problems and then gradually understand more difficult problems.
This is too difficult for me.	If I keep trying, this will get easier.
I give up. I'll never be able to do this.	I will have another go.
I got some wrong. That means I'm not good at this.	I got some questions right, so I am beginning to understand this.

Number and place value

Learning objectives

- Read, write, order and compare numbers to at least 1 000 000 and determine the value of each digit
- Compose, decompose and regroup numbers using standard and non-standard partitioning
- Count forwards or backwards in steps of powers of 10 for any given number up to 1 000 000
- Round any number up to 1 000 000 to the nearest 10, 100, 1000, 10 000 and 100 000
- Understand what is meant by a number sequence, find term-to-term rules and work out missing terms in number sequences
- Read Roman numerals to 1000 (M) and recognise years written in Roman numerals

Key words

place value   million   round   powers of 10
sequence   ascending   descending   term
term-to-term rule

Unit introduction

Teaching guidance

Display three rows of blank digit cards, like the ones shown here. Ask the children to copy them into their books.

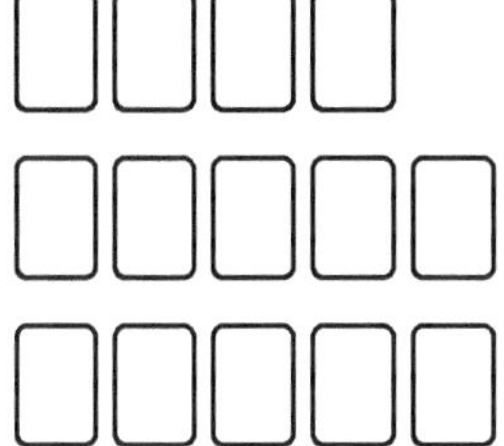

Tell the class that you are going to give them clues and that they are going to use these to work out what each number is. Emphasise that you will read out each clue once only, so they need to listen carefully.

Start with a 4-digit number. Give the class four clues, one for each digit. For example, for the number 9706, the clues could be:
- *Divide the number of hours in a day by 4. Write the answer in the ones place.*
- *The digit in the thousands place is 100 times smaller than 900.*
- *Halve the number of days in two weeks to find the digit in the hundreds place.*
- *Subtract the hundreds digit from the thousands digit. The digit in the tens place is 2 less than that.*

Once the children have identified the number, ask them to find a different way of representing it. For example, they could use a part–whole model, expanded notation or write it in words.

Repeat the process for a 5-digit number. Here are some examples of clues – adapt them for your number or think of your own.
- *Find the third/smallest odd number. Write it in the … place.*
- *Find the sum of the digits in the … and … place, then divide the answer by 2.*
- *Subtract 2 from the digit in the … place.*
- *Find the product of 287 × 3 × 0.*
- *Add 3 to the difference between 11 and 9.*
- *Subtract 10 from the number of days in two weeks.*
- *Divide 400 by 100 to get the digit in the … place.*
- *The remaining digit is 14 less than the sum of the digits in the … and … place.*

Then ask the children to choose another 5-digit number and to write their own set of challenging clues. They can then exchange their clues and try to work out each other's numbers.

Numbers above 9999

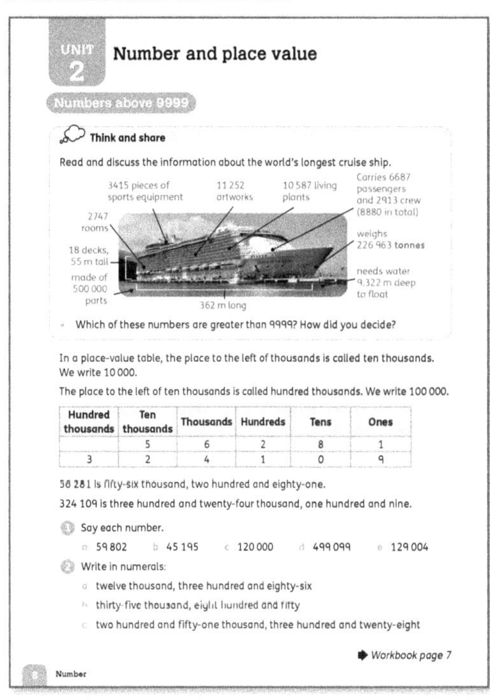

Materials

Place-value table with columns for 10 000s, 1000s, 100s, 10s and 1s; place-value cards (ten thousands, thousands, hundreds, tens and ones, page 21) or 0–9 digit cards

Warm-up

As a mental warm-up activity, do some of the Counting backwards and forwards' activities from the 'Place value and number sense' section of the Activity bank (page 22).

Focus

- <u>Think and share:</u> Revise the concept of *place value* with the class by working through the question about the cruise ship on **Pupil book 5 page 8**. Give the children time to discuss the photograph and to say the numbers.
- Check that they can identify the numbers that are greater than 9999 (11 252 artworks, 10 587 living plants, 226 963 tonnes, 500 000 parts) and let them share how they did this.
- Read the information and talk about the extended place-value table. Make sure the children understand that *place* refers to the position of any digit in a number and that *value* refers to how much the digit is worth in the number.
- Ask questions to make sure the children understand how the place-value table works and are able to partition numbers and write them in expanded form.
- Partitioning means saying how many of each place value there are in the number. (For example: 12 352 = 1 ten thousand, 2 thousands, 3 hundreds, 5 tens and 2 ones.)
- Expanded form means writing the partitioned number as multiples of each place value. (For example: 12 352 = (1 × 10 000) + (2 × 1000) + (3 × 100) + (5 × 10) + (2 × 1).)
- Remind the class that place value allows them to write any number using just the digits from 0 to 9.
- Let the children read through the numbers in question 1 before asking some children to each say a number aloud.
- Let the children complete question 2 in pairs if they need support. Check that they have written the numbers correctly and that they can identify the number in the ten thousands place.

Follow-up

Use **Workbook 5 page 7** to consolidate understanding of place value in numbers with up to six digits. The children should be able to complete question 1 independently. Let the children work in pairs to complete question 2 and question 3 and then have them work with another pair to assess each other's work.

Challenge

Let the children take part in a place-value scavenger hunt. Ask them to find large numbers in the media (or online). Ask them to list the numbers they find and write how each number is used. (For example: 8849 m – the height of Mount Everest; 280 000 – the number of passengers passing through Dubai International Airport in one day; 14 181 – the number of alligators and crocodiles kept as pets in Canada.)

Support

Give the children a set of place-value cards or several sets of 0–9 digit cards. Say several numbers up to 99 999 and ask the children to lay out the cards to show that number. Ask them to say the number once they have made it to reinforce number names.

Interesting mistakes

- Some children may say the digits of the number in order instead of the correct number name. (For example, they may say 'five four three' instead of 'five hundred and forty-three'.) Point this out and encourage them to say numbers correctly by identifying the value of each digit and saying the numbers in expanded form.
- Some children find reading numbers and writing a number they have heard difficult. Provide lots of practice to reinforce the idea that we read numbers in groups of three from left to right. This can be confusing if the children are used to working with numbers by looking at the digits in order of increasing place value (ones, then tens, then hundreds, etc.). Encourage children to add by partitioning the numbers and adding the larger place values first. (For example: 234 + 435 = 200 + 400 + 30 + 30 + 4 + 5 = 600 + 60 + 9 = 669). This method can be more helpful than column algorithms for children who are still trying to learn how to read and write numbers.

Answers for Pupil book 5 page 8

1 a fifty-nine thousand, eight hundred and two
b forty-five thousand, one hundred and ninety-five
c one hundred and twenty thousand
d four hundred and ninety-nine thousand and ninety-nine
e one hundred and twenty-nine thousand and four

2 a 12 386   b 35 850   c 251 328

Answers for Workbook 5 page 7

1 12 350: one TTh, two Th, three H, five T, no O
58 016: five TTh, eight Th, no H, one T, six O
22 483: two TTh, two Th, four H, eight T, three O
18 095: one TTh, eight Th, no H, nine T, five O
123 407: one HTh, two TTh, three Th, four H, no T, seven O
908 400: nine HTh, no TTh, eight Th, four H, no T, no O
325 125: three HTh, two TTh, five Th, one H, two T, five O

2 257 995, 876 125, 127 555
899 245, 899 765, 465 255, 350 425, 250 105, 525 000, 499 300, 552 475, 908 544

3 Individual answers.

What is a million?

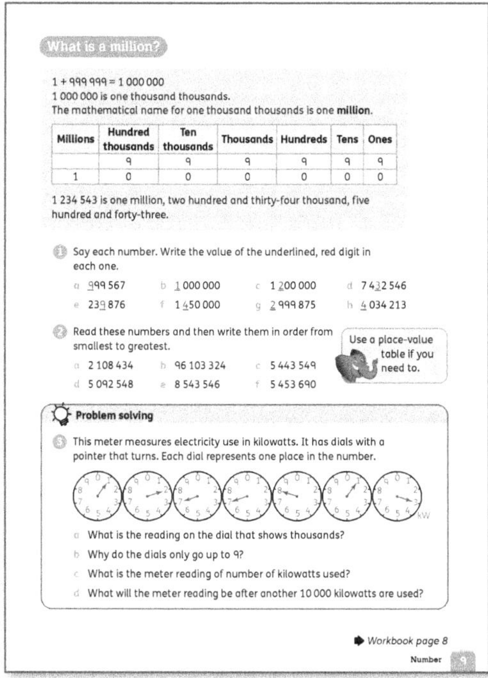

80 and 7 to make 254 487. Include some examples with nothing covered in one or more rows so the children have to say numbers that include 0 as a place holder.

- It is important for the children to say the number (for example: *two hundred and fifty-four thousand, four hundred and eighty-seven*) so that they link place value with the counting system.
- Once you have covered a few numbers and let the children say them, you can say a number and ask different children to come up and stick cards onto the table to represent the number.
- Turn to **Workbook 5 page 8** and let the children work through the questions to check that they are confident working with numbers up to 999 999.
- Read through the information at the top of **Pupil book 5 page 9** to consolidate the ideas before asking the children to complete questions 1–3.
- The children can do question 1 orally in small groups before writing the value of the underlined digit.
- <u>Problem solving:</u> Read through question 3 with the class and make sure they know that the dials record how much electricity is used. If possible, show them the school electricity meter or a photograph of it. Note that modern meters may be digital.

Challenge

- Draw a row of six or seven empty boxes in a row on the board. Ask the children to make a copy of this on paper.
- Pick a card from a set of digit cards and ask the children to select a box to write the number. Then return the card to the pack and shuffle. Repeat until the children have filled all the boxes.
- Once the children have placed a digit, they cannot move it. The aim is to make the largest possible number when all the boxes have been filled.
- Discuss with the children the strategies they used to make decisions about where to place digits.
- Repeat, but this time the children try to make the smallest number.
- Ask the children to write each of these numbers in numerals:
  nine hundred and twenty thousand and seventy-three
  one million, two hundred and seven thousand, four hundred and forty-one
  five million, two hundred thousand and fifty
- Ask the children questions in context, such as: *The reading on an electricity meter is 56 432. What will the reading be after another 10 000 units are used? (66 432) Another 1000? (57 432) Another 15 000? (71 432)*

Materials

Large place-value table (page 21); six cards with adhesive tack attached to cover numbers on the place-value table; place-value cards; sets of 0–9 digit cards

Warm-up

Select any place value or counting activity from the 'Place value and number sense' section of the Activity bank (pages 22–23) as a starter for this lesson.

Focus

- Ask the children to say what they think a *million* is and let them give examples of numbers in the millions.
- Count on beyond 100 000 and then extend the place-value table into the millions. Point out that 10 hundred thousands make 1 000 000. Count on in hundred thousands to teach the vocabulary.
- Make sure the children understand that 6-digit numbers have six places: hundred thousands, ten thousands, thousands, hundreds, tens and ones. Demonstrate the value of each digit using place-value cards and/or other concrete apparatus.
- Display a table like the one below.
- Use cards to cover a number in each row of the table and have the children say the numbers aloud. For example, you could cover 200 000, 50 000, 4000, 400,

100 000	200 000	300 000	400 000	500 000	600 000	700 000	800 000	900 000
10 000	20 000	30 000	40 000	50 000	60 000	70 000	80 000	90 000
1000	2000	3000	4000	5000	6000	7000	8000	9000
100	200	300	400	500	600	700	800	900
10	20	30	40	50	60	70	80	90
1	2	3	4	5	6	7	8	9

Support

- Use place-value cards to make an 'incomplete' number. For example: display 900 000, 20 000, 1000, 40 and 3. Ask the children what card is missing for the number *nine hundred and twenty-one thousand, three hundred and forty-three.* (Answer: 300)
- Repeat this for several numbers with five or more digits to reinforce place value in numbers with ten thousands and hundred thousands.

Interesting mistakes

Some children may need additional support to write and/or say numbers in which zero is a place holder (for example, 234 4056, 1 200 304, etc.). Spend time asking the children to enter numbers into a place-value table and to say the numbers in expanded form, emphasising where there is a zero in a column and what it means.

Answers for Pupil book 5 page 9

1 **a** nine hundred thousand
b one million
c two hundred thousand
d thirty thousand
e nine thousand
f four hundred thousand
g two million
h four million

2 a, d, c, f, e, b: 1 108 534, 5 092 548, 5 443 549, 5 453 690, 8 543 546, 9 6103 324

3 **a** 7
b When a dial reaches 0, the dial to its left moves on one.
c 1 277 813 kW
d 1 287 813 kW

Answers for Workbook 5 page 8

1 **a** six hundred **b** nine
c sixty **d** four hundred
e fifty thousand **f** fifty thousand
g eighty **h** seven hundred thousand

2

| g | d | a | | e | f | b | h | | c |

0 ... 500 000 ... 1 000 000

Represent numbers in different ways

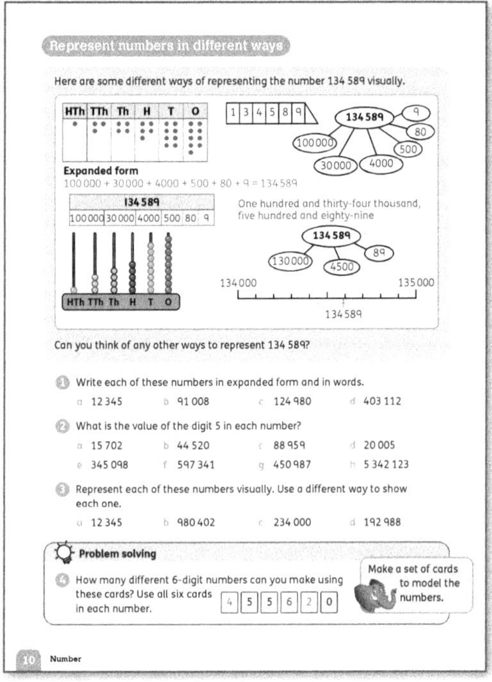

Materials

Poster paper and marker pens; card and scissors; number lines for support

Warm-up

- Let the children work in pairs to read through and then discuss the different ways of representing numbers shown at the top of **Pupil book 5 page 10**. Most of these representations should be familiar to them, but you may need to revise the terms *partitioning* and *expanded form*.
- Let the children share other methods they know of representing numbers (for example, using place-value counters, or cubes, rods and blocks).

Focus

- The children can work on their own to complete question 1. Let them check each other's work.
- Do question 2 orally, letting the children take turns to give the values.
- Let the children discuss question 3 and then work in groups to represent the numbers on a large sheet of poster paper.
- Problem solving: If the children need to, let them use card and scissors to make their own digit cards to model the problem in question 4.

Support

- Help the children to locate large numbers on a number line. Point out the start and end points and then work systematically through the smaller divisions.
- Provide a number of different number lines for additional practice. There are many interactive versions available online.
- The children may also need support with non-standard partitioning (for example, partitioning

134 589 as 130 000, 4500 and 89), as they may not see the value of breaking up a number in different ways. It may make more sense to the children to use non-standard partitioning in the context of comparing or finding the sum or difference between two numbers.

Answers for Pupil book 5 page 10

1 a 10 000 + 2000 + 300 + 40 + 5 = 12 345;
   twelve thousand, three hundred and forty-five
b 90 000 + 1000 + 8 = 91 008;
   ninety-one thousand and eight
c 100 000 + 20 000 + 4000 + 900 + 80 = 124 980;
   one hundred and twenty-four thousand, nine hundred and eighty
d 400 000 + 3000 + 100 + 10 + 2 = 403 112;
   four hundred and three thousand, one hundred and twelve

2 a five thousand   **b** five hundred
   c fifty   **d** five
   e five thousand   **f** five hundred thousand
   g fifty thousand   **h** five million

3 a–d Individual answers, for example using part-whole models, number lines, abacuses and expanded form.

4 Note for teachers: Students will probably use trial and improvement to solve this problem. The short way is to use permutations.
There are five different choices for the first digit, then five choices for the second digit, then four choices for the third etc, so the number of permutations of 6 digits is 5 × 5 × 4 × 3 × 2 × 1 = 600.
However, since 5 is repeated, the number of permutations is 600 ÷ 2 = 300.

Materials
A set of 0–9 digit cards for each child; beads or counters in different colours to model abacuses (if needed)

Warm-up
- As a mental warm-up activity, ask the children to work in pairs. They each shuffle their digit cards and place five cards on their desk. They then turn over their cards and use them to make a 5-digit number.
- The pairs then compare their numbers and the child who made the greater number gets a point.
- They repeat this five times, shuffling and dealing new cards each time.

Focus
- Ask the children to look at the four representations of numbers at the top of **Pupil book 5 page 11**. Once they have decided which is the greatest number (D 12 493), let them share their reasoning with the class.
- Work through the examples of using place value to compare numbers. The children have compared numbers before so the aim here is to show them how to develop this skill to compare numbers with more places.
- Revise the use of the < and > symbols, then let the children work independently to complete question 1. Check the answers as a class and discuss any interesting mistakes.

Answers for Pupil book 5 page 11

1 a <   **b** >   **c** <
   d <   **e** <   **f** >
   g <   **h** <   **i** <

Compare numbers

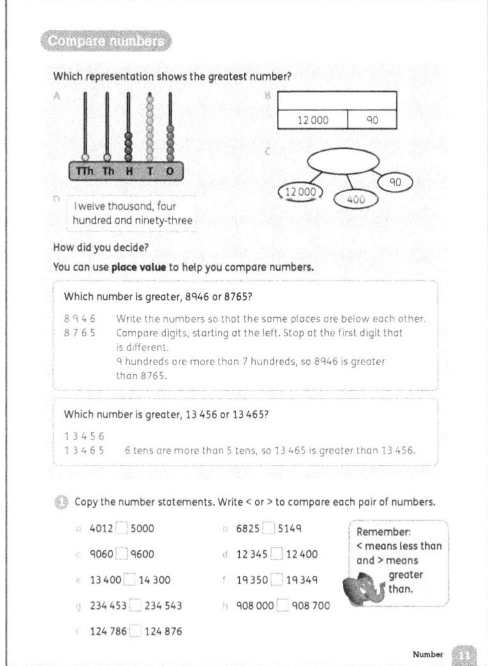

Compare and order numbers

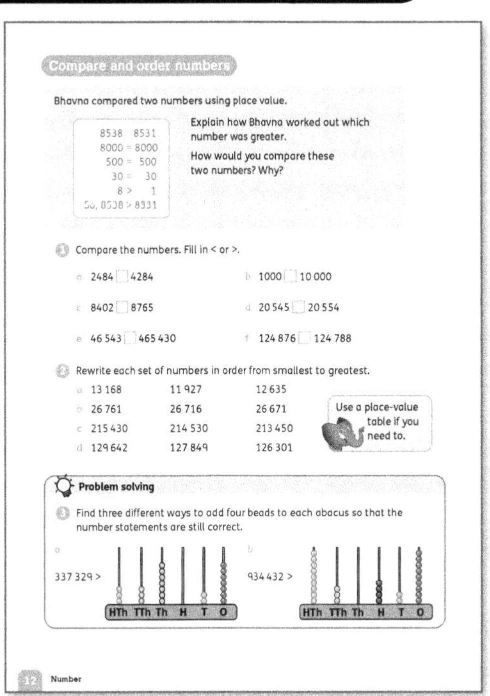

Warm-up

Use the 'Compare 3-digit numbers' or 'Compare 5- and 6-digit numbers' activity in 'Place value and number sense' (page 23) as a mental starter for this lesson.

Focus

- Let the children work in pairs to discuss the example at the top of **Pupil book 5 page 12** and to say how they would do the same comparison. The question asks the children to explain how Bhavna worked out which number was greater. (Bhavna compared the place values of each digit in order from the thousands to the ones. Other possible ways of comparing the numbers include showing them on a number line.)
- They can then complete question 1 and question 2 on their own as consolidation.
- <u>Problem solving:</u> Let the children work in pairs to complete question 3. They can model the numbers using beads or counters if they need to.

Support

Some children may need additional support to order a set of numbers. Give these children some blank cards and a blank number line. The children can copy the numbers in each set onto the cards and move the cards around to order them on the number line.

Interesting mistakes

The children may mix up the < and > signs. Some modern research suggests that children may understand the meaning of the signs if they think about the mathematical development of these signs rather than trying to remember 'crocodile mouths' or other methods.

- Point out that the equals sign is made of two lines that are the same distance apart because the numbers (or values) on either side of the sign are equal (for example, 23 = 23).
- We can also think of the < sign as two lines. There is more space between the lines on the right because the number to the right is greater than the one to the left (for example, 12 < 13).
- Similarly, the > sign has more space between the lines on the left as the number to the left is greater (for example, 13 > 12).

Answers for Pupil book 5 page 12

1 a < b < c <
d < e < f >

2 a 11 927, 12 635, 13 168
b 26 671, 26 716, 26 761
c 213 450, 214 530, 215 430
d 126 301, 127 849, 129 642

3 Individual answers, for example:
First image: shows 337 329 > 337 027
You could add *up to* three beads to the 'H' and one to the 'Ones'.
Second image: shows 934 432 > 930 429
You could add *up to* four beads to the 'Th' and up to three to the 'Ones'.

Round numbers

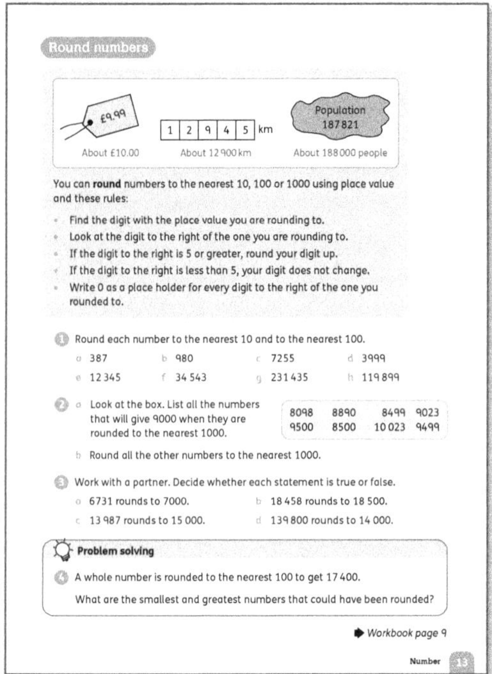

Warm-up

Use one of the 'Place value and number sense' activities from the Activity bank (pages 22–23) as a mental warm-up.

Focus

Use a number talk (pages 17–18) to revise and discuss the concept of *rounding* numbers. Give the class the following problem to consider:

> Maria and Zara each pick three numbers from a box without looking at them.
>
> - Maria picks 80, 120 and 170.
> - Zara picks 90, 140 and 110.
>
> If a number is closer to 100 than to 200, it scores 5 points.
>
> If a number is closer to 200 than to 100, it scores 10 points.
>
> *What is each girl's score?* (Maria scores 20; Zara scores 15)
>
> *How did you decide?* (Possible answer: Find the difference between the number and 100 and between the number and 200, then compare the answers to find the smaller difference.)
>
> *What is the score for 150? Why?* (10 points because you round up 50 to the next 100.)

- Encourage the children to think about the problem and let them indicate when they have an answer to share. Spend some time talking about how they decided and what they would do with 150 to make sure they remember that the digit 5 in the rounding position means round up.

- Use the examples and explanation at the top of **Pupil book 5 page 13** to reinforce the rules for rounding to different places. This lesson focuses on rounding to 10, 100 and 1000. The children will round to other places in the next lesson.
- Let the children complete question 1 and question 2. Check their work before you move on.
- The children work in pairs to discuss question 3. Ask them to explain how they decided if each statement is true or false. Then ask them to change the false statements to make them true.
- <u>Problem solving:</u> Use question 4 to assess whether the children can apply what they have learnt. If they can solve this problem, they have a good understanding of rounding.

Follow-up

- Ask the children to complete **Workbook 5 page 9** in class or as a homework activity to check that they can confidently round numbers to the nearest 10, 100 and 1000.
- Tell children: *Zef thinks that 3.009 is greater than 3.1 because 9 is greater than 1.* Ask them to convince you that Zef is not correct.
- Write the digits 1, 2, 4, 5 and 9 on the board. Ask children to use all of these digits to make the number with 3 decimal places that is closest to 30.

Support

Use number lines with children who need support with rounding numbers, so they can see whether a number is closer to one leading number or the next one. For example, on a 1000 to 2000 number line with 1500 marked halfway, they can see that 1495 is closer to 1000 than 2000. You do not need to mark all the intervals on the number lines, just key numbers. You can use blank number lines and label the intervals with the numbers you are working with.

Answers for Pupil book 5 page 13

1 **a** 380; 400   **b** 980; 1000   **c** 7260; 7300
   d 4000; 4000 **e** 12 350; 12 300 **f** 34 540; 34 500
   g 231 440; 231 400
   h 119 890; 119 900
2 **a** 8890, 9023, 8500, 9499
   b 8000, 8000, 10 000, 10 000
3 **a** true   **b** true
   c false   **d** false
4 17 350; 17 449

Answers for Workbook 5 page 9 

1 253, 252, 249, 248, 245
2 1467, 1456, 1483, 1532, 1522
3

Number	To the nearest 10	To the nearest 100	To the nearest 1000
1369	1370	1400	1000
3481	3480	3500	3000
11 402	11 400	11 400	11 000
18 492	18 490	18 500	18 000
26 445	26 450	26 400	26 000
42 359	42 360	42 400	42 000
41 385	41 390	41 400	41 000
88 884	88 880	88 900	89 000
89 912	89 910	89 900	90 000
87 783	87 780	87 800	88 000

4 **a** 1295, 1304
   b 1250, 1349
   c Individual answers.

More rounding

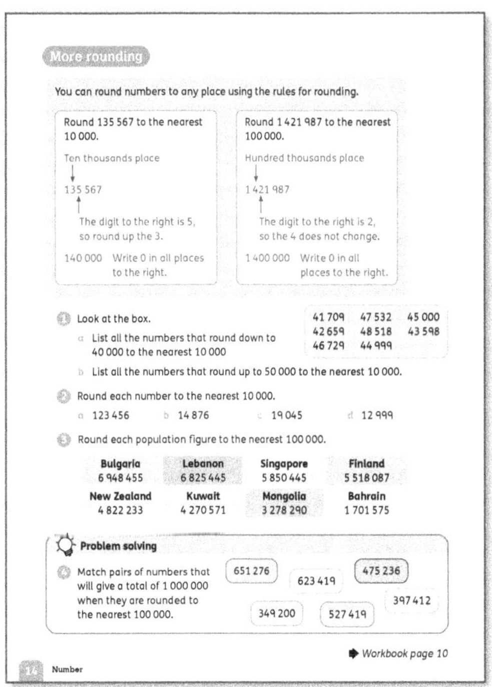

You can round numbers to any place using the rules for rounding.

Round 135 567 to the nearest 10 000.

Ten thousands place

135 567

The digit to the right is 5, so round up the 3.

140 000   Write 0 in all places to the right.

Round 1 421 987 to the nearest 100 000.

Hundred thousands place

1 421 987

The digit to the right is 2, so the 4 does not change.

1 400 000   Write 0 in all places to the right.

1 Look at the box.

41 709   47 532   45 000
42 659   48 518   43 598
46 729   44 999

  a List all the numbers that round down to 40 000 to the nearest 10 000
  b List all the numbers that round up to 50 000 to the nearest 10 000.

2 Round each number to the nearest 10 000.

  a 123 456   **b** 14 876   **c** 19 045   **d** 12 999

3 Round each population figure to the nearest 100 000.

Bulgaria	Lebanon	Singapore	Finland
6 948 455	6 825 445	5 850 445	5 518 087
New Zealand	**Kuwait**	**Mongolia**	**Bahrain**
4 822 233	4 270 571	3 278 290	1 701 575

Problem solving

4 Match pairs of numbers that will give a total of 1 000 000 when they are rounded to the nearest 100 000.

651 276   623 419   475 236   349 200   527 419   397 412

➡ *Workbook page 10*

14   Number

Materials

Atlases or a world map

Warm-up

Use one of the 'Place value and number sense' activities from the Activity bank (pages 22–23) as a mental warm-up.

Focus

This lesson builds on the previous lesson. Explain that you can use the rules for rounding to round numbers to any place (including decimal places). Work through the two examples at the top of **Pupil book 5 page 14** with the class.

- The children can do question 1 independently.
- Remind the children that if they are rounding to the ten thousands place, they need to look at the digit in the thousands place. Give them time to write the answers to question 2 and then check these as a class.
- For question 3, display a large world map or direct the children to use atlases. Have them find all the countries mentioned in the table. Then ask them to round the numbers to the nearest hundred thousand.
- <u>Problem solving:</u> Let the children try to solve question 4 on their own. They can then share how they approached this problem with the class.

Follow-up

Use **Workbook 5 page 10** as an informal assessment task. Let the children work independently to complete the rounding activities. Share the answers with the class and let the children check their own work.

Discuss possible answers to question 1b. Ask each child to write a statement assessing their own ability to round numbers and a suggestion for how to improve if they think they need to do so.

Challenge

Let the children do a survey to find out how and when people use rounded numbers and estimated values at home and at work. People who work with measurements are often very good at estimating. For example, a builder may be able to look at a load of bricks or tiles and estimate how many there are. The children can prepare a short talk or video for the class about how people use rounding and estimating.

Answers for Pupil book 5 page 14

1 a 41 709, 42 659, 43 598, 44 999
   b 47 532, 45 000, 48 518, 46 729

2 a 120 000   b 10 000   c 20 000   d 10 000

3 Bulgaria   7 000 000;   Lebanon   6 800 000;
   Singapore   5 900 000;   Finland   5 500 000;
   New Zealand 4 800 000;   Kuwait   4 300 000;
   Mongolia   3 300 000;   Bahrain   1 700 000

4 623 419 and 394 200; 623 419 and 397 412;
   475 236 and 527 419

Answers for Workbook 5 page 10

1 a 10 000: 12 345, 9800
      20 000: 22 630, 17 908
      30 000: 28 636, 31 258
      40 000: 42 987, 38 500
      50 000: 49 823, 54 599
      60 000: 61 298, 58 323
      70 000: 69 555, 70 950
      80 000: 83 500, 75 000
      90 000: 89 765, 91 600
   b Individual answers.

2 250 000 and 349 999; 750 000 and 849 999;
   1 150 000 and 1 249 999; 3 550 000 and
   3 649 999; 1 850 000 and 1 949 999

Number sequences and rules

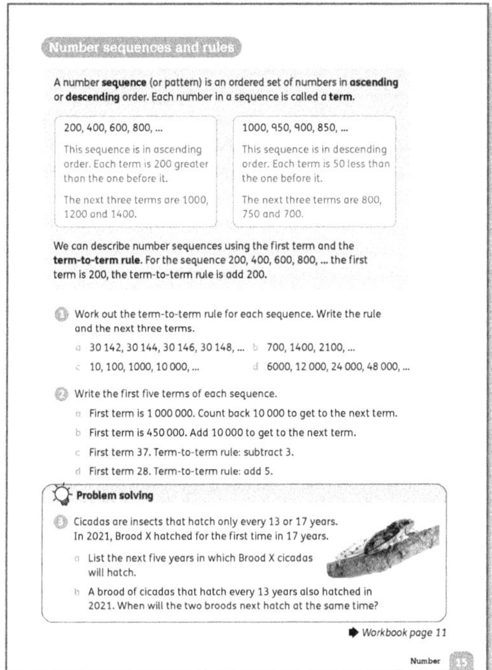

Materials

Number chart; calculators (if needed)

Warm-up

- Do some different skip-counting activities as a mental starter for this unit. Vary the counting interval and the starting numbers. Include counting forwards and backwards.
- Alternatively, display any number chart and play games in which the children have to find numbers by counting backwards and/or forwards in given amounts.

Focus

- Let the children experiment with drawing increasing or decreasing sequences of shape patterns. Spend some time discussing how many shapes are needed to make each new pattern in the sequence. Introduce the idea of rules for sequences and encourage the children to express these rules mathematically as far as possible.
- Use the children's sequences to revise the idea that each pattern in the sequence is a *term*, and to

make sure that the children understand the words *ascending* and *descending*.

- Turn to **Pupil book 5 page 15** and work through the examples with the class. Point out that the *term-to-term rule* is based on the difference between terms and tells us what we have to do to get to the next term.
- Let the children work in pairs, using a calculator if necessary, to describe and continue the number *sequences* in questions 1 and 2.
- <u>Problem solving</u>: Encourage the children to find out more about cicadas. Note that only some species of cicada have long intervals between hatching times; others breed annually. The 17-year cicadas in the USA hatched in 2021. The children can search online to find out more. Question 3b requires them to apply what they know about common multiples to find the answer. Listing the years is not an efficient strategy here.

Follow-up

Use **Workbook 5 page 11** as an additional activity to consolidate completing number sequences and following rules to build number sequences.

Challenge

You can find a range of games involving sequences and missing numbers online. Search for number sequences and term-to-term rules at an appropriate level.

Support

The focus here is on sequences and the difference between terms, rather than calculation skills, so it makes sense to allow the children to use a calculator to check their work and to generate sequences of their own.

Answers for Pupil book 5 page 15

1 a add 2; 30 150, 30 152, 30 154
   b add 700; 2800, 3500, 4200
   c multiply by 10: 100 000, 1 000 000, 10 000 000
   d multiply by 2: 96 000, 192 000, 384 000

2 a 1 000 000, 990 000, 980 000, 970 000, 960 000
   b 450 000, 460 000, 470 000, 480 000, 490 000
   c 37, 34, 31, 28, 25
   d 28, 33, 38, 43, 48

3 a 2038, 2055, 2072, 2089, 2106
   b 2242; the lowest common multiple of 13 and
      17 is 13 × 17 = 221.

   So the first time they both hatch again in the
   same year is 2021 + 221 = 2244

Answers for Workbook 5 page 11

1 a 590, 540, **490**, **440**, 390, **340**
   b 10, 20, 40, **80**, **160**, 320, **640**
   c 5157, **5147**, **5137**, 5127, 5117
   d 1093, **1085**, 1077, 1069, 1061, **1053**, **1045**
   e 4122, 4223, 4324, **4425**, **4526**, **4627**

2 1024 → 1014 → 1004 → 994 → 984 → 1084 → 1184
   → 1284 → 1384 → 2384 → 3384 → 4384 → 5384
   → 5284 → 5184 → 5084 → 4984 → 4884

Powers of 10

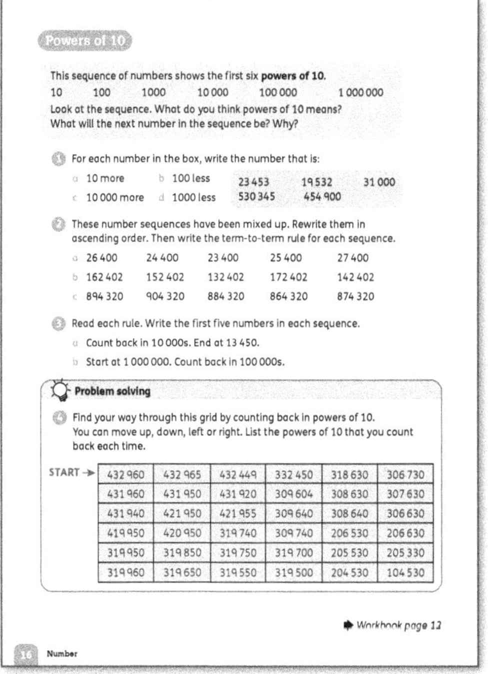

Powers of 10

This sequence of numbers shows the first six **powers of 10.**

10   100   1000   10 000   100 000   1 000 000

Look at the sequence. What do you think powers of 10 means?
What will the next number in the sequence be? Why?

1 For each number in the box, write the number that is:
   a 10 more   b 100 less
   c 10 000 more   d 1000 less

   23 453   19 532   31 000
   530 345   454 900

2 These number sequences have been mixed up. Rewrite them in ascending order. Then write the term-to-term rule for each sequence.
   a 26 400   24 400   23 400   25 400   27 400
   b 162 402   152 402   132 402   172 402   142 402
   c 894 320   904 320   884 320   864 320   874 320

3 Read each rule. Write the first five numbers in each sequence.
   a Count back in 10 000s. End at 13 450.
   b Start at 1 000 000. Count back in 100 000s.

Problem solving

4 Find your way through this grid by counting back in powers of 10. You can move up, down, left or right. List the powers of 10 that you count back each time.

START →					
432 960	432 965	432 449	332 450	318 630	306 730
431 960	431 950	431 920	309 604	308 630	307 630
431 940	421 950	421 955	309 640	308 640	306 630
419 950	420 950	319 740	309 740	206 530	206 630
319 950	319 850	319 750	319 700	205 530	205 330
319 960	319 650	319 550	319 500	204 530	104 530

➡ *Workbook page 12*

16 Number

Materials

Small pieces of paper or card (if needed); small counters

Warm-up

Select any of the activities 'Count forwards and backwards in tens, hundreds and thousands' from the 'Place value and number sense' section of the Activity bank (page 23) as a mental starter for this lesson.

Focus

- Discuss the questions at the top of **Pupil book 5 page 16** with the class. Let them share their understanding of the term *powers of 10*. (The power of 10 is the number of tens multiplied together in a product. The next number of the sequence is 10 000 000 because it is 1 000 000 × 10 or 10 × 10 × 10 × 10 × 10 × 10 × 10.) If you think it is appropriate, you can introduce the notation 10^1, 10^2, 10^3 and so on, but the children do not need to use this notation at this stage.
- Discuss with the class how they will present their answers to question 1. Try to elicit that an efficient method would be to use a table.
- Allow any children who need additional support with question 2 to write the numbers on pieces of paper or card so that they can physically arrange them.
- For question 3, the children can work in pairs to discuss the rules before they write the answers.
- <u>Problem solving</u>: For question 4, let the children use a small counter to move from number to number so they can keep track without writing in their Pupil book.

Follow-up

Let the children work independently to complete **Workbook 5 page 12**. Observe them as they work to make sure they are able to do the questions. Discuss the strategies that the children use to add or subtract powers of 10 as a class.

Answers for Pupil book 5 page 16

1 a 23 463, 19 542, 31 010, 530 355, 454 910

   b 23 353, 19 432, 30 900, 530 245, 454 800

   c 33 453, 29 532, 41 000, 540 345, 464 900

   d 22 453, 18 532, 30 000, 529 345, 453 900

2 a 23 400, 24 400, 25 400, 26 400, 27 400; add 1000

   b 132 402, 142 402, 152 402, 162 402, 172 402; add 10 000

   c 864 320, 874 320, 884 320, 894 320, 904 320; add 10 000

3 a 53 450, 43 450, 33 450, 23 450, 13 450

   b 1 000 000, 900 000, 800 000, 700 000, 600 000

4 Start: 432 960, – 1000, 431 960, – 100, 431 950, – 10 000, 421 950, – 1000, 420 950, – 10 000, 419 950, – 100 000, 319 950, – 100, 319 850, – 100, 319 750, – 10, 319 740, – 10 000, 309 740, – 100, 309 640, – 1000, 308 640, – 10, 308 630, – 1000, 307 630, – 1000, 306 630, – 100 000, 206 630, – 100, 206 530, – 1000, 205 530, – 1000, 204 530, – 100 000, 104 530: End

Answers for Workbook 5 page 12

1 a 896, 996, 1096, 1196, 1296, 1396

   b 8900, 9900, 10 900, 11 900, 12 900, 13 900

   c 91 876, 81 876, 71 876, 61 876, 51 876, 41 876

   d 3699, 13 699, 23 699, 33 699, 43 699, 53 699

   e 12 087, 11 087, 10 087, 9087, 8087, 7087

   f 923 087, 913 087, 903 807, 893 807, 883 807, 873 807

2 a 20 546, 21 546, 22 546, 23 546, 24 546
   20 446, 21 446, 22 446, 23 446, 24 446
   20 346, 21 346, 22 346, 23 346, 24 346
   20 246, 21 246, 22 246, 23 246, 24 246
   20 146, 21 146, 22 146, 23 146, 24 146

   b Across: – 100 100; down: 1100
   445 778, 345 678, 245 578, 145 478, 45 378
   446 878, 346 778, 246 678, 146 578, 46 478
   447 978, 347 878, 247 778, 147 678, 47 578
   449 078, 348 978, 248 878, 148 778, 48 678
   450 178, 350 078, 249 978, 149 878, 49 778

Count in powers of 10

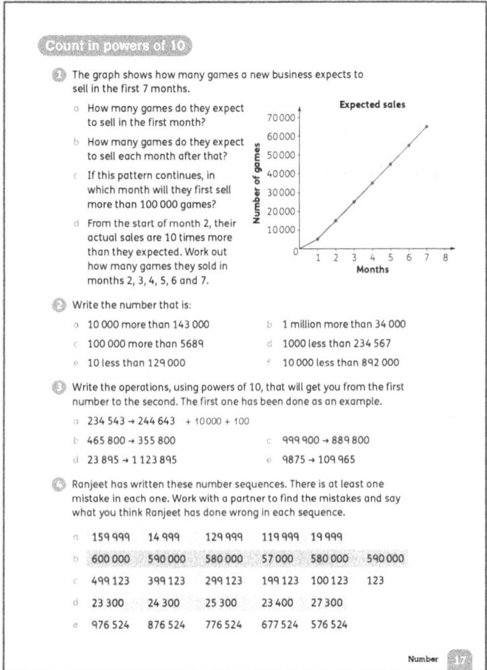

Materials
Number charts and counters; number lines (if needed)

Warm-up
Use 'Count forwards and backwards in tens, hundreds and thousands' from the 'Place value and number sense' section of the Activity bank (page 23) as a mental starter for this lesson.

Focus
Turn to **Pupil book 5 page 17** and discuss questions 1–4 before the children complete them independently or in pairs. You may need to revise how to read and interpret a line graph with the class.

Interesting mistakes
Bridging across barriers such as 100 or 1000 can cause difficulties when skip counting.

- For example, when counting backwards in hundreds from 1200, some children may say: '1200, 1100, 1900, 1800, 1700, ...' and then realise at some point that they have already said some of the numbers.
- Alternatively, the children may omit the barrier number (in this case, 1000) and say: '1200, 1100, 900, 800, ...'

Placing counters on number charts or drawing jumps on a number line can help the children see what happens when they reach the barrier number. Sustained practice will help them become more confident with this concept.

Answers for Pupil book 5 page 17

1 a 5000   b 10 000

c In the middle of month 11 (half-way between months 10 and 11)

d 150 000, 250 000, 350 000, 450 000, 550 000, 650 000

2 a 153 000   b 1 034 000   c 105 689

d 233 567   e 128 990   f 882 000

3 a Provided as an example.
(operations can be done in any order)

b − 10 000 − 100 000

c − 100 − 10 000 − 100 000

d + 100 000 + 1 000 000

e − 10 + 100 + 100 000

4 a The term-to-term rule is 'subtract 10 000'
14 999 should be 149 999
19 999 should be 109 999

b The term-to-term rule is 'subtract 10 000'
57 000 should be 570 000
The second 580 000 should be 560 000
The second 590 000 should be 550 000

c The term-to-term rule is 'subtract 100 000'
100 123 should be 99 123
123 should be − 877

d The term-to-term rule is 'add 1000'
23 400 should be 26 300

e The term-to-term rule is 'subtract 100 000'
677 524 should be 676 524

Roman numerals

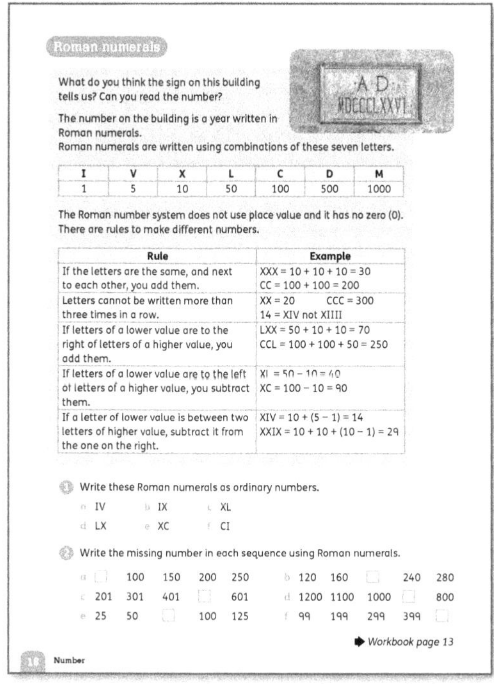

I	V	X	L	C	D	M
1	5	10	50	100	500	1000

Rule	Example
If the letters are the same, and next to each other, you add them.	XXX = 10 + 10 + 10 = 30 CC = 100 + 100 = 200
Letters cannot be written more than three times in a row.	XX = 20 CCC = 300 14 = XIV not XIIII
If letters of a lower value are to the right of letters of a higher value, you add them.	LXX = 50 + 10 + 10 = 70 CCL = 100 + 100 + 50 = 250
If letters of a lower value are to the left of letters of a higher value, you subtract them.	XI = 50 − 10 = 40 XC = 100 − 10 = 90
If a letter of lower value is between two letters of higher value, subtract it from the one on the right.	XIV = 10 + (5 − 1) = 14 XXIX = 10 + 10 + (10 − 1) = 29

Materials

Sets of flashcards and smaller cards with the Roman numbers I, V, X, L, C, D and M (one set for each group of pupils); photos of dates written in Roman numerals (optional)

Warm-up

Select one of the 'Compare 3-digit number' or 'Compare 5- and 6-digit numbers' activities from the 'Place value and number sense' section of the Activity bank (page 23) as a mental warm-up for this lesson.

Focus

Roman numerals to 100 were covered in Level 4.

- Remind the class that Roman numerals were the system of numbers used by ancient Romans and that they were first used about 3000 years ago.
- Remind the children that our number system is called the Hindu–Arabic system. This system uses digits from 0 to 9 and place value to write numbers. The Roman numeral system used combinations of seven letters (numerals) in different ways to write numbers. The Roman numeral system had no 0 and did not use place value.
- Use the flashcards to show the children how to build numbers using Roman numerals. Start with I, V and X. Explain that sometimes we need to add numerals and sometimes we need to subtract numerals to work out the number.
- Give the groups sets of cards and let them investigate how to build Roman numbers by applying the rules given on **Pupil book 5 page 18**. They are asked to say what the sign on the building in the picture means. (This gives the date when it was built: 1876 CE.)
- Once the children have explored the rules, let them complete question 1. For question 2, remind them that they need to work out the rule for each sequence to find the missing number.

Follow-up

Let the children work in pairs to complete the questions on **Workbook 5 page 13**. Allow them to refer back to the rules for building Roman numerals in the Pupil book if necessary. Once the children have completed the questions, you could let them find an online tool for converting between Roman and Hindu–Arabic numbers to check their work.

Answers for Pupil book 5 page 18

1 a 4   b 9   c 40

d 60   e 90   f 101

2 a L   b CC   c DI   d CM

e LXXV   f ID

Answers for Workbook 5 page 13

1 a XIII, XIV, XV, XVI

b XIX, XX, XL, XLIX, L, LX, LXX

c LXXI, LXXXI, LXXXII

d LXVII, LXXXVII, LXXXVIII

2 a <   b >   c <   d =

e <   f <   g =   h <

i >   j <

3 a LVI   b CXXIV   c LI

d MMXIV   e MXX

Work with Roman numerals

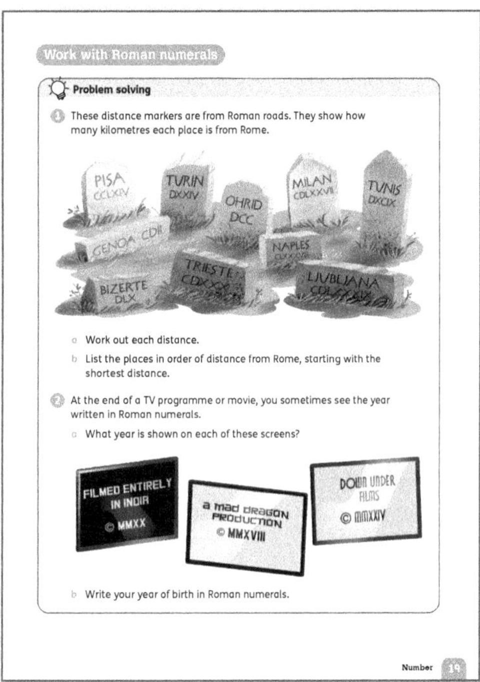

Materials

Atlases or large map showing Italy and the other countries around the Mediterranean Sea; pictures of the dates shown at the end of TV programmes or movies

Warm-up

As a mental starter for this lesson, display some Roman numbers such as:

X   XXIV   D   L   LX   DM   M

Ask the children to work out the Roman numeral that is 10 more, 10 less, 100 more and 100 less than each one.

Focus

Display the map or ask the children to use atlases. Turn to **Pupil book 5 page 19** and let the children find the cities shown on the distance markers. Most of the cities are in Italy, but Tunis and Bizerte are in Tunisia, Ohrid is in Macedonia and Ljubljana is in Slovenia.

- <u>Problem solving:</u> For question 1, let the children work on their own or in pairs to work out and order the distances.
- Show the children photos of dates written in Roman numerals, for example in the credits at the end of TV programmes or movies, if possible. Let them try to work out the answers to question 2 on their own.

Challenge

- Ask the children: *What is the biggest number you can write using the seven Roman numerals you have learnt?*
- Ask the children how to explore how to write bigger numbers in Roman numerals.

Interesting mistakes

The children may list the values of each letter separately or they may treat the letters as individual values rather than as a 'number'. For example, they may think that XIX means 10, 1 and 10 rather than representing the number 19.

- Remind them that they need to think of the 'letters' as digits that make up a number. (The Romans only had seven digits, not the ten that we have.) The Romans used different rules for making the numbers, not the base 10 place-value rules of our number system.
- Continue to provide support and allow the children to check the rules for making numbers as they work. You could display these in the classroom.

Answers for Pupil book 5 page 19

1 a Pisa: 264; Turin: 524; Milan: 477; Tunis: 599; Genoa: 402; Ohrid: 700; Naples: 188; Bizerte: 560; Trieste: 430; Ljubljana: 489

   b Naples, Pisa, Genoa, Trieste, Milan, Ljubljana, Turin, Bizerte, Tunis, Ohrid

2 a 2020, 2018, 2024

   b Individual answers.

Support

Give the children a Roman numeral jigsaw puzzle to solve. Use a grid of Roman numerals from 1 to 100 like the one shown at the bottom of the page. Cut out differently shaped pieces (for example, as shown by the different shading here). Ask the children to reassemble the grid and paste it into their books to use as reference.

I	II	III	IV	V	VI	VII	VIII	IX	X
XI	XII	XIII	XIV	XV	XVI	XVII	XVIII	XIX	XX
XXI	XXII	XXIII	XXIV	XXV	XXVI	XXVII	XXVIII	XXIX	XXX
XXXI	XXXII	XXXIII	XXXIV	XXXV	XXXVI	XXXVII	XXXVIII	XXXIX	XL
XLI	XLII	XLIII	XLIV	XLV	XLVI	XLVII	XLVIII	XLIX	L
LI	LII	LIII	LIV	LV	LVI	LVII	LVIII	LIX	LX
LXI	LXII	LXIII	LXIV	LXV	LXVI	LXVII	LXVIII	LXIX	LXX
LXXI	LXXII	LXXIII	LXXIV	LXXV	LXXVI	LXXVII	LXXVIII	LXXIX	LXXX
LXXXI	LXXXII	LXXXIII	LXXXIV	LXXXV	LXXXVI	LXXXVII	LXXXVIII	LXXXIX	XC
XCI	XCII	XCIII	XCIV	XCV	XCVI	XCVII	XCVIII	XCIX	C

Ask some or all of these questions to assess how well the children have understood the concepts in this unit.

- *What are the ten digits we use for representing all numbers?* (0, 1, 2, 3, 4, 5, 6, 7, 8, 9)
- *What makes it possible for us to represent large numbers, using only the 10 digits?* (Place value: the position of a digit determines its value.)
- *Put the correct sign, < or >, between these numbers.* (Give the children a pair of numbers.)
- *How do you order a set of numbers? Which part of each number do you look at to help you? Why?* (Look at the highest value place value first. If these are the same, look at the next highest place value and so on.)
- *What is the value of each digit in this number?* (Give the children some 5-, 6- and 7-digit numbers.)
- *How do you say this number?* (Write down some numbers in numerals for the children to say.)
- *Draw a part–whole model to show 1 234 509.* (part–whole model with whole 1 234 509 and parts 1 000 000, 200 000, 30 000, 4000, 500 and 9)
- *Write this number in expanded notation.* (Give the children some numbers.)
- *What is wrong with this expanded notation: 3 445 098 = 3 000 000 + 400 000 + 50 000 + 90 + 8?* (40 000 is missing and 50 000 should be 5000)
- *What rule was used to make this sequence?* (Provide some different number sequences.)
- *Is there any other way of carrying on this sequence? How?*
- *What comes next in this sequence? Why?*
- *Write these numbers in order of increasing size: 230 954, 230 949, 230 994, 230 998, 230 940.* (230 940, 230 949, 230 954, 230 994, 230 998)
- *What is 10 more/10 less/100 more/100 less/1000 more/1000 less than (a given number)?*
- *What is 2375 rounded to the nearest 10?* (2380)
- *What is 423 544 rounded to the nearest 1000?* (424 000)
- *Write these numbers using Roman numerals: 100, 250, 400, 512, 900, 930.* (C, CCL, CD, DXII, CM, CMXXX)
- *What number is represented by each Roman numeral?* (Give some Roman numerals with values between 1 and 1000.)
- *Find three Roman numbers that are the same front to back and back to front.* (For example: XXX, CXC, MCM.)
- *Naresh says that DLX and DXL are the same number because Roman numerals don't use place value. What mistake has he made?* (He has not noticed that the order of the letters is different. A smaller value written before a larger value means that the smaller value is subtracted from the larger value. DLX = 560 and DXL = 540)

UNIT 3 Properties of shapes

Learning objectives

- Revise the names and properties of polygons
- Use side lengths and angle sizes to classify polygons as regular or irregular
- Identify 3D shapes including cubes and cuboids from 2D representations (nets)
- Identify, describe and sketch 3D shapes in different orientations
- Know that angles are measured in degrees: estimate and compare acute, obtuse and reflex angles
- Compare angles, estimate and measure angles in degrees and draw angles of a given size
- Know that angles at a point total one whole turn (360°) and that angles at a point on a straight line total half a turn (180°), and use these facts to calculate missing angles
- Identify other multiples of 90°
- Recognise and describe spatial patterns and solve related problems
- Recognise the use of symbols to represent unknown quantities in calculations

Key words

polygon   vertex, vertices   trapezium   heptagon   nonagon   net   angle   acute angle   obtuse angle   reflex angle   scalene triangle   isosceles triangle   equilateral triangle   square number

Unit introduction

Materials

Large sheet of paper or card; pencils; crayons and marker pens

Teaching guidance

Put the children in groups. Explain that they are going to make an A–Z poster of maths vocabulary used to talk about shapes and their properties. Give each group the materials they will need.

- Give the groups time to choose the words they will include on their poster. They should also include diagrams or definitions to show what the words mean.
- Note that they may need support to find words for some of the more difficult letters. Here are some ideas:
  G – geometry, geometric shape or gradient
  M – measure, middle or model
  U – undecagon (an 11-sided polygon)
  W – wedge
  XYZ – labels for the vertices of a triangle
  WXYZ – labels for the vertices of a quadrilateral.
  Note that there are no appropriate words for J.
- Have a class discussion about what to do if the children cannot find a word for a letter.

Revisit polygons

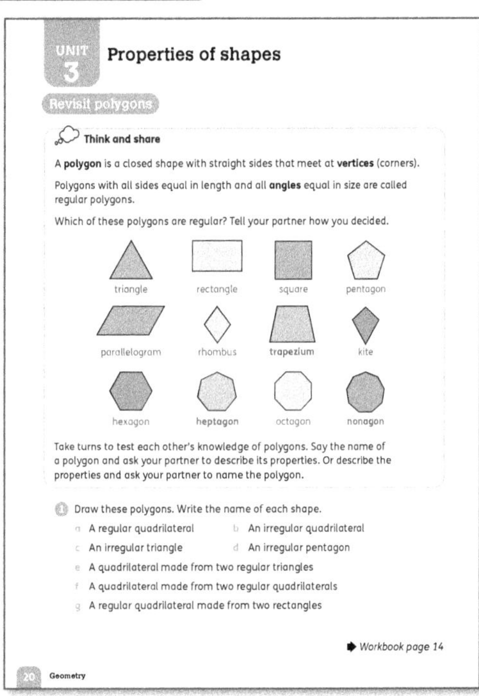

Materials

Pictures of polygons in different orientations; rulers, pencils, shape stencils; squared paper (if needed); geoboards and elastic bands (if available)

Warm-up

Select activities that reinforce number facts and mental calculation skills from the Activity bank (pages 22–28). The children will not be doing any calculations for a few lessons.

Focus

- <u>Think and share:</u> Turn to **Pupil book 5 page 20**. Let the children work in pairs to revise the names and properties of *polygons* (such as *trapezium, heptagon* and *nonagon*. All of the shapes given are regular

apart from the rectangle, the parallelogram, the trapezium and the kite., and the meaning of 'regular' polygon (all sides and angles equal).

- Display some pictures of shapes of different sizes and in different orientations and ask the children to name them. Encourage them to use the correct names. (For example, the children should use the word 'rhombus' and not the incorrect term 'diamond'. They should give the correct names for types of quadrilaterals, rather than just calling them quadrilaterals.) Ask the children to identify regular and irregular polygons, and to identify equal side lengths and equal angles in shapes.
- Ask the children to divide the shapes into two groups and to say what properties they used to group them. For example, they might make a group of shapes with four sides and a group of shapes with any other number of sides or a group of regular and a group of irregular polygons.
- Repeat this, asking the children to make three and then four groups to encourage them to look at different properties of the shapes.
- For question 1, the children should use rulers, shape stencils, a pair of compasses or any other equipment they have to draw the shapes as accurately as possible. Children who need additional support might find it easier to work on squared paper.

Follow-up

Ask the children to complete **Workbook 5 page 14**. Remind the children that if they have half a shape, the other part should also be half. This may make a symmetrical shape, but not always. If you have geoboards and elastic bands, let the children use these to help them complete the shapes.

Challenge

Let the children use the squares they have drawn (or new ones) to make a three-piece tangram puzzle. They can cut out the pieces and use them to investigate other polygons they can make by joining them in different ways. They should sketch the shapes they make.

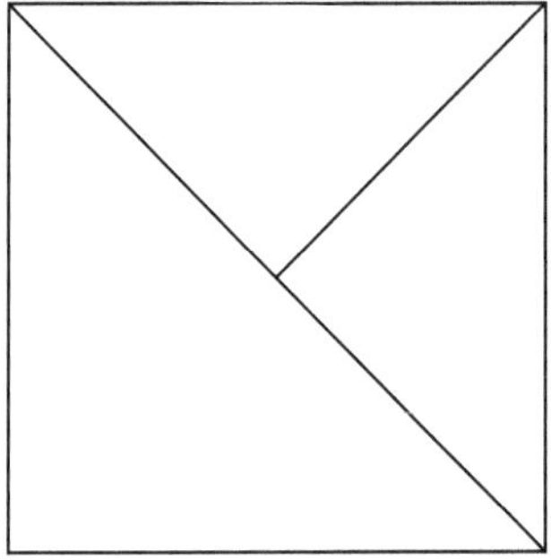

Interesting mistakes

The children may think that a polygon is regular if it has equal sides and ignore the condition of equal angles. For example, they may say that a rhombus is regular. Encourage the children to fold the shapes to match up angles and/or use symmetry to determine whether a shape is regular or irregular.

Answers for Pupil book 5 page 20

1
  a Individual drawings; square
  b Individual drawings; possible answers e.g. rectangle, parallelogram, rhombus, trapezium, kite
  c Individual drawings; possible answers e.g. scalene triangle, right-angled triangle, isosceles triangle
  d Individual drawings; pentagon
  e Individual drawings; rhombus
  f Individual drawings; rectangle
  g Individual drawings; square

Answers for Workbook 5 page 14

1
  a square   **b** rectangle   **c** square
  d pentagon   **e** octagon   **f** hexagon
  g regular octagon   **h** octagon

3D shapes and nets

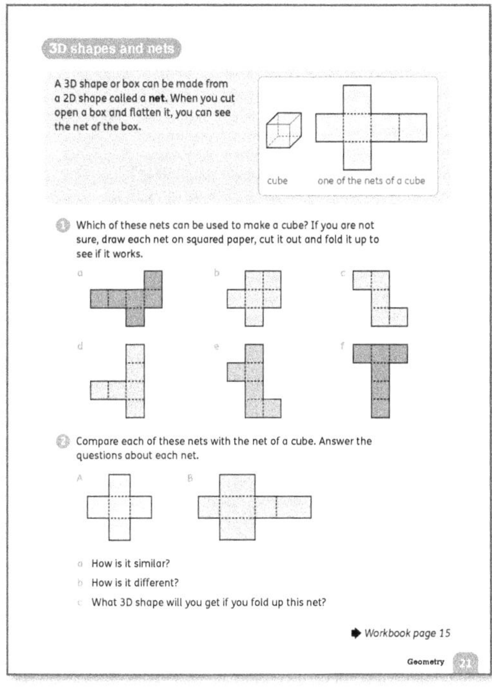

Materials

A selection of different 3D shapes (which could include everyday items such as food boxes); flashcards with the names of 3D shapes; squared paper; thin card; glue or sticky tape; scissors or craft knives; geoboards or squared dotted paper

Warm-up

Select activities that reinforce number facts and mental calculation skills from the Activity bank (pages 22–28) as mental starters for this lesson.

- Show the children a variety of 3D shapes. Ask them to name the shapes they recognise.
- Put flashcards with 3D shape names on a table and ask the children to place each shape next to the appropriate card. Ask: *Can you think of any common objects that you could add to each set?*

- Hold up a shape for the children to see. Ask the children to point to a face, an edge and a *vertex* (they learned these terms in Level 3). Ask questions such as: *How many faces does this solid have? What shape is this face?*
- Hide a 3D shape behind a screen, for example, an open book. Describe the shape. For example, tell the children how many faces it has and what shape the faces are. Ask the children to guess what the hidden shape is.
- Repeat, but this time ask a child to select and hide a 3D shape and describe it for the rest of the class.
- Alternatively, have the children ask questions about the shape with either 'yes' or 'no' answers. (For example, 'Does it have eight faces?') After each question, the children can guess what the shape is.

Focus

- Turn to **Pupil book 5 page 21** and read the information about nets together. Let the children trace with a finger the line that they would cut out to make the net.
- The children can do question 1 in pairs or small groups. If they need to, let them physically make and fold the nets.
- Do question 2 orally with the class.

Follow-up

The children can use the net on **Workbook 5 page 15** to make a cube. You could scan or photocopy the template or the children could trace it. Supply each group with thin card, scissors or craft knives and glue or sticky tape.

Support

Some children may need additional support to draw nets accurately. Help them with using a ruler, positioning tabs, cutting out and gluing as required. You could show the class an online video demonstration of how to build shapes from nets.

Answers for Pupil book 5 page 21

1 a, e and f are nets of a cube

2 These are some suggestions. Children's answers may vary.
Net A
  a It has square faces
  b There are only 5 faces but a cube has 6
  c An open-top box
Net B
  a It has 6 faces
  b Some faces are rectangular, whereas all the faces of a cube are identical squares.
  c A cuboid

Answers for Workbook 5 page 15

1 Individual cubes.

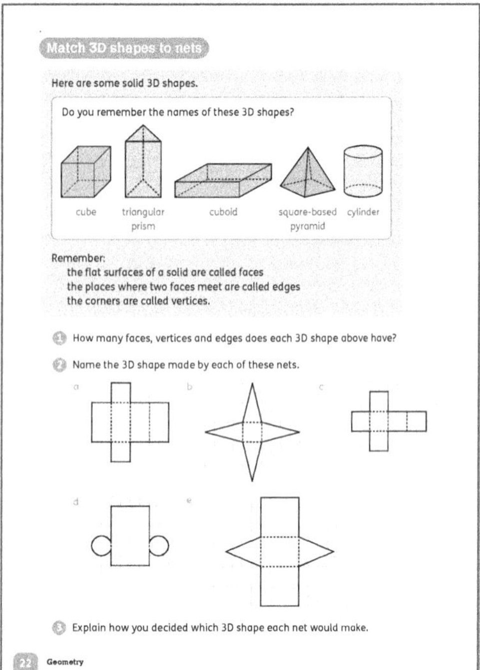

Materials

3D shapes, including cuboid, square-based pyramid, cube, cylinder and triangular prism; squared paper and scissors (if needed)

Warm-up

Select activities that reinforce number facts and mental calculation skills from the Activity bank (pages 22–28) as mental starters for this lesson.

Focus

- Work through **Pupil book 5 page 22** with the class to revise the correct terminology for talking about solids.
- Do questions 1–3 orally as a class.
- It may be useful to draw up a table to record the children's answers as you work through question 1, with columns for shape, faces, vertices and edges. For a cylinder, we usually count the curved surface as a 'face' as it is made from a flat shape, so a cylinder has 3 faces and 2 edges but no vertices.
- Compare the shapes and numbers of faces on the nets with 3D shapes, to help the children identify the shape that each net makes. You may also find it helpful to copy, cut out and fold the net to demonstrate.

Interesting mistakes

Some children may not use the correct terminology when talking about the properties of 3D shapes. Regularly use the correct terminology with them by playing games in which the children describe shapes or listen to descriptions in order to identify shapes.

Challenge

- Let the children work in pairs or small groups to investigate regular (Platonic) solids. There are only five of these: tetrahedron, cube, octahedron, dodecahedron and icosahedron.
- The children can find out what the regular solids are, why they are called Platonic solids and make models

of them to do a presentation. The children could build these shapes from nets or using modelling clay.
- They could also investigate combining these solids to make models for the classroom. For example, sticking a tetrahedron onto each face of an icosahedron produces a solid called a stellated icosahedron (shown here).

Support

- Make a set of vocabulary cards and a set of matching diagram cards of 3D shapes for the children to play games that involve matching the word to the diagram.
- Use the cards to play memory games (Pelmanism) or card games such as 'Snap', in which the children say 'snap' when the matching word and diagram are played one after the other.

Answers for Pupil book 5 page 22

1 cube: 6 faces, 8 vertices, 12 edges
triangular prism: 5 faces, 6 vertices, 9 edges
cuboid: 6 faces, 8 vertices, 12 edges
square-based pyramid: 5 faces, 5 vertices, 8 edges
cylinder: 3 faces, 0 vertices, 2 edges

2 a cuboid   b square-based pyramid
   c cube   d cylinder
   e triangular prism

3 Individual answers.

Draw 3D shapes

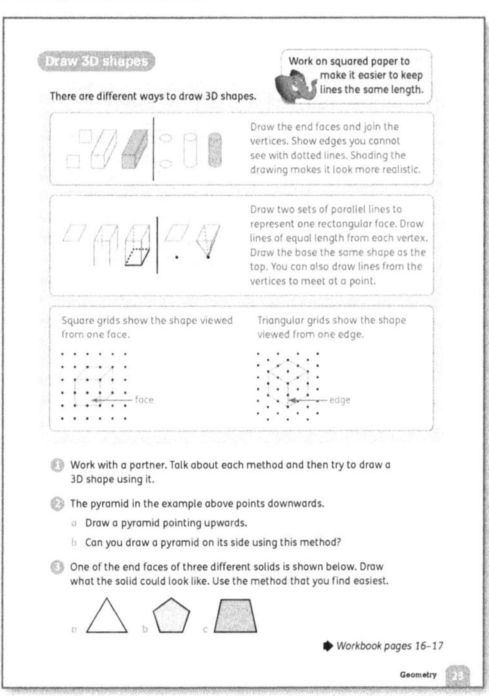

Materials

Pencils and rulers; squared paper; squared and triangular dotted paper (you can find both online)

Warm-up

Select activities that reinforce number facts and mental calculation skills from the Activity bank (pages 22–28) as mental starters for this lesson.

Focus

It is best to use a practical approach to the work on drawing 3D shapes.

- Let the children work in pairs to read through the methods shown on **Pupil book 5 page 23**.
- Provide squared and triangular dotted paper for question 1 so the children can try out each method as they work through the examples.
- Before moving on to the next activity, have the children use **Workbook 5 page 16** to consolidate this learning and to show that they can draw 3D shapes given the two end faces.
- Let the children try question 2 on **Pupil book 5 page 23** and then discuss what they did and what worked well or didn't work well.
- Use question 3 to check that the children can draw a reasonable diagram to represent a 3D shape.

Follow-up

Use **Workbook 5 page 17** to consolidate the work on drawing 3D shapes using the different methods. The children can do the questions in class or as a homework task.

Challenge

Let the children investigate the use of perspective in art and technical diagrams. They can do their own research and prepare a short talk or slide show to report back to the class.

Answers for Pupil book 5 page 23

1 Individual drawings.

2 a Possible drawing: **b** Possible drawing:

3 Possible drawings:

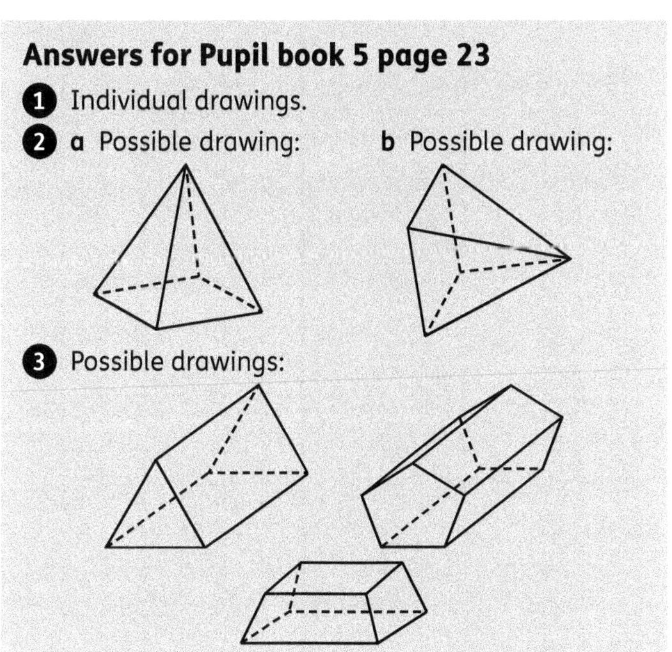

Answers for Workbook 5 page 16

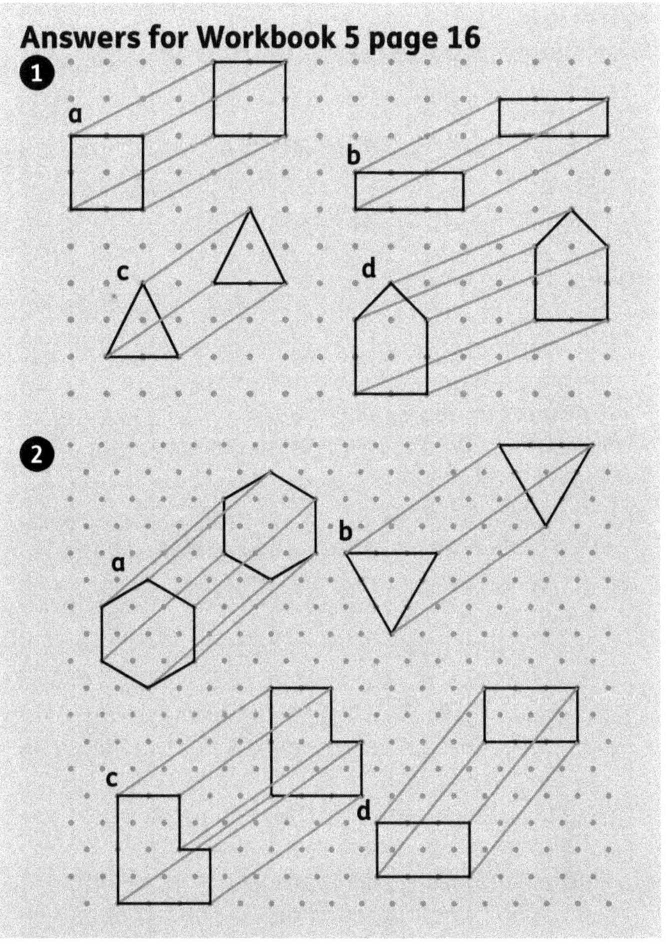

Answers for Workbook 5 page 17

1 **a–d** Individual drawings.
2 **a–d** Individual drawings.

Classify angles

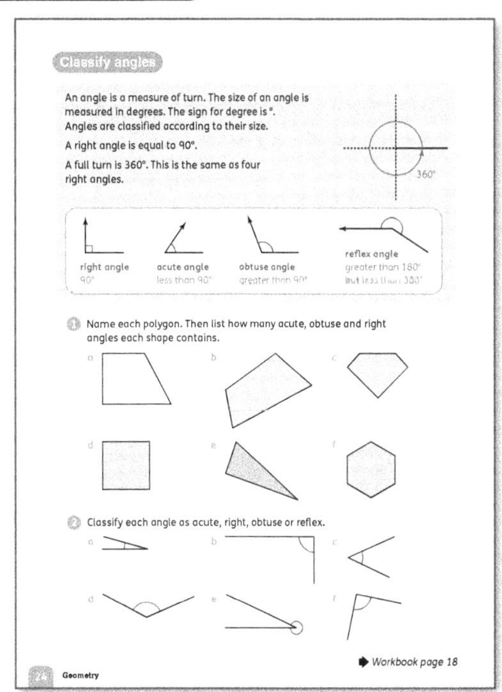

Materials

Flashcards with angles of different sizes drawn on them

Warm-up

Select one of the 'Angles' activities from the 'Geometry and measures' section of the Activity bank (pages 28–29) as a mental starter for this lesson.

Focus

- Tell the class they are going to work with angles. Ask them for a definition of an *angle* and agree on the best definition, for example a measure of turn, measured in degrees.
- Show the children the corner of a piece of paper (or a page in a book) and ask them to say what angle is shown (right angle).
- Ask the children to look around the classroom and identify right angles and angles that are bigger or smaller than right angles.
- Hand out flashcards showing different angles to groups of five children. Ask the children to arrange themselves in a line showing the angles in order of size.
- Encourage the children to talk about how they compared the angles.
- Hold up a flashcard showing an angle. Ask the children to draw a smaller or larger angle.
- Hold up two flashcards showing angles of different sizes. Ask the children to draw an angle that is between the two angles shown.
- Turn to **Pupil book 5 page 24**. Revise the degree symbol, and the number of degrees in a full turn and a right angle (quarter turn). Revise the terms *right angle*, *acute angle* and *obtuse angle*, and teach the term *reflex angle* (greater than 180° and less than 360°).
- Let the children work on their own to complete question 1 and question 2.

Follow-up

Let the children work in pairs to identify the angles in the questions on **Workbook 5 page 18**.

Answers for Pupil book 5 page 24

1
- **a** trapezium; 2 right angles, 1 acute, 1 obtuse
- **b** quadrilateral; 0 right angles, 2 acute, 2 obtuse
- **c** pentagon; 3 right angles, 0 acute, 2 obtuse
- **d** square; 4 right angles, 0 acute, 0 obtuse
- **e** right-angled triangle; 1 right angle, 2 acute, 0 obtuse
- **f** hexagon; 0 right angles, 0 acute, 6 obtuse

2
- **a** acute   **b** right   **c** acute
- **d** obtuse   **e** reflex   **f** right

Answers for Workbook 5 page 18

1 Possible answers (there are more):

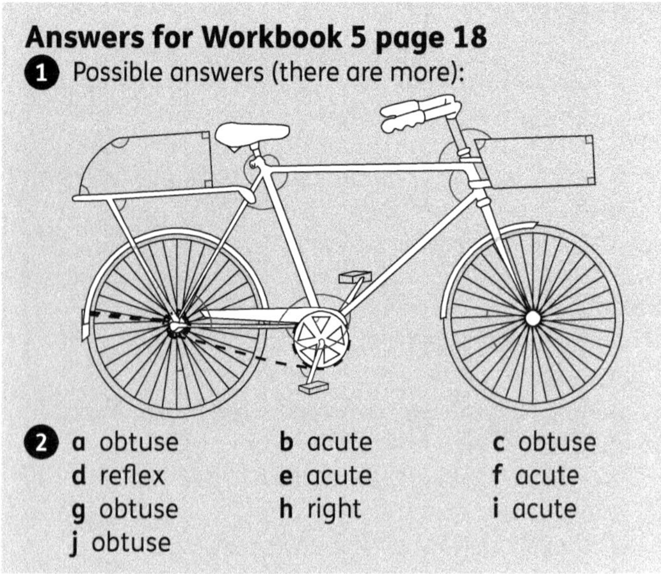

2
- **a** obtuse   **b** acute   **c** obtuse
- **d** reflex   **e** acute   **f** acute
- **g** obtuse   **h** right   **i** acute
- **j** obtuse

Measure angles in degrees

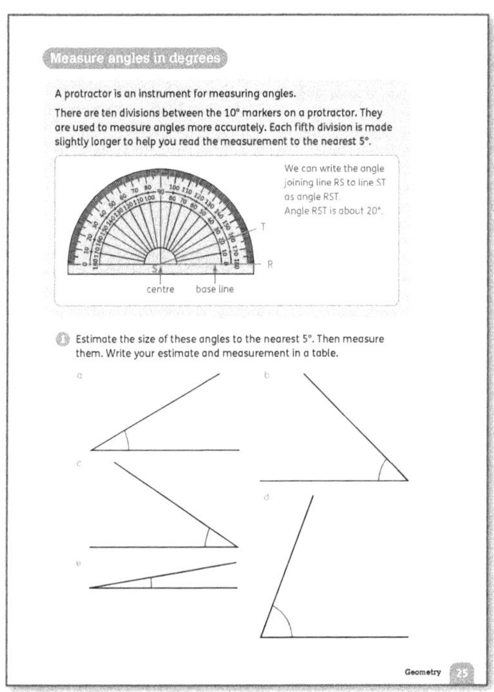

Previously, the children have considered angles as turns (although they know they are measured in degrees) and they have classified and ordered them by size without actually measuring them.

Now the children will learn how to read and use a protractor. If you introduce the protractor as just another (curved) measuring scale, the children are more likely to find it easy to use.

Materials

Protractors (half circle or full circle); video showing how to measure angles (if possible)

Warm-up

Select activities that reinforce number facts and mental calculation skills from the Activity bank (pages 22–28) as mental starters for this lesson.

Focus

- Teach the children the correct way to place the protractor and read the scale. If possible, demonstrate this using a real protractor on an overhead projector or whiteboard, or find a demonstration video or animation to show the children.
- Remind the children that they can use either of the scales on the protractor, provided they always start with 0 on the base line. (Point out that this is not so different from many rulers, which have from 0 to 30 cm on one side and 0 to 300 mm in the opposite direction on the other side.) Demonstrate measuring angles facing in opposite directions, like 1a and 1c in the Pupil book, using the correct scale.
- If the children do not have protractors, give them a printed protractor. If you can print these onto acetate sheets, they can use them as they are. If not, print them on card and the children will need to cut out the protractor as well as the inner semi-circle so that they can see the angle lines. They can also use a pin to mark the centre point. You can find printable protractors like this one online.

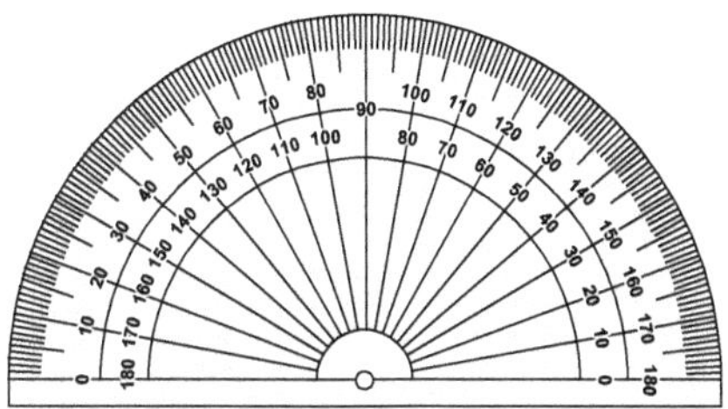

- Turn to **Pupil book 5 page 25**. Do the estimating part of question 1 orally with the class to make sure that they think about the angles and what size they are likely to be. Then ask the children to measure the angles.

Interesting mistakes

The children may not line up the protractor correctly or use the wrong scale. Encourage the children to think about whether their reading makes sense. For example, if they have read 165° for an acute angle, something must be wrong – either they have lined up the protractor incorrectly or they have used the wrong scale.

Answers for Pupil book 5 page 25

1 **a** Individual answers; 30°.

   b Individual answers; 45°.

   c Individual answers; 34°.

   d Individual answers; 70°.

   e Individual answers; 10°.

Draw and measure angles

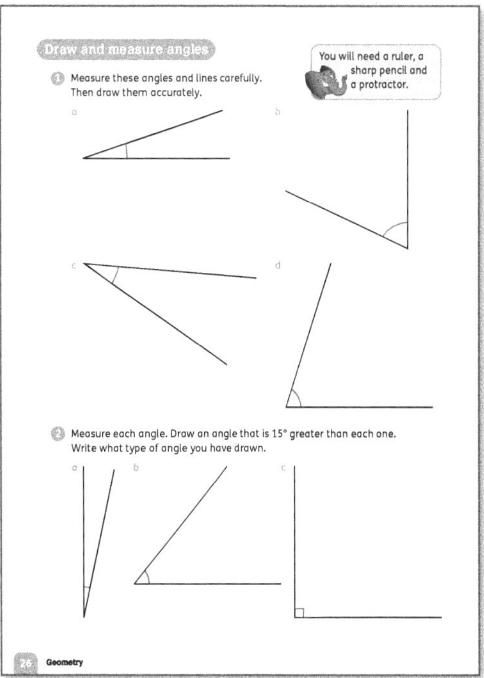

Materials

Pencils, rulers and protractors

Warm-up

In this lesson, the children will be adding 15 to different angles, so do some mental addition and subtraction activities in which the children have to give the number that is 15 more or less than a given number. Start with multiples of 5 but also include some other numbers in the range 0 to 180.

Focus

- Turn to **Pupil book 5 page 26**. Discuss with the class how they can make an accurate copy of each angle in question 1. Ask them which units they will use to measure the lines accurately (millimetres) and which units they will use to measure the angles (degrees). Discuss how they will assess their completed diagrams for accuracy.
- The children can work on their own to do question 2. They can then check each other's work by measuring the drawn angles.

Support

Some children may need additional support to place the protractor and read the angle. Prepare a worksheet with some angles of different sizes with a protractor positioned on each angle. The children can write the size of each angle on the sheet. You can download worksheets or prepare your own. For the first few, indicate with an arrow where they need to read the angle. For example:

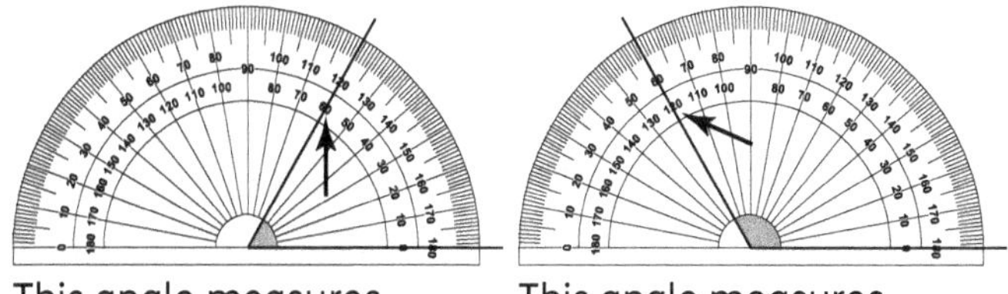

This angle measures_______   This angle measures_______

Interesting mistakes

The children may read the protractor in the wrong direction. Encourage them to use their finger to move from the zero line on the angle around the scale to the other arm of the angle. This will help them to see which scale they should use.

Answers for Pupil book 5 page 26

1 a 20°; Individual drawings.
   b 65°; Individual drawings.
   c 30°; Individual drawings.
   d 72°; Individual drawings.
2 a 12°; Individual drawings of 27° angle; acute
   b 52°; Individual drawings of 67° angle; acute
   c 90°; Individual drawings of 105° angle; obtuse

Angles in triangles and quadrilaterals

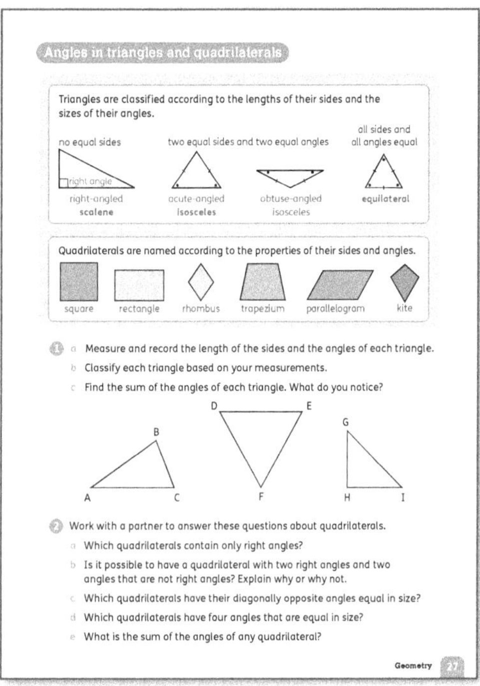

Materials

Pencils, rulers and protractors

Warm-up

Select activities that reinforce number facts and mental calculation skills from the Activity bank (pages 22–28) as mental starters for this lesson.

Focus

This lesson is largely investigative. In the activities on **Pupil book 5 page 27**, the children will revise what they have already learnt about triangles (such as *scalene*, *isosceles* and *equilateral*) and quadrilaterals. They will also explore the angle properties of shapes in more detail, using the measuring skills from the previous lesson.

- Let the children work independently to complete question 1. Encourage them to check each other's measurements and discuss the question in pairs or small groups.
- Let the children work in pairs to complete the discussion tasks involving quadrilaterals in question 2.

Challenge

In places that get lots of snow in winter, it is better to have roofs with acute angles at the top. Ask students to suggest why this is and to draw some houses with roofs suitable for snowy climates. You could find some images of houses in Scandinavian countries to show children, which they can compare with their answers.

Answers for Pupil book 5 page 27

1 a AB = 3.2 cm, AC = 3.4 cm, BC = 2.0 cm;
     DE = 3.5, cm, EF = 3.5 cm, DF = 3.5 cm;
     GH = 2.3 cm, HI =2.3 cm, GI = 3.2 cm
   b ABC is acute-angled scalene; DEF is equilateral; GHI is right-angled isosceles
   c They all sum to 180°.
2 a Square and rectangle
   b Yes. The only quadrilateral that fits the description is a trapezium with two right angles.
   c square, rectangle, rhombus, parallelogram
   d square, rectangle
   e 360°

Compare angles

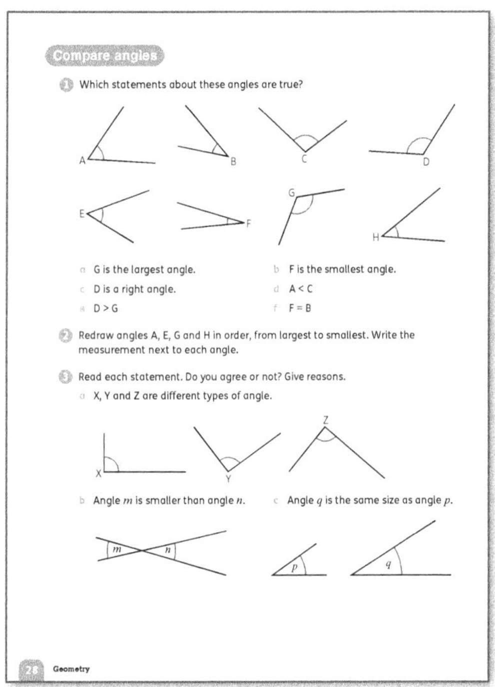

Materials

Pencils, rulers and protractors; pictures of different styles of houses found in different climates (with roofs of different angles including one flat roof), if possible

Warm-up

Select either 'Compare 3-digit numbers' or 'Compare 5- and 6-digit numbers' from the 'Place value and number sense' section of the Activity bank (page 23) as a mental starter for this lesson.

Focus

The aim of this lesson is to consolidate classifying, comparing and measuring angles. The activities address some of the misconceptions that the children might have. For example, they might think that the lengths of the arms of an angle affect its size, or that only angles marked with a square angle marker are right angles.

- Turn to **Pupil book 5 page 28**. Let the children work in pairs or small groups to do question 1 orally. They will need to measure the angles to help them decide. Take feedback from the class when they have completed the question.
- The children can then work independently to complete question 2. They need to measure, order and redraw the angles.
- Give the children time to read and think about the statements in question 3. Then let different children share their ideas with the class. Focus on the reasons they give to support their answers.
- Then show the class pictures of some different roof styles found in different climates. Explain some of the reasons for roofs with different angles. For example, high, acute-angled roofs are suitable for colder, wetter climates as snow and rainwater run off these roofs. Steep roofs help to retain heat to keep the house warm. They make the house feel hotter so they are not suitable for warm places. Flat roofs are only suitable in very dry climates, and often they are not completely flat to allow water to run off when it does rain.
- Let the children describe the angle of the roof in each picture and say what kind of climate they think this type of roof is suited to.

Challenge

Let the children investigate local roof styles and prepare a short talk on which are most suited to the local climate, giving reasons for their answers. If local architecture doesn't provide interesting examples, you could ask the children to investigate other places.

- Some interesting designs are found in Bermuda, where white, stepped roofs are built to collect rainwater.
- Traditional Indonesian houses often have two slopes to allow for cooling and to get rid of rainwater.
- Turf roofs are designed to keep homes cool and also to collect water.
- Flat roofs are used in desert regions for many reasons: they keep out and reflect the sun; they are easy to build; the people in desert regions do not need to worry about rain or snow; people might sleep outside on the roof during very hot nights.

Answers for Pupil book 5 page 28

1   a true   b true   c false
         d true   e false   f false

2   G 120°, A 59°, E 51°, H 42°

3   a no; they are all right-angles
         b no; they are the same size
         c yes; they are the same size

Angles on a straight line

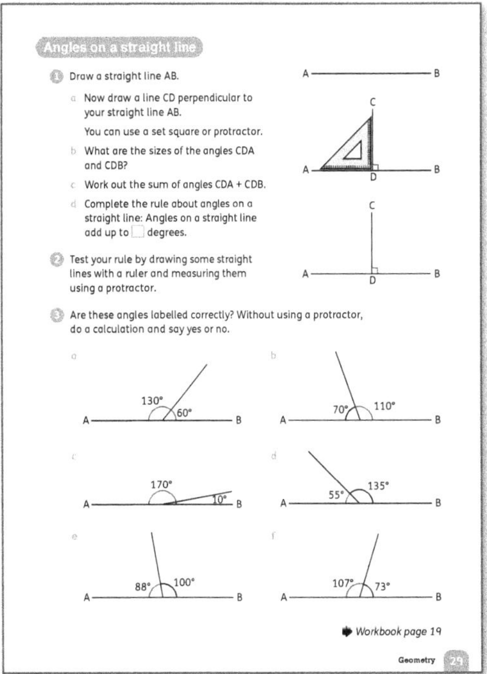

Materials

Pencils, rulers; set squares (if possible); protractors

Warm-up

As the children will be calculating to find angles that total 180°, do some mental addition and subtraction activities in the range 0–180 as a mental starter for this lesson.

Focus

- Turn to **Pupil book 5 page 29** and let the children work independently or in pairs to do the investigation in question 1. Then discuss what they have learnt from this investigation as a class. They should be able to tell you that a straight line is two quarter turns, which are the same as two right angles. 90 + 90 = 180, so a straight line must be 180°.
- For question 2, let the children draw some lines in different orientations and use a protractor to show they are all 180°.
- For question 3, it is important that the children realise that AB is a straight line and that they can calculate to say whether the angles given are correct or not. They should not measure to do this. Ask: *Does it matter how many angles there are at a point on the line?* (No, it does not.) *Do two angles that are not right angles drawn at a point on the line still add up to 180°?* (Yes, they do.)

Follow-up

In the activity on **Workbook 5 page 19**, the children have to use the given angle in each diagram to work out the size of the unknown angle. Try to use the term 'unknown angle' rather than 'missing angle'. (The angle is there, it is just not labelled, so the terminology 'missing angle' can be confusing.) Let the children work out the sizes of the angles and then check their answers as a class.

Calculate unknown angles

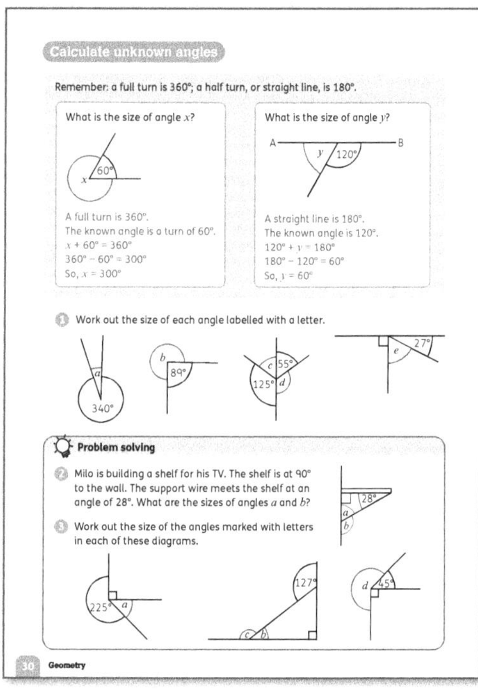

This is an important lesson as it forms the basis for later work that the children will do in algebra and geometry.

Materials
Circular spinner with movable pointer

Warm-up
Use a circular spinner with a pointer (or a clock face) to revise quarter, half and full turns, and the number of degrees in each. Ask the children to show where the pointer will be after a 90° turn clockwise/anticlockwise, a half turn etc. Then show a three-quarter turn. *How many degrees is this?* (180° + 90° or 360° − 90° = 270°). What if the spinner makes a full turn and then an extra quarter turn? (360 + 90 = 450°). Repeat for other multiples of 90°.

Focus
- Explain that we use what we know in mathematics to work out what we do not know. In this case, we know some angle facts and we can use them to work out other angle facts.
- Turn to **Pupil book 5 page 30**. Work through the two examples carefully with the class, asking questions as you do so, and explaining the process. For example, ask: *If we have already turned 60°, how many more do we need to turn to complete a full turn? How can we write a number sentence to show this?* and so on.
- It is important to point out to the children that they cannot measure to solve unknown angle problems like these as the diagrams are often not drawn accurately. An angle marked 60° on a diagram may not measure 60° on a protractor.
- The children can work in pairs to work out the size of the unknown angles in question 1. At this stage they need to show their working, but they do not need to solve formal equations.
- Problem solving: Before the children do question 2, ask them what they know about the sum of angles in any triangle (they always add up to 180°). The children can then do questions 2 and 3.

Challenge
Give the children a set of clues for the size of a given number of angles round a point. For example:
- *Angles a to e are five angles round a point. Together these angles make a full turn.*
- *Angle a is $\frac{1}{3}$ of a right angle.*
- *Angle b is $\frac{1}{8}$ of the full turn.*
- *c is twice the sum of a and b.*
- *d is half the size of e.*
- *Work out the size of each angle.*

You can then challenge the children to make up their own set of clues for angles about a point or on a straight line. They can exchange these and try to solve each other's clues.

Support
To help the children understand that the angle sum of a triangle is always 180°, let them draw some different triangles, cut them out and tear off the angles. They can then put the angles together against a point on a line to see that they 'fill' the straight line and are therefore equivalent to 180°.

Shape patterns and sequences

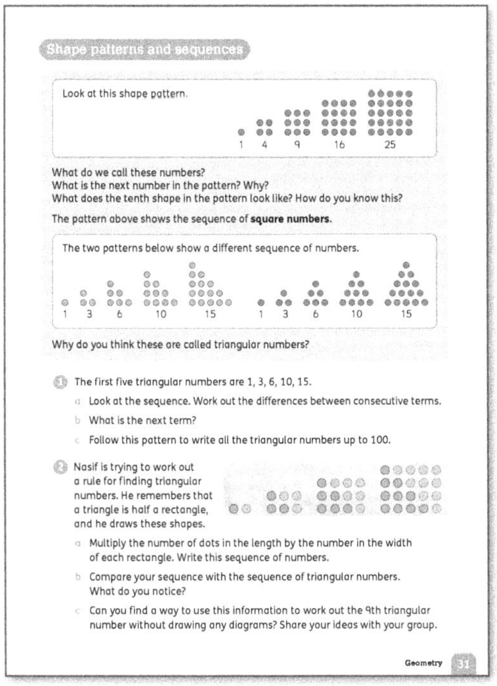
Shape patterns and sequences

Look at this shape pattern.

1 4 9 16 25

What do we call these numbers?
What is the next number in the pattern? Why?
What does the tenth shape in the pattern look like? How do you know this?

The pattern above shows the sequence of **square numbers.**

The two patterns below show a different sequence of numbers.

1 3 6 10 15 1 3 6 10 15

Why do you think these are called triangular numbers?

1. The first five triangular numbers are 1, 3, 6, 10, 15.
   a Look at the sequence. Work out the differences between consecutive terms.
   b What is the next term?
   c Follow this pattern to write all the triangular numbers up to 100.
2. Nasif is trying to work out a rule for finding triangular numbers. He remembers that a triangle is half a rectangle, and he draws these shapes.
   a Multiply the number of dots in the length by the number in the width of each rectangle. Write this sequence of numbers.
   b Compare your sequence with the sequence of triangular numbers. What do you notice?
   c Can you find a way to use this information to work out the 9th triangular number without drawing any diagrams? Share your ideas with your group.

Geometry 31

This lesson builds on the work that the children have already done on number sequences and what they learnt about square numbers in Level 4.

Materials
Counters for each group to make shape patterns (if needed)

Warm-up
Skip count in different sized jumps from different starting numbers, or starting at 0 add 1, then 2, then 3 and so on as a mental warm-up for this lesson.

Focus
- Look at the *square number* pattern at the top of **Pupil book 5 page 31**. Give the children time to read and think about the questions before taking feedback. Focus on the reasons for their answers. The children are asked why they think the numbers in the second sequence are called triangular numbers. (They are called triangular numbers because you can draw isosceles or equilateral triangles using each number of dots.)
- Let the children discuss the questions in question 1 in pairs or groups before they write any answers, then take feedback.
- Let the children spend time working through question 2 on their own before they share their ideas with their group.

Challenge
- Explain to the children that a composite number is a number with more than two factors, so all numbers that are not 1 or prime are composite numbers. Every composite number can be shown as a rectangular array, so they are also sometimes called 'rectangular numbers'.
- Allow the children to investigate different ways of showing the first ten rectangular numbers (4, 6, 8, 9, 10, 12, 14, 15, 16, 18) as rectangles. Let them produce a short report summarising what they discovered.

Interesting mistakes
Some children may not relate the difference between terms to the term numbers. They may say things like: 'It increases by one more each time'. Encourage them to draw up a table so that they can see the connection between the difference and the term number.

Term number	1	2	3	4
Term	1	3	6	10
Difference		+2	+3	+4

Answers for Pupil book 5 page 31
1. a 2, 3, 4, 5 b 21
   c 1, 3, 6, 10, 15, 21, 28, 36, 45, 55, 66, 78, 91
2. a 2, 6, 12, 20
   b The answers are double the corresponding triangular numbers.
   c It is half of the number of dots in a rectangle that is 9 tall and 10 wide.

End-of-unit check
Ask some or all of these questions to assess how well the children have understood the concepts in this unit.

- Show the children different shape names (hexagon, quadrilateral, pentagon, and so on) and ask: *How many sides does this shape have? How many vertices?* (hexagon: 6 sides, 6 vertices; quadrilateral: 4 sides, 4 vertices; pentagon: 5 sides, 5 vertices)
- Show the children a set of different polygons and ask: *How could you group these shapes into sets? What are the criteria for each of your sets?* (For example: Number of sides, regular and irregular.)
- Display or name a 3D shape and ask: *How many faces/edges/vertices does this shape have?*
- Show a 2D or 3D shape and ask: *How can you describe this shape to me?* (For example: 2D shapes: number of sides, number of angles, number of equal sides/angles, regular/irregular. 3D shapes: number of sides, number of faces, number of vertices, shapes of faces, number of equal sides.)
- *Draw a net for a triangular prism/pyramid.* (See the nets on **Pupil book 5 page 22**.)
- Ask the children to draw different 3D shapes using different methods. Give instructions such as: *Here are two end faces of a 3D shape. Use them to draw the 3D shape.*
- Show different triangles and ask: *What is the correct name for this triangle? What properties are used to name it?* (Right-angled triangle: one right angle. Acute-angled isosceles triangle: two equal sides, two equal angles, all angles acute. Obtuse-angled isosceles triangle: two equal sides, two equal angles, one angle obtuse. Equilateral triangle: all sides and all angles equal.)

- Display a number of different angles, including reflex angles, and ask: *What type of angle is this? Why?* (For example: The angle is less than 90°, so it is an acute angle. The angle is 90°, so it is a right angle. The angle is greater than 90° and less than 180°, so it is an obtuse angle. The angle is greater than 180°, so it is a reflex angle.)
- *How can you tell if a triangle is isosceles?* (It has two equal sides and two equal angles.)
- Show a triangle with two angle sizes given (adding up to 90° or close to 90°) and ask: *Is this triangle right-angled? How can you tell?* (If the two given angles add up to 90°, the triangle is right-angled, because the angles in a triangle add up to 180° and 180 – 90 = 90. If the two given angles do not add up to 90°, the triangle is not right-angled, because the angles in a triangle add up to 180°.)
- Show different angles and ask: *How can you estimate the size of this angle?* (Children can use whether the angle is less than or greater than 90°/180° and whether it is close to a multiple of 90° or close to, for example, half of 90°.)
- *What important tips would you give someone to help them use a protractor properly?* (For example: Make sure that the protractor is lined up correctly, so the line on the protractor between 0° and 180° is along one of the lines of the angle. Read the size of the angle starting from the 0° that lies on the line of the angle. Think about whether the reading makes sense, for example is the value read for an acute angle less than or greater than 90°?)
- Display a set of angles and ask: *Is each angle acute, obtuse, right angle or reflex?* (Acute angles are less than 90°, right angles are equal to 90°, obtuse angles are greater than 90° and less than 180°, and reflex angles are greater than 180°.)
- *How can you find this unknown angle without measuring? How did you do it?* (The children need to give the correct reason, for example: Angles on a straight line add up to 180°. Angles round a point add up to 360°. The angles in a triangle add up to 180°.)
- Provide a diagram with some of the angles incorrectly labelled (that is, not possible) and ask the children to find the mistakes, giving reasons for their answers. For example:

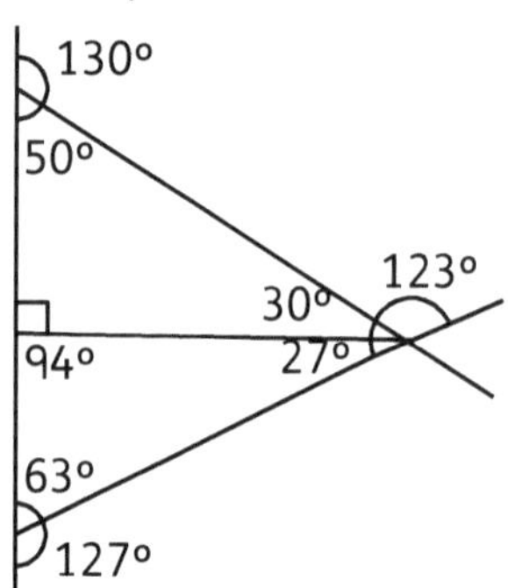

(90° and 94° do not add up to 180°. 63° and 127° do not add up to 180°. 94°, 27° and 63° do not add up to 180°.) Note that different answers are possible depending on which angles the children say are incorrect. For example, if they say that 94° is not correct, then the other two angles in that triangle are correct. If they assume that 94° is correct and change the 90° angle to 86°, then one of the other angles in the triangle is not possible. Let the children make their decisions and explain or justify them to you or to their groups.

UNIT 4 Addition and subtraction

<table>
<tr><td>

Learning objectives

- Add and subtract mentally with increasingly larger numbers

- Use appropriate laws of arithmetic to simplify calculations

- Add and subtract numbers with more than four digits, including using formal written methods

- Use rounding to check answers to calculations and determine, in the context of a problem, levels of accuracy

- Solve addition and subtraction multi-step problems in context, deciding which operations and methods to use and why

</td><td>

Key words

inverse operation   multiple   area   estimate   approximate

</td></tr>
</table>

Unit introduction

Teaching guidance

- Start with a discussion of different strategies for adding and subtracting mentally. Ask the children what they do when they add or subtract in their heads.
- The strategies could include: doubling; adding in parts; using known facts for smaller numbers to add larger numbers; compensation. List the strategies as the children say them.

- Display an addition pyramid like the one below. Explain that the number in each block is the sum of the two numbers below it.

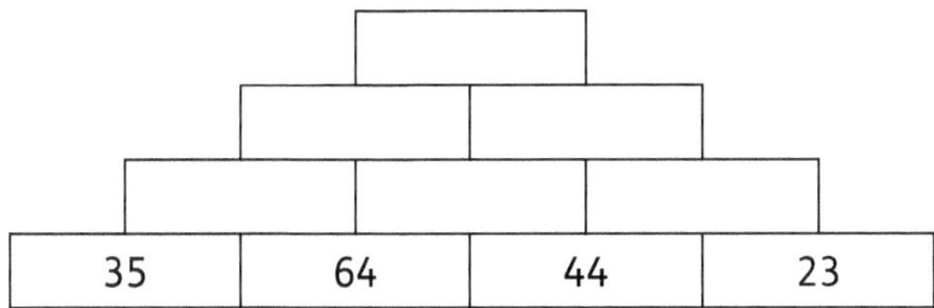

- Give the children time to complete the pyramid. Then ask:
  - *How did you think and work to find the answers?*
  - *When did the sums start to become more challenging? What made them more challenging?*
  - *Did you use different strategies for different numbers? Which strategies did you use? Why?*
- Next, display the pyramid below. Tell the children that, as before, each block is the sum of the two blocks below it.

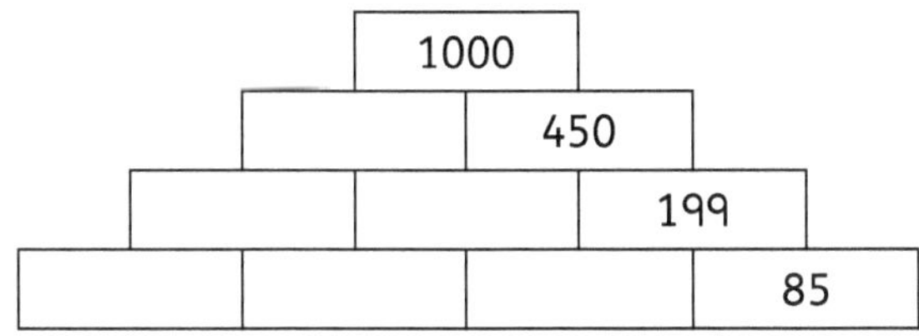

- Give the children time to complete the pyramid. Then discuss what they did to find each answer and why they needed to do it.
- Next, ask the children why it is important to be able to calculate mentally in daily life. Ask them to give some examples of when they or other people do this.

Mental calculation

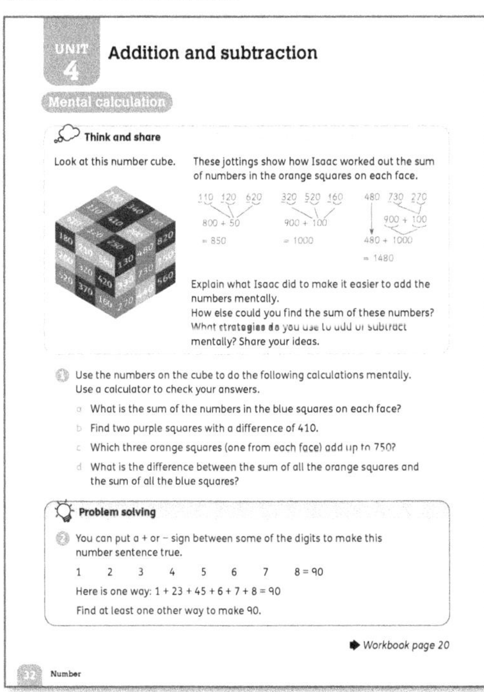

Materials
Calculators; place-value tables (page 21)

Warm-up
This lesson deals with the addition of *multiple* numbers. Do a mental starter in which the children add several small numbers mentally.

Focus
- <u>Think and share</u>: Turn to **Pupil book 5 page 32**. Remind the children that we do jottings to keep track of our mental calculations. Jottings are informal notes that we make for ourselves.
- Let the children work in pairs or small groups to discuss the jottings and to answer the questions. The children should notice that Isaac has added the hundreds and tens digits separately and then added these totals.
- Take feedback as a class. Add any new strategies that the children suggest for adding and subtracting mentally to the list you made in the 'Unit introduction' activity. Keep this list on display in the classroom.
- Let the children work independently to do the calculations in question 1, then ask them to check their own work using a calculator. Remind them that they can make jottings if they need to. You may need to adapt this activity for children who cannot distinguish the colours on the cube face.
- <u>Problem solving</u>: Question 2 is an example of a task where jottings are useful. The children can copy the numbers and draw lines to join them and make notes to keep track of the totals.

Follow-up
Let the children complete the activities on **Workbook 5 page 20** independently. Once they have done this, check the answers as a class and ask the children to share how they worked and what mental strategies they used. If there are any interesting mistakes, encourage the children to discuss these and what went wrong.

Support
- Some children may find it difficult to add three (or more) numbers such as 40 + 70 + 60 mentally. Encourage them to look at how to rearrange numbers to make it easier (40 + 60 + 70 = 100 + 70 = 170).
- Similarly, the children may be able to recall and use basic facts such as 6 + 7, but not realise that they can use these facts to help with additions such as 16 + 7 or 60 + 70. You can help by having a number talk (pages 17–18) about extending number facts that involve small numbers to larger numbers. You can model this on place-value tables if needed.

Answers for Pupil book 5 page 32
1 a Front face: 1050; top face: 1000; side face: 920; total: 2970
   b 820 and 410   c 110, 160, 480
   d 360
2 Individual answers, for example,
   1 − 2 + 34 + 56 −7 + 8 = 90

Count on to add

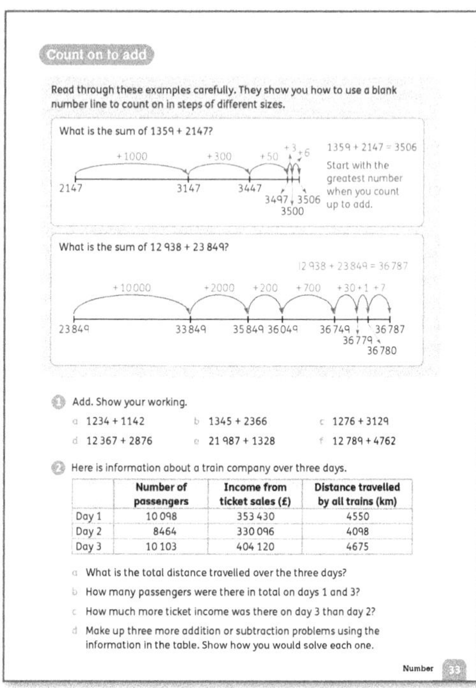

Warm-up
Select any 'Place value and number sense' activities from the Activity bank (pages 22–23) as a mental starter for this lesson.

Focus
- Ask the class: *What is the sum of 955 and 176?* Explain that we can use a technique called counting on.
- Draw a number line on the board. Write 955 on the left.
- Show the class how to count up in steps to add 176.
- Talk about each jump as you do it. For example:

First, I add 100.

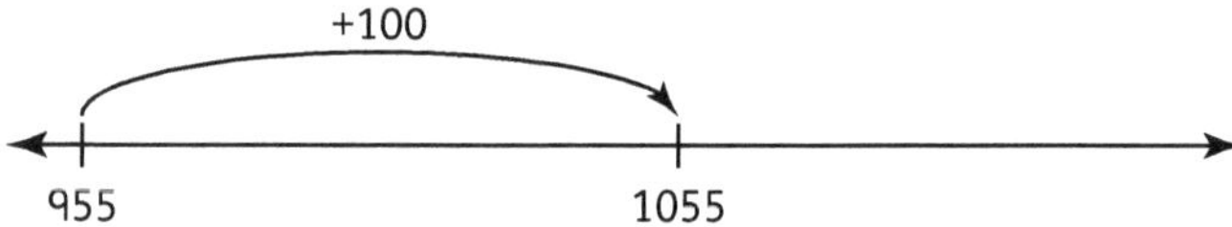

Next, I count on 70 in two steps.

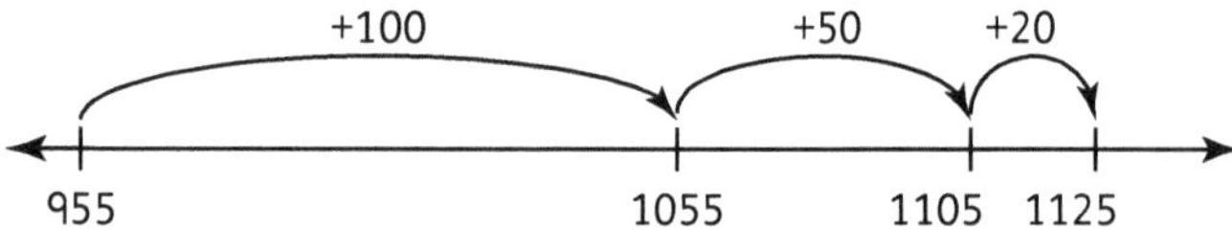

Finally, I count on 6 in two steps.

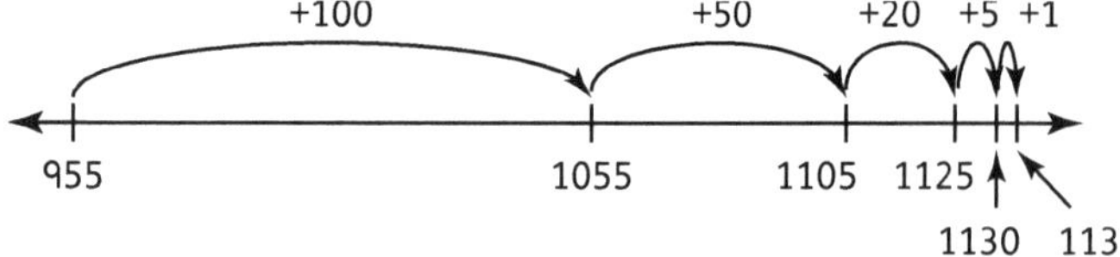

Ask: *Would you count on in the same way?* Let the children suggest another way to do the same calculation.
- Work through another example, asking the children to choose the steps (and explain their choices) before moving to **Pupil book 5 page 33**.
- Let the children work in pairs to read and discuss the examples at the top of the page.
- Note that question 1 does not specify that the children have to use number lines and count on to do the calculations. Some children might find the column addition method more efficient, so it would be counter-productive to ask them to use a less efficient method.
- The children can work independently to solve the problems in question 2 and to make up their own. Share some of the children's problems with the class.

Count in steps to subtract

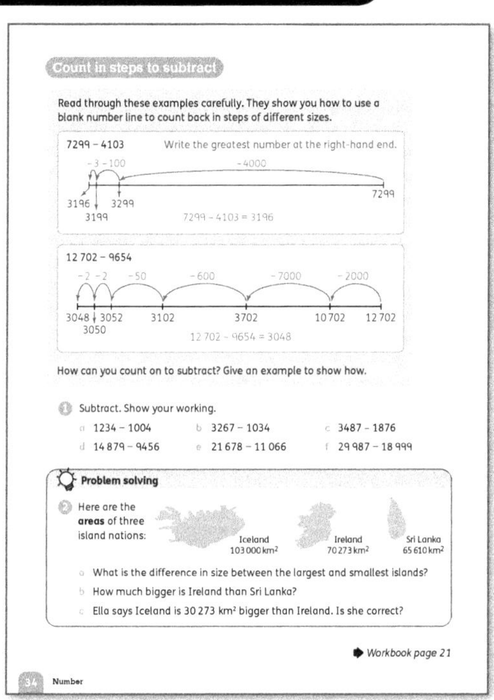

Warm-up
Select 'Mental addition and subtraction' from the 'Mental problem solving' section of the Activity Bank (pages 24–25) as a mental warm-up for this lesson.

Focus
- The question in the box asks the children to use an example to explain how you can count on to subtract. Discuss their suggestions with the class. (Instead of counting back in steps, you can count forward: for example, to subtract 167 from 2152 work out 167 **+ 3** = 170; 170 **+ 30** = 200; 200 **+ 800** = 1000; 1000 **+ 1152** = 2152. Then add: 1152 + 800 + 30 + 3 = 1985.)

- Draw a number line on the board and say to the class: *I have $195. I give away $172. How can we work out what I have left?*
- Draw a blank number line. Write 195 at the right-hand end of the line. Ask the class: *Why have I written 195 here and not on the left?* (Because it is the greater number so it must go at the right-hand end of the line.)
- Discuss how to find the answer by counting back. As with the additions in the previous lesson, draw each jump and explain what you are doing.

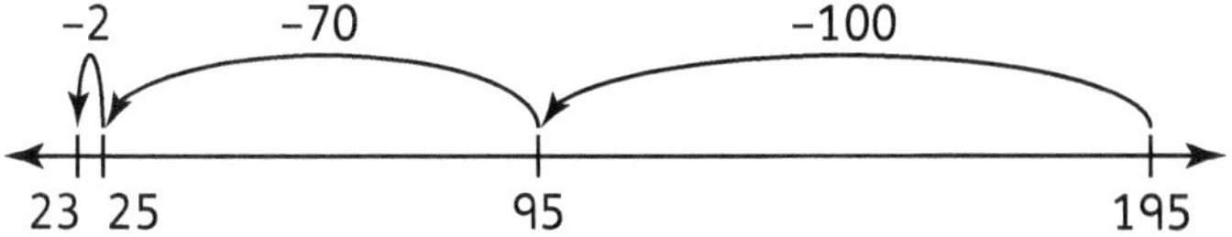

- Explain that we count back to 'take away' the smaller number from the larger. Point out that 23 is the difference between the two numbers.
- Redraw the number line. This time, write 172 on the left. Discuss how you can count up to find the difference between two numbers. Again, draw and explain the jumps, for example:

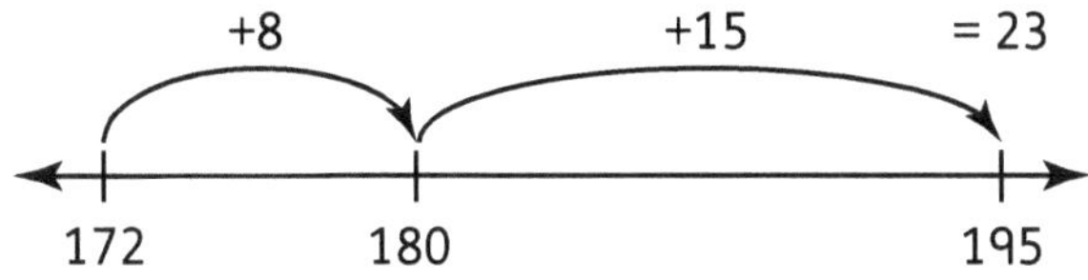

- Ask: *What jumps would you do?* Let the children explain their reasoning. (For example, they may jump in smaller increments to avoid having to add 15, or they may start with a jump of 20 (to get to 192) and then jump 3 to get to 195).
- Next, work through the counting back examples at the top of **Pupil book 5 page 34** with the class. After each one, ask the children to find the difference by counting on.
- As with the addition activities, the children are not told what method to use for question 1. Children who are confident using column subtraction may prefer to use that method.
- Problem solving: Before the children do question 2, remind the class that *area* is measured in square units and that km² means square kilometres.

Follow-up

Use **Workbook 5 page 21** to assess the children's understanding of counting in steps to add and subtract. The number lines force the children to count in the given steps, which demonstrate effective strategies.

Answers for Pupil book 5 page 34

1   **a** 230   **b** 2233   **c** 1611
    d 5423   **e** 10612   **f** 10988

2   **a** 37390 km²   **b** 4663 km²
    c No. Iceland is 32727 km² bigger than Ireland

Answers for Workbook 5 page 21

1   **a** 565   **b** 754   **c** 951   **d** 2121
   e 4207   **f** 8246   **g** 17   **h** 92
   i 190   **j** 397   **k** 447   **l** 1152

Interesting mistakes

- The children may more easily recognise some problems as subtraction than others. For example, they may realise that if there are 250 passengers and 45 get off the boat, they can subtract to find out how many are left. However, they may not recognise problems that involve comparing two amounts or finding the missing part of a whole as subtractions.
- Encourage the children to draw bar models (page 21) to represent situations, as this can help them to visualise the problem and more easily see what operation is required. Continue to discuss mistakes and ask the children to explain their thinking so that you can address any misconceptions.

Find pairs

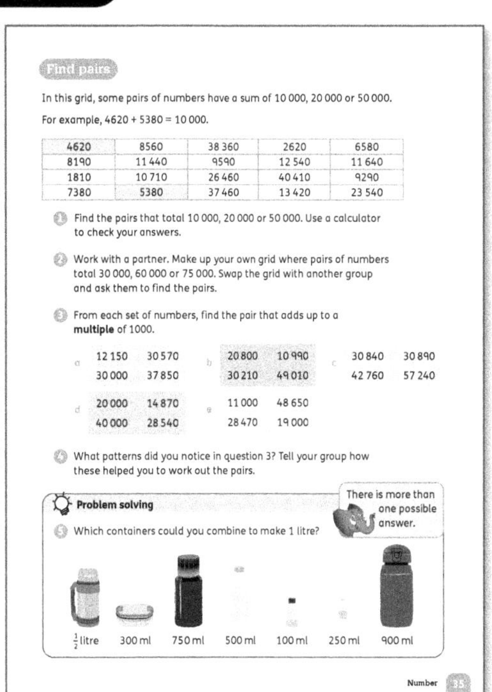

In this lesson, the children need to use the addition and subtraction facts they already know to make complementary pairs. Encourage the children to think about what they are doing and get them to discuss and share strategies for doing this.

Materials

Calculator; small counters

Warm-up

Revise adding and subtracting pairs of numbers to 100 as a mental starter for this lesson.

Focus

- Turn to **Pupil book 5 page 35**. Let the children work in pairs to complete question 1. They may find it helpful to cover numbers with small counters as they make and list the pairs so they reduce the number of choices each time. Let the children use a calculator to check their pairs.
- For question 2, discuss with the class how many blocks they will have in their grid. The grid in the book has 20 blocks, but they could have 16 to make the task more manageable.
- Let the children work in pairs to do question 3. Note that there is only one pair to find in each set.
- For question 4, give the children time to share their ideas in groups and then take feedback from the class.
- <u>Problem solving:</u> The children can work on question 5 independently.

Challenge

Ask the children to use the capacities of the containers shown in question 5 to make up some more problems. For extra challenge, ask them to make up two-step problems.

Support

If the children need support to make pairs to 100 and 1000, show them how to use jottings and place value to find the pairs. For example:

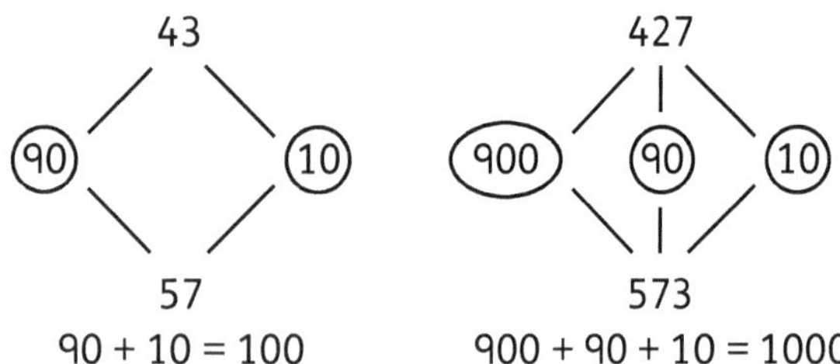

$$90 + 10 = 100 \qquad 900 + 90 + 10 = 1000$$

Interesting mistakes

Some children may still not be able to recall pairs to 10 (or 9). Help by providing continued and regular practice. Play games in which the children lay out number cards and take pairs that make 9 or 10 (or even 100) to increase their fluency.

1

Pairs that add up to 10 000	Pairs that add up to 20 000	Pairs that add up to 50 000
4620 + 5380	8560 + 11 440	38 360 + 11 640
8190 + 1810	6580 + 13 420	9590 + 40 410
7380 + 2620	10 710 + 9290	12 540 + 37 460

2 Individual answers.

3 **a** 12 150 + 37 850
   b 10 990 + 49 010
   c 42 760 + 57 240
   d 20 000 + 40 000
   e 11 000 + 19 000

4 Possible answers include:
- Look for two numbers which were already multiples of 1000.
- Look for two numbers where the digits in the tens column added to 100.
- Look for two numbers where the digits in the hundreds column added to 1000.

5 $\frac{1}{2}$ litre and 500 ml
   100 ml and 900 ml
   750 ml and 250 ml

Add larger numbers

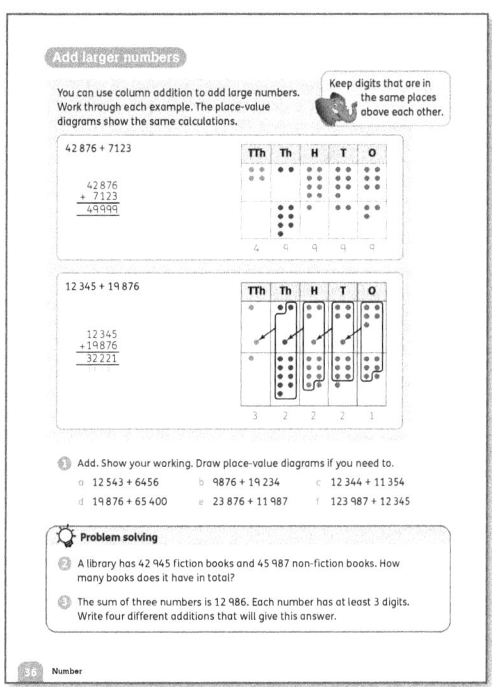

The children have already learnt how to add in columns. This lesson extends their skills to work in a higher number range.

Materials

Place-value tables (page 21) and counters; squared paper

Warm-up

Choose any 'Calculation skills' activities from the Activity bank (pages 25–28) as a mental starter for this lesson.

Focus

- Give the children time to talk in groups about what they remember about written methods for addition. Take feedback from the groups.
- Ask questions such as: *When do you need to carry to the next column? Why? Do you ever carry more than 1? Why do you write the 1 that you are carrying at the bottom of the column?* and so on.

- Work through the examples on **Pupil book 5 page 36** with the class. The visual representations remain important, particularly for children who still need support with place value.
- Explain that, as we work with larger and larger numbers, it is more efficient to use compact written methods. There is nothing 'wrong' with the other methods; they just take longer.
- Tell the children that they can use the place-value table and counters to model calculations if they need to. The children can work independently on question 1.
- Problem solving: Encourage the children to draw a bar model (page 21) for question 2.
- For question 3, point out that 'at least 3' means that there must be three digits in each number, but there can be more. Again, encourage the children to draw bar models or number lines to find the three numbers in each addition.

Interesting mistakes

The children must pay attention to the correct place value of each digit; otherwise their calculations can become thoroughly confused. This is especially important when adding or subtracting two numbers that have a different number of digits (for example, a 2-digit and a 3-digit number, or a 3-digit and 4-digit number). If necessary, they can make labels above the columns TTh, TH, H, T, O or they can work on squared paper to keep digits in the same place aligned in a column.

Answers for Pupil book 5 page 36

1   **a** 18 999   **b** 29 110   **c** 23 698
    d 85 276   **e** 35 863   **f** 136 332

2   88 932   **3**   Individual answers.

Subtract larger numbers

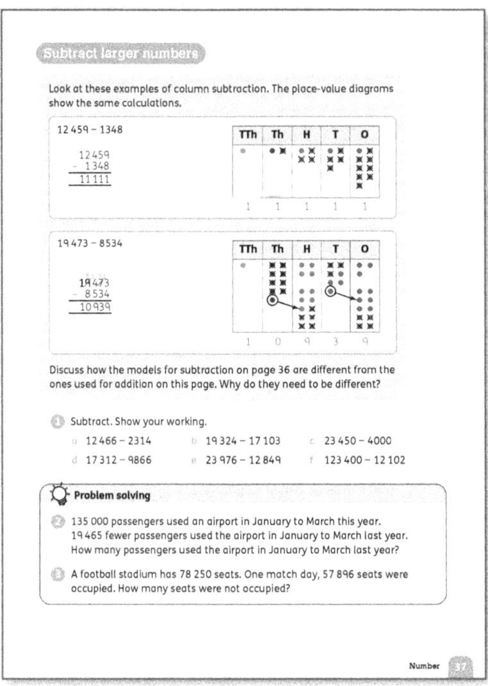

The children have already learnt how to subtract in columns. This lesson extends their skills to working in a higher number range.

Materials

Place-value tables (page 21) and counters; squared paper

Warm-up

Choose any 'Calculation skills' activity from the Activity bank (pages 25–28) as a mental starter for this lesson.

Focus

- Remind the children that they have already used written subtraction. Ask them to remember and share what they found easy and what they found challenging about this method. Allow other children to suggest how to overcome the challenges that are mentioned.
- Turn to **Pupil book 5 page 37**. Ask the children to read through the examples and compare the written method for each subtraction with the visual representation. Discuss with the class how the models for subtraction are different from those for addition and the reasons why they are different. (The diagrams need only one row because digits are crossed out from the starting number of digits. When subtracting, the arrows go from left to right because they show that a digit in a column is exchanged for 10 digits in the place-value column to the right. When adding, the arrows go from right to left to show that 10 digits in a column can be exchanged for 1 digit in the place-value column to the left.)
- Encourage the children to use written column methods for question 1, but allow them to model the subtractions with place-value tables and counters if they need to.
- Problem solving: Read the problems in questions 2 and 3 with the class and discuss the language and how they know what they need to find out. Ask them to draw diagrams to represent the problem.

Challenge

Give the children a list of topics, such as:
- Areas of different countries
- Population figures
- The area covered by forests on each continent
- Distances between cities
- The values of different countries' imports/exports
- Passenger numbers on rail or air transport
- Crowd numbers attending sport or cultural events.

Ask the children to choose a topic and do some research to find a set of numbers they can use in their own subtraction problems (or combined addition and subtraction problems). You can limit the number range to the millions, but if the children are confident working with even larger numbers, allow them to do so.

Support

Search for interactive games online that practise addition and subtraction.

Interesting mistakes

Children sometimes subtract the smaller digit from the larger one, for example in 57 – 39 they subtract 7 from 9.

- Modelling the subtractions on a place-value table (page 21) with counters can help the children to see how subtraction works and to understand that they have to subtract the 'bottom' number from the 'top' number, even when a digit is greater than the one above it. You can also use base-ten blocks (pages 20–21) to help with this.
- Avoid saying that we cannot subtract a bigger number from a smaller one (as subtractions such as 5 – 7 = –2, which have negative answers, are mathematically possible and normal). We cannot have negative digits in a number, so we don't do this in column subtraction.

Answers for Pupil book 5 page 37

1   **a** 10 152   **b** 2221   **c** 19 450
    d 7446   **e** 11 127   **f** 111 298
2   115 535   **3**   20 354

Inverse operations

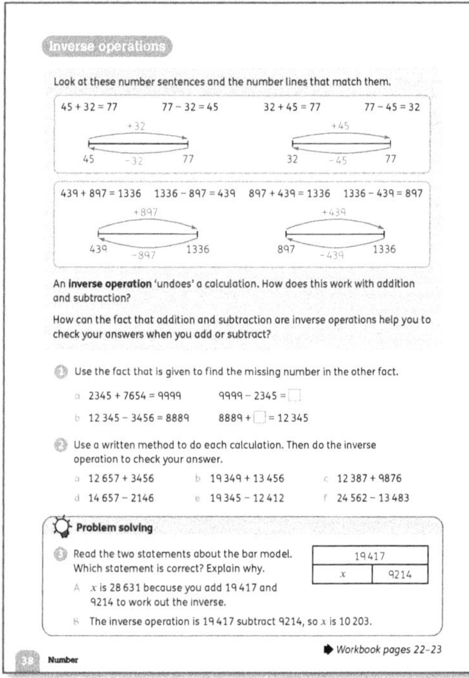

Materials

Calculators; abacuses; base–ten blocks (pages 20–21); counters

Warm-up

To give the children a break from number and calculation work, choose any activity from the 'Geometry and measures' section of the Activity bank (pages 28–29) as a mental warm-up for this lesson and to revise geometry vocabulary.

Focus

- Play a quick movement game. Take the class into the school playground and tell them to spread out. Ask them to hop forward a number of times. Then, ask them how they can hop back to their starting point. Elicit that if they go forward three hops, they need to go back three hops to get to the starting point. Repeat a few times, sometimes starting with hopping back and then hopping forward to get to the starting point.
- Back in the classroom, turn to **Pupil book 5 page 38**. Ask the children how the number lines are similar to the hops they did. Let them work in groups to discuss the questions and then let different children share the ideas from their group with the rest of the class.
- Check that the children understand the concept of an *inverse operation* by giving them two numbers to add. The questions in the example box ask the children to explain how an inverse operation undoes a calculation and helps you to check the answer to an addition or subtraction. (To check an addition, you can do a subtraction: for example, to check that 259 + 374 = 633 you can find 633 – 374 and check that it gives 159 (or find 633 – 259 and check that it gives 374). To check a subtraction, you can do an addition: for example, to check that 736 – 489 = 247 you can find 247 + 489 and check that it gives 736.) They can use abacuses, base-ten blocks or counters to model the operation if they need to. Then ask them to check their answer using subtraction. Ask them to explain what they did (for example, to check 23 + 29 = 52, subtract 23 from 52 and check the answer is 29).
- The children can work independently to complete question 1 and question 2.
- <u>Problem solving</u>: Let the children think about their answers to question 3. Then have them share their ideas with a partner or in small groups.

Follow-up

Use **Workbook 5 page 22** as an informal assessment task to check that the children can use inverse operations and column addition and subtraction confidently. For question 2, they can show their working next to the diagrams or in their books.

On **Workbook 5 page 23**, tell the class that we often make mistakes when we do calculations. Let the children work on their own or in pairs to find and circle the mistakes in the calculations. Spend some time discussing their answers to question 2.

Answers for Pupil book 5 page 38

1   **a** 7654   **b** 3456
2   **a** 16 113   **b** 32 805   **c** 22 263
    d 12 511   **e** 6933   **f** 11 079
3   B is correct. The bar model shows that $x + 9214 = 19\,417$, so using inverse operations gives $x = 19\,417 - 9214 = 10\,203$.

Answers for Workbook 5 page 22

1 a

```
   3155
  −1420
   1735
```

b

```
   7863        3667
  −4196       +4196
   3667        7863
```

c

```
  12410       16219
  +3809       −3809
  16219       12410
```

d

```
  17217       21325
  +4108       −4108
  21325       17217
```

2 a 3300 km **b** 3060 km
c 3400 km **d** 2399 km

Answers for Workbook 5 page 23

1 a 2681 **b** 21 304 **c** 3647 **d** 3443
e 50 613 **f** 14 865 **g** 28 785 **h** 34 439

2 Individual answers, for example, they forgot to carry a one when adding, they subtracted the smaller digit from the larger digit, instead of the bottom number from the top number.

Estimate and give approximate answers

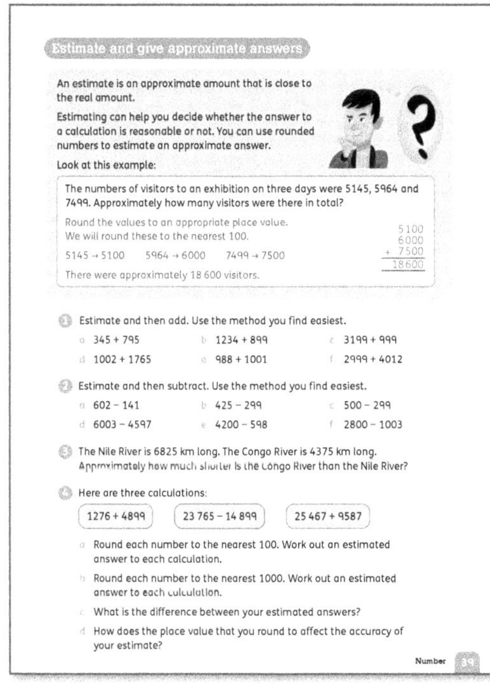

Materials
Photographs or pictures showing large numbers of items; flashcards with the words *estimate* and *approximate* on them; several supermarket receipts with the totals cut off

Warm-up
As a mental starter, display a picture of a large number of items and ask the children to estimate how many items there are. Let them explain how they made their estimates.

Focus
- Display the words *estimate* and *approximate*.
- Ask the children what it means if a plumber, builder or electrician gives you an estimate for a job you want them to do. (In this context, an estimate is an approximate total cost for a job, usually with a breakdown of the individual costs that make up the total.) Then tell the class that the letters ETA stand for 'estimated time of arrival'. Discuss what this means. (It is the expected time that a person or means of transport will arrive if the journey goes as planned.)
- Explain that in maths we round numbers and use those to calculate an estimated answer to a calculation.
- Give each group a set of three or four till receipts with the totals cut off. Ask them to calculate an estimate for each total and to place the receipts in ascending order of total cost. Let the children share how they worked to do this. Focus on estimating and not on accurately rounding decimals at this stage. Discuss a suitable level of accuracy for the estimated totals if you want to make sure that you have enough money to pay (for example, to the nearest dollar rather than to the nearest cent).
- Explain that we estimate amounts in maths to check that our answers are reasonable.
- Turn to **Pupil book 5 page 39**. Work through the example with the class. Then let them complete questions 1–3. You may wish to tell the children what to round to (for example, nearest 100 or nearest 1000), or you could let them decide this for themselves. Discuss a suitable level of accuracy for the answer to question 3. *Do we need to know the difference in the lengths of the two rivers to the nearest kilometre? Would the nearest 10 km or nearest 100 km be accurate enough?* (The answer to the nearest 100 km is accurate enough because the lengths are so long.)
- Question 4 involves thinking critically about how the place we round to changes the estimated answer. Let the children discuss the questions in groups and then take feedback as a class.

Answers for Pupil book 5 page 39

1
a 350 + 800 = 1150; 1140
b 1200 + 900 = 2100; 2133
c 3200 + 1000 = 4200; 4198
d 1000 + 1800 = 2800; 2767
e 1000 + 1000 = 2000; 1989
f 3000 + 4000 = 7000; 7011

2
a 600 − 140 = 460; 461
b 430 − 300 = 130; 126
c 500 − 300 = 200; 201
d 6000 − 4600 = 1400; 1406
e 4200 − 600 = 3600; 3602
f 2800 − 1000 = 1800; 1797

3 6800 − 4400 = 2400 km

4
a 6200, 8900, 35 100
b 6000, 9000, 35 000
c 200, 100, 100
d Possible answer: If you round to a smaller number (for example, 100 rather than 100), your estimate will be more accurate.

Estimate when adding and subtracting

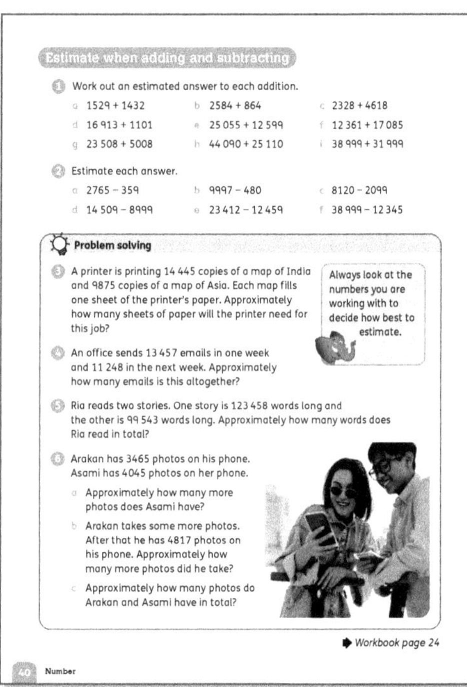

Materials
Calculators; blank number lines

Warm-up
Select any suitable 'Rounding and estimating' activity from the Activity bank (page 24) as a starter for this lesson.

Focus
- **Turn to Pupil book 5 page 40**. Explain that we can use estimating to check that our answers are reasonable. This is particularly important when we use a calculator. If we do not have an idea of an approximate answer, we may not realise that we have made mistakes.
- For question 1 and question 2, let the children work on their own to round the numbers and find an approximate answer. Point out that you are not asking them to work out the actual answers here.
- <u>Problem solving:</u> Let the children work in pairs on question 3. They can then use a calculator to find the actual answer. Encourage them to discuss with their partner whether their calculated answer is reasonable, by comparing it with their estimate.
- The children can work on their own or in pairs to solve the word problems in questions 4–6. Make sure they write an estimate before they find the actual solution and use it to check that their answer is reasonable.

Follow-up
Let the children complete the estimates and calculations on **Workbook 5 page 24** independently. Then let them compare answers and check each other's work.

Interesting mistakes
- The children may not see the value of estimating and leave out that part of any instruction.
- Continue to provide activities (such as question 1 and question 2), where the estimate is the required answer and they do not have to do the actual calculation.
- Insist that the children record their estimates before they start any problem-solving activities.

Challenge
Tell the children: *A journalist rounded the number of people who attended an art exhibition to the nearest 100 and reported that 5300 people attended. What are all the possible numbers of people who attended? Show how you worked this out.* (any number between 5250 and 5349)

Support
Remind the children who need support with rounding that they can use number lines to help them (see page 39).

Answers for Pupil book 5 page 40

Possible answers are:

1
a 1500 + 1400 = 2900
b 3000 + 900 = 3900
c 2000 + 5000 = 7000
d 17 000 + 1000 = 18 000
e 25 000 + 13 000 = 38 000
f 12 000 + 17 000 = 29 000
g 24 000 + 5000 = 29 000
h 44 000 + 25 000 = 69 000
i 39 000 + 32 000 = 71 000

2
a 2800 − 400 = 2400   b 10 000 − 500 = 9500
c 8100 − 2100 = 6000   d 14 500 − 9000 = 5500
e 23 000 − 12 000 = 11 000
f 39 000 − 12 000 = 27 000

3 14 000 + 10 000 = 24 000

4 13 500 + 11 200 = 24 700

5 123 000 + 100 000 = 223 000

6
a 4000 − 3500 = 500   b 4800 − 3500 = 1300
c 4800 + 4000 = 8800

Answers for Workbook 5 page 24

1 10 000 + 10 000; 20 000; 24 023
10 000 + 10 000; 20 000; 26 099
20 000 + 20 000; 40 000; 37 653
20 000 – 10 000; 10 000; 10 944
20 000 – 10 000; 10 000; 8778
50 000 – 10 000; 40 000; 40 645

2 Individual answers; 59 000 km²

3 Mauritius 320 000; Sri Lanka 420 000

Multi-step problems

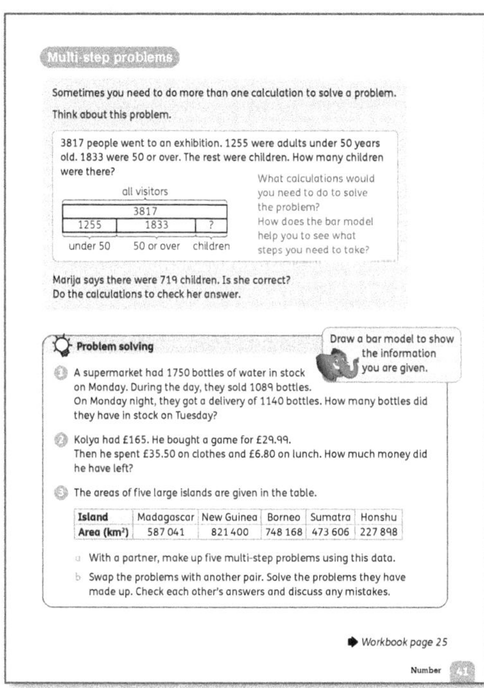

Materials

Rulers and marker pens; calculators; atlas or world map

Warm-up

Use any suitable 'Mental problem solving' activity from the Activity bank (pages 24–25) as a warm-up for this lesson.

Focus

- Place the children in pairs. Give them time to read the problem at the top of **Pupil book 5 page 41**, study the diagram and answer the questions. Have a class discussion about the results. (There were a total of 1255 + 1833 = 3088 adults. So, there were 3817 – 3088 = 729 children. Therefore, Marija is not correct. She could have checked her answer by addition: 1255 + 1833 + 719 = 3807. The total should be 3817, so her answer is not correct.)
- Explain that bar models are useful for showing problems. In multi-step problems, we usually need more than one bar.
- Turn to **Workbook 5 page 25**. Let the children work independently to complete the bar models and solve the multi-step problems. This activity gives the

children experience of different types of bar models and allows them to see how they can work with two bars to show steps in a problem. It is helpful to do this kind of work before asking the children to draw their own bar models.

- Check the answers as a class and let the children compare the bar models they drew to show the last problem.
- Problem solving: Let the children work through question 1 and question 2 on **Pupil book 5 page 41** in class or as homework.
- Give the children time to work in pairs on question 3. They can write their problems on cards or sheets of paper and then exchange them with other pairs to solve. Discuss the activity afterwards. Ask the children what they found easy, what they found challenging and what interesting mistakes they made when trying to solve each other's problems.

Challenge

Ask the children to write a suitable word problem to match this set of calculations.

£19.47 − £3.50 = £15.97
£4.25 + £1.99 = £6.24
£15.97 + £6.24 = £22.21

Interesting mistakes

- The children might use bar models incorrectly, which can lead to confusion. For example, the children might represent equal quantities using unequal bars. For example, they might show a comparison problem like this:

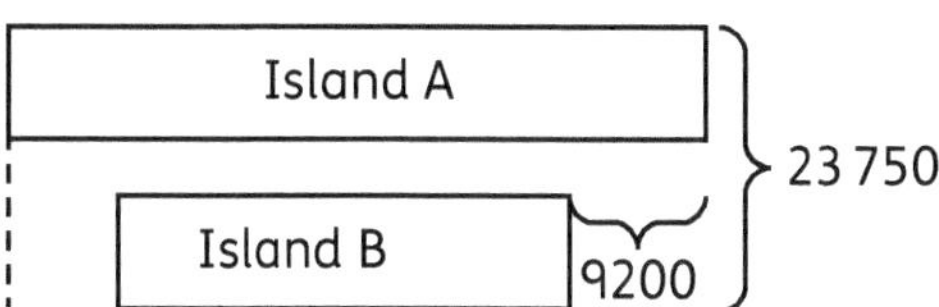

- One way of addressing this is to tell the children always to draw a vertical line on the left and to start all bars in the problem against that line. This will ensure that the difference will be represented correctly as a gap at the other end. Drawing the bars so they touch each other will also help the children to compare the bars more easily.

Answers for Pupil book 5 page 41

1 1801 bottles   **2** £92.71

3 Individual answers.

Answers for Workbook 5 page 25

1 £470   **2** 935 and 1665

3 9815   **4** 21 940

Assess computation skills by asking the children to complete addition and subtraction calculations involving 4-, 5- and 6-digit numbers.

Give the children some problems that combine addition and subtraction to check that they can solve multi-step problems.

You may also like to ask some or all of these questions to assess how well the children have understood the concepts of this unit.

- *What is the sum of/difference between 366 and 178/941 and 199/253 and 187?* (366 and 178: 544/188; 941 and 199: 1140/742; 253 and 187: 440/66)
- *How do you decide whether to add or subtract to find an answer?* (For example: Words such as total or sum may mean add. Words such as difference, less than or how much bigger may mean subtract.)
- *What is the total of 472 and 129/588 and 259/1376 and 3269?* (601/847/4645)
- *What do you need to add to* (give a number) *to make 10 000? How did you work this out?* (For example: Use number bonds to 9 and 10. The ones digits add up to 10 and the tens, hundreds and thousands digits add up to 9. Count back on a number line.)
- *Find the sum of* (give two numbers).
- *How much greater is* (give a number) *than* (a smaller number)?
- *How can you draw a diagram to show this subtraction?* (Give two numbers.) (For example: a place-value table with counters.)

- *What do I need to add to* (give a number) *to get 2400?*
- *My total is 12 514. What two 4-digit numbers could I have added to get this total?* (For example: 4213 and 8301.)
- *I know that 12 369 + 5897 = 18 266. What is 18 266 less 12 370? How did you work this out?* (5896. It is 1 less than 18 266 less 12 369.)
- *Show me how you would add/subtract these two numbers.* (Give two numbers.)
- *Approximately how much is 12 345 and 3 876?* (For example: 12 000 + 4000 = 16 000.)
- *How do we use estimating in daily life?* (For example: Adding up items in a shop, deciding whether we can afford something, estimating size of a room for decorating.)
- *Draw a bar model to show this problem:*
  Khai had to complete a number of steps to get to the next level in a maths game. He completed half of the steps in class and then completed 5 more steps in the bus on the way home. That left him with 15 steps to complete. How many steps did he need to complete altogether to get to the next level? (40)

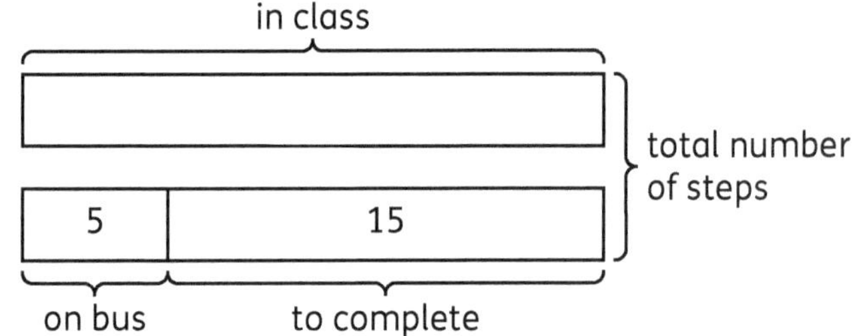

UNIT 5
Decimals and percentages

Learning objectives

- Extend the place-value table to include thousandths, and relate these to tenths and hundredths

- Understand and explain the value of any digit in a decimal with up to three decimal places

- Compose, decompose and regroup numbers with up to three decimal places using standard and non-standard partitioning

- Round decimals (two decimal places) to the nearest whole number and to one decimal place

- Read, write, order and compare numbers with up to three decimal places

- Recognise the per cent symbol (%) and understand that per cent relates to 'number of parts per hundred'

- Write a percentage as a fraction with a denominator of 100 and as a decimal

- Use waffle diagrams to show percentages of a whole (100 parts) and answer questions about the proportions shown

- Read and write decimal numbers as fractions

- Solve problems using decimals, fractions and equivalent percentages

Key words

thousandth   percentage   whole   decimal
equivalent fraction   denominator
ascending   descending

Materials
0–9 digit cards

Teaching guidance
- Remind the class that they worked with decimals in Level 4. Ask the children to define a *decimal*. Refine their suggestions to create a definition that works for the class.
- Draw a row of six boxes on the board, putting a decimal point before the last two boxes. Ask the children to copy these boxes.

- Tell the children you are going to pick a card from a set of 0–9 digit cards and then return it to the pack and shuffle the cards before picking again.
- The children write each digit in a box as it is picked. Once positioned, a digit cannot be moved.
- The aim is to create the largest number when all six boxes have been filled.
- After one round, discuss with the children the strategies they used.
- Repeat a few times. Then change to making the smallest possible number.
- Discuss how the children's strategies change when they have to make a small number.

Revisit decimal place value

[Reproduction of Pupil book 5 page 42]

UNIT 5 — Decimals and percentages
Revisit decimal place value

Think and share

Each grid represents one **whole**. What does one part of each grid represent?

What is the shaded part of each grid as a fraction? What is it as a **decimal**? What is the unshaded part of each grid as a decimal?

The place-value table shows that $\frac{25}{100}$ is made up of 2 tenths and 5 hundredths.

$\frac{1}{100}$ is made up of 0 tenths and 1 hundredth so we write a 0 in the tenths column as a place holder.

Ones	tenths	hundredths
0	2	5
0	0	1

1. For each diagram, write a decimal to represent the shaded part and another decimal to represent the unshaded part.

2. Look at the digit 9 in each number.

47.9   9.13   0.39   3.96   90.6

a. In which number does the digit 9 have the highest value?
b. In which two numbers does the digit 9 have the same value?

➜ Workbook page 26

42 Number

Materials
Calculators; place-value cards (page 21), or 0–9 digit cards (plus decimal point cards) for each child; place-value table that includes hundredths (page 21); crayons or marker pens

Warm-up
- Play an adapted version of the 'Wipe out calculator game' in the Activity bank (page 24) with numbers that include two decimal places.
- Give the children a set of place-value cards or several sets of 0–9 digit cards and a decimal point card.
- Say a decimal number and ask the children to lay out the cards to show that number. Ask them to say the value of some of the digits in the number.
- Repeat with more decimal numbers.

Focus
- <u>Think and share:</u> Turn to **Pupil book 5 page 42**. Let the children work in pairs to read the information and answer the questions about parts of a *whole*. Take feedback as a class and resolve any misconceptions. (In each grid, one part represents one tenth ($\frac{1}{10}$) or 0.1. $\frac{4}{10}$ or 0.4 is shaded in the top grid and $\frac{1}{4}$ or 0.25 is shaded in the second grid. 0.6 of the top grid is unshaded and 0.75 of the bottom grid is unshaded.)
- The children should be able to complete question 1 independently. Check their work to make sure that they can.
- The children can do question 2 orally in pairs. Alternatively, do this question with the class, giving the children time to think about the answers before asking for contributions.

Follow-up
Use **Workbook 5 page 26** to check that the children can represent decimals in a diagram and write decimals with two decimal places. Observe the children as they work to see whether they count squares or whether they use complementary pairs to 100 to work out the unshaded decimal. Once they have completed the questions, ask different children to say how they worked out the decimals.

Interesting mistakes
There are many common misconceptions associated with decimals.
- Some children think that the digits after the decimal point work in the same way as those before. For example, they might say 34.56 as 'thirty-four point fifty-six'. This leads to incorrect comparison of numbers – for example, the children might say that '34.56 is larger than 34.7 because fifty-six is bigger than seven'. Use place-value tables or notation cards to emphasise the value of the digits in a number and regularly practise how to say numbers correctly.
- Another common error relates to making mistakes in sequences of decimals. For example, the children might continue the sequence 0.6, 0.7, 0.8, 0.9 with 0.10, which they think represents ten tenths. Discuss with the children what ten tenths means, using diagrams to support understanding. Use a place-value table to show that 0.1 and 0.10 are the same number.

Answers for Pupil book 5 page 42
1 a 0.06; 0.94   b 0.57; 0.43   c 0.22; 0.78
d 0.02; 0.98   e 0.13; 0.87   f 0.78; 0.22
2 a 90.6   b 47.9 and 3.96

Thousandths

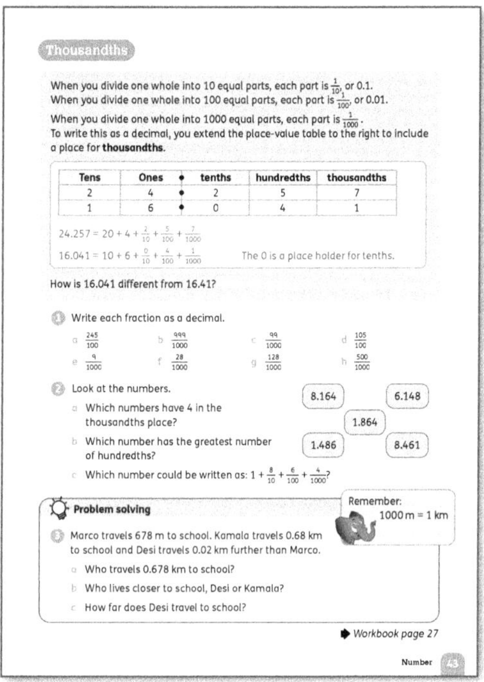

Materials

- Four strips showing one whole, tenths, hundredths and thousandths (see below); 1000 charts (large square divided into 1000 small squares, search for this online); blank number lines with ten divisions; crayons or marker pens

Warm-up

Do some activities involving multiplication and division by 10 and 100 as a mental warm-up for this lesson. Include questions such as: *What number is 10 times greater/smaller than …? What number is 100 times greater/smaller than …?*

Focus

Display a strip showing one whole:

- Ask the class what each piece represents if you divide the strip into 10 equal parts like this:

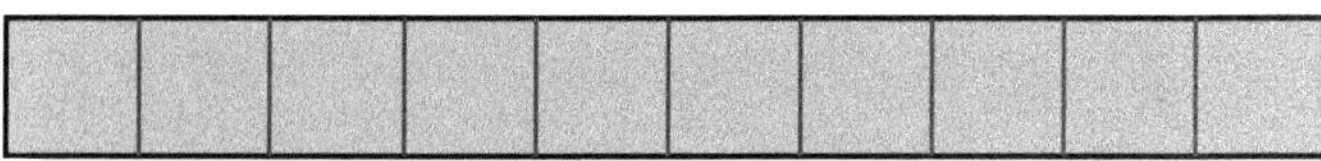

- They should be able to tell you that each part is $\frac{1}{10}$ of the whole. Remind them that $\frac{1}{10} = 0.1$
- Ask: *What do we get if we divide each of the 10 parts into 10 parts?* (We get hundredths. Each part is 1 of 100 equal parts or $\frac{1}{100}$.) *How can we write $\frac{1}{100}$ as a decimal?* (0.01)

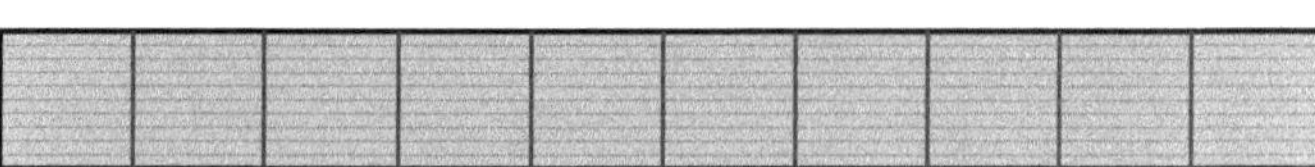

- Then ask: *What if we divide each of the hundredths into 10 equal parts?* (We get 1000 small blocks and each block is $\frac{1}{1000}$ of the whole.)
- Ask: *How can we write $\frac{1}{1000}$ as a decimal?* (0.001)

Next, show the children a 1000 chart with some blocks shaded. Use different colours for the tenths, hundredths and thousandths. In this example, $\frac{376}{1000}$ blocks are shaded.

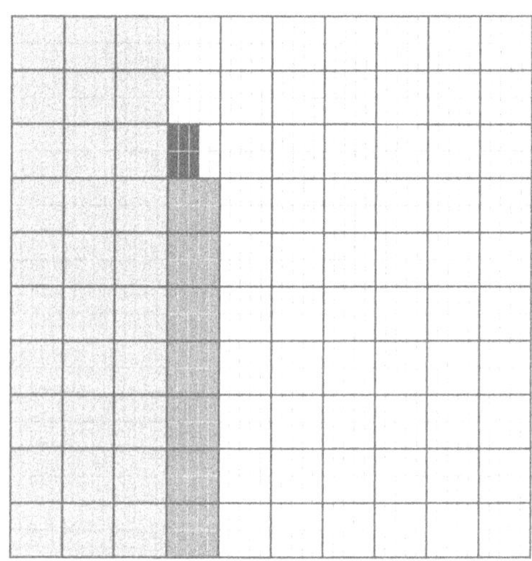

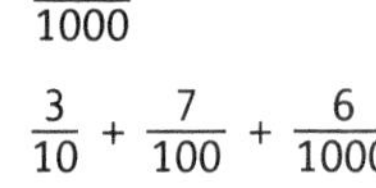

$$\frac{376}{1000}$$

$$\frac{3}{10} + \frac{7}{100} + \frac{6}{1000}$$

- Point out that $\frac{376}{1000} = \frac{3}{10}$ (the long strips) $+ \frac{7}{100}$ (the squares) and $\frac{6}{1000}$ (the smallest blocks). We can write this number as 0.376
- Point to the 1000 chart and show that 0.376 is more than $\frac{3}{10}$ but less than $\frac{4}{10}$. This means it is between 0.3 and 0.4 on a number line.
- Draw the tenths number line on the board and show the position of 0.376, between 0.3 and 0.4.

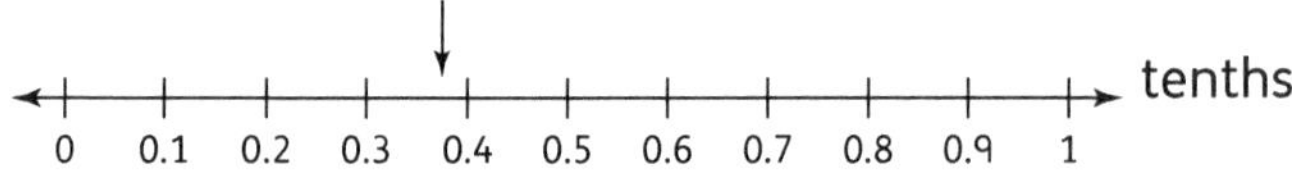

- If we zoom in, we get the hundredths number line, which shows the divisions between 0.3 and 0.4. Show the class that 0.376 is between 0.37 and 0.38.

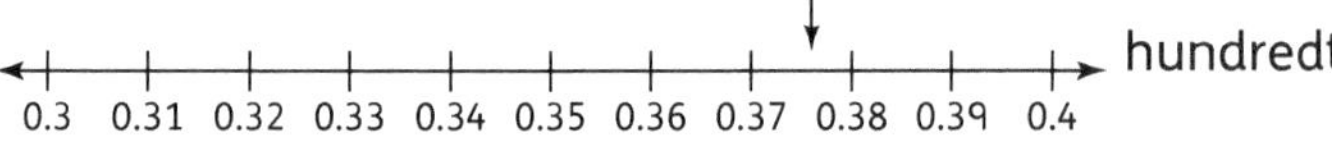

- If we zoom in further, we get the thousandths
  number line, which shows the divisions between 0.37
  and 0.38. We can show the exact position of 0.376 on
  this number line.

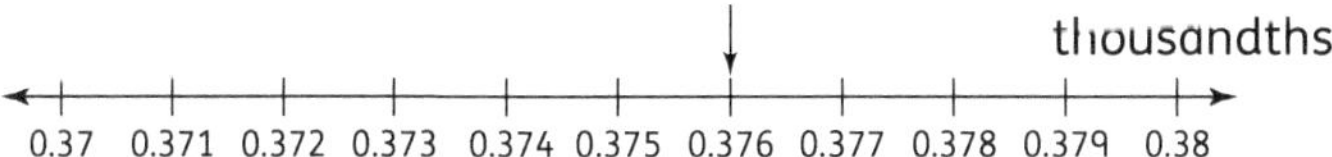

Give each child a 1000 chart, and three blank number lines.

- Ask the children to choose a number of thousandths
  to shade on their square. (Or tell them a number to
  shade, for example: $\frac{641}{1000}$.)
- Then ask them to write the fraction and decimal next
  to it (for example, $\frac{641}{1000}$ and 0.641).
- They can then mark the position of the number on a
  number line showing tenths, then hundredths, and
  then thousandths.

Turn to **Pupil book 5 page 43**. Read through the text and
examples with the class to make sure that the children
understand what *thousandths* are and how they are
represented in a place-value table. The question in the
example box asks the children to explain how 16.041
is different from 16.41. ($16.041 = 16 + \frac{4}{100} + \frac{1}{1000}$, whereas
$16.41 = \frac{4}{10} + \frac{1}{100}$.)

- Spend some time dealing with 0 as a place holder in
  decimals. Demonstrate to the class that zeros can
  be left off the end of decimals. For example: *We can
  write $\frac{500}{1000}$ as 0.500, but it is more efficient to write 0.5.*
  Refer back to the strips or grids to show that $\frac{500}{1000}$ is
  equivalent to $\frac{5}{10}$ so it can be written as 0.5.
- The children can work independently on question 1
  to write each fraction as a decimal.
- Question 2 can be done orally with the class or as a
  written activity.
- <u>Problem solving</u>: Let the children discuss question 3
  before they solve the problem. They should work out
  by themselves that they can express 678 m as $\frac{678}{1000}$ km
  or 0.678 km.

Follow-up

Use **Workbook 5 page 27** to consolidate work on
thousandths. The questions use an area representation
for thousandths. The children can work independently or
in pairs.

Challenge

Let the children do their own research and prepare an
information sheet showing how decimals with three
places (thousandths) are used in daily life. Possible
examples include: race times to thousands of a
second; processing speeds of computers (gigahertz);
measurements in medicine (pH of blood gases, for
example); engineering precision (measurements taken
with a micrometer); baseball batting averages.

Answers for Pupil book 5 page 43

1 a 2.45   b 0.999   c 0.099   d 1.05
    e 0.009   f 0.028   g 0.128   h 0.5
2 a 8.164 and 1.864   b 1.486   c 1.864
3 a Marco   b Kamala   c 698 m or 0.698 km

Answers for Workbook 5 page 27

1 a 0.064   b 0.032   c 0.02
    d 0.08   e 0.016   f 0.03
2 a top right   b middle right
    c bottom left   d middle (beneath seating area)
3 0.621

Represent decimals in different ways

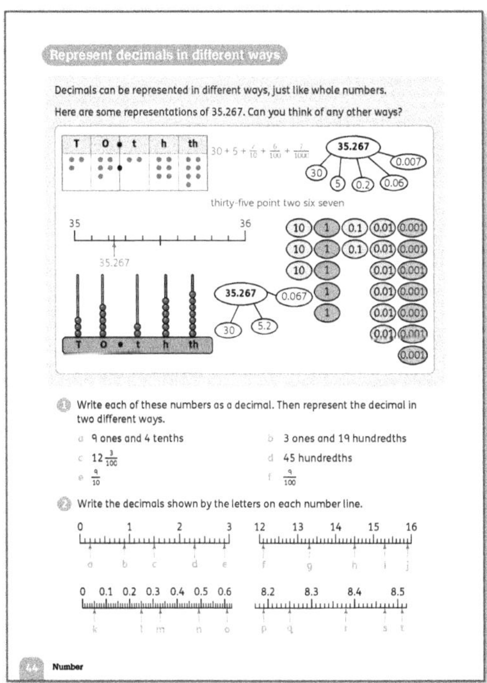

It is important that the children learn to treat
decimals (and other fractions) as numbers rather
than separate elements of maths. This lesson
focuses on representing and partitioning decimals
in different ways.

Materials

Marker pens; place-value tables (page 21); place-value
counters

Warm-up

Use the 'Write the value of the given digit' activity
from the 'Place value and number sense' section of
the Activity bank (page 22) as a mental starter for
this lesson.

Focus

- Turn to **Pupil book 5 page 44** and let the children
  discuss the representations of decimals and suggest
  any others they can think of.
- For question 1, let the children write the numbers and
  show them in different ways. They can then compare
  their representations with others in their group.
  Encourage them to use a range of representations.
- Remind the children to look at the number lines and
  work out what the divisions represent before they
  complete question 2.

Challenge

- Give the children some pairs of decimals (for example, 3.4 and 3.5 or 10.25 and 10.26). Ask them to write five decimals with three decimal places that would be between their two decimals on a number line.
- Ask the children to discuss these questions in their groups. *Is $\frac{1}{1000}$ greater than or less than $\frac{1}{100}$?* (Less than) *Explain how you know.* (For example, there are more thousandths than hundredths in one whole, so thousandths must be smaller than hundredths.) *How many thousandths are there in 2 wholes?* (2000) *Why?* (1000 thousandths make one whole) *How many thousandths are there in one hundredth?* (10) *Why is $\frac{5}{1000}$ not equal to 0.5?* (Because there are no tenths and no hundredths in $\frac{5}{1000}$, so you use 0 placeholders, and write 5 in the thousandths place, 0.005)

Support

Let the children physically model decimals with counters and place-value tables to help them to draw their representations. Decomposing and regrouping the numbers will help them gain confidence and a better understanding of place value to thousandths.

Answers for Pupil book 5 page 44

1   a 9.4   b 3.19   c 12.03
    d 0.45   e 0.9   f 0.09

2   a = 0.2, b = 0.9, c = 1.5, d = 2.3, e = 2.9
    f = 12.1, g = 13.3, h = 14.5, i = 15.3, j = 15.9
    k = 0.05, l = 0.25, m = 0.33, n = 0.49, o = 0.61
    p = 8.19, q = 8.25, r = 8.38, s = 8.47, t = 8.51

Compare and order decimals

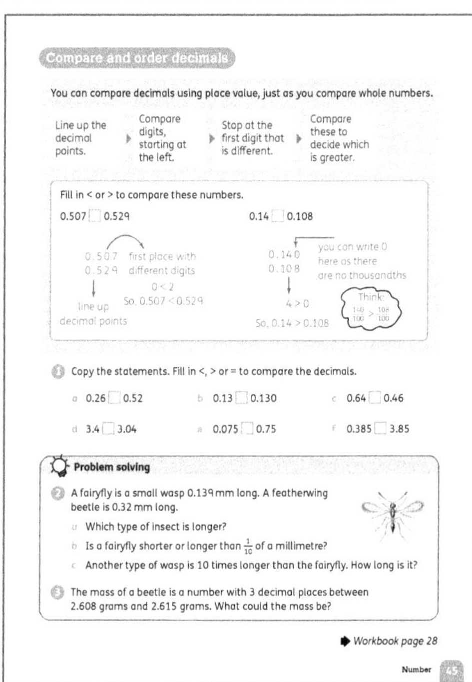

Materials

Flashcards showing numbers with two and three decimal places

Warm-up

Use 'Compare 3-digit numbers' or 'Compare 5-and 6-digit numbers' (page 23) as a starter for this lesson.

Focus

- Remind the class that we can compare numbers using the =, < and > symbols. Write statements on the board such as 10.31 > 10.19. Ask the children to say whether each statement is true or false and how they decided.
- Ask the class what it means to put numbers in *ascending* or *descending* order.
- Ask five or six children to come to the front of the class and give each child a flashcard showing a decimal with two decimal places each.
- Ask the rest of the class to give instructions to the children so that they stand in a line with the cards in ascending order. Give another child a card and ask them to stand in the correct position in the line.
- Repeat the activity using decimals with three decimal places.
- Repeat again, with different decimals and descending order.
- Turn to **Pupil book 5 page 45**. Let the children read the steps in the flow diagram and the two worked examples. Ask the children if they would need to show each step of the process in their jottings. (The children are likely to work more efficiently than this as they gain understanding and experience.)
- The children can work on question 1 in pairs.
- <u>Problem solving:</u> Read through the problems in question 2 and question 3 and discuss the questions before asking the children to work out the answers.

Follow-up

Use **Workbook 5 page 28** as an informal assessment task to check that the children are able to use number lines to show the position of decimals and to compare and order decimals.

Interesting mistakes

- The children may mistakenly apply rules for whole numbers when comparing and ordering decimals. For example, they may say that '0.065 > 0.21 because 65 > 21' or that '3.06 > 3.5 because it has more digits'.
- Similarly, they may mistakenly apply what they know about fractions. For example, because $\frac{1}{405} > \frac{1}{450}$, the children may say that '0.405 > 0.450'.
- Other mistakes may arise from thinking that zeros to the right of a decimal number change the value. For example, the children may say that '0.3 < 0.300 because 3 < 300'.
- The children may also think that decimals with only tenths are greater than decimals that include thousandths because tenths are bigger than thousandths.

If any children make any of these mistakes, have a number talk (pages 17–18) and ask the children to show visually how they arrived at their conclusion. The visual representation will often help them to see why they made the mistake. Shading the decimal parts of a number on 1000 charts (see page 68) will also help the children to understand the comparisons.

Answers for Pupil book 5 page 45

1 **a** 0.26 < 0.52 **b** 0.13 = 0.130
c 0.64 > 0.46 **d** 3.4 > 3.04
e 0.075 < 0.75 **f** 0.385 < 3.85

2 **a** a featherwing beetle
b longer **c** 1.39 mm

3 2.609 g, 2.610 g 2.611 g, 2.612 g, 2.613 g, 2.614 g

Answers for Workbook 5 page 28

1

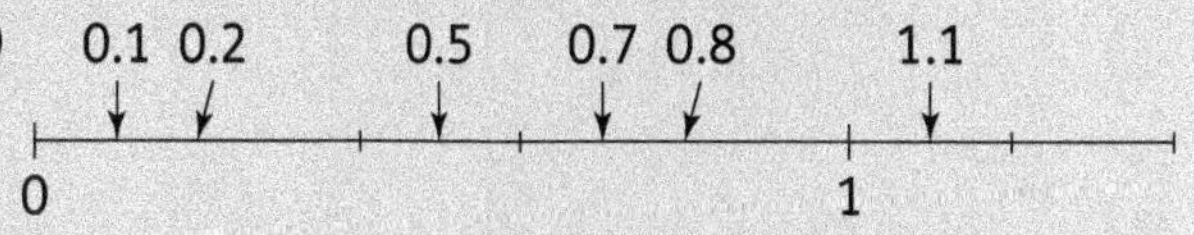
0.1
0.2
0.5
0.7
0.8
1.1
0
1

2

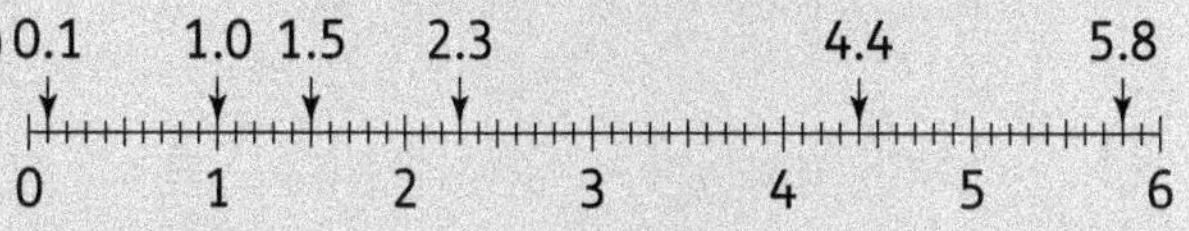
0.1
1.0
1.5
2.3
4.4
5.8
0
1
2
3
4
5
6

3

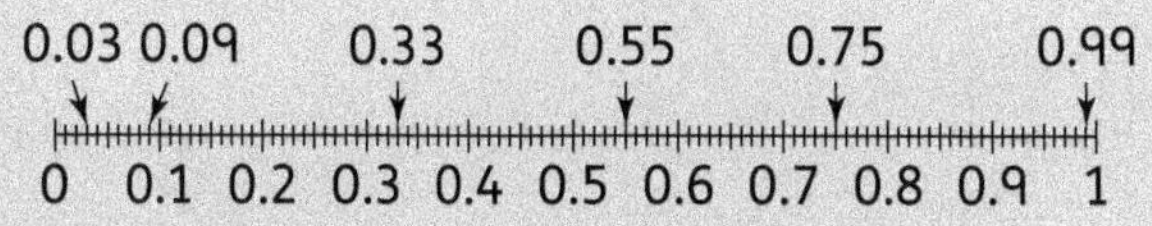
0.03
0.09
0.33
0.55
0.75
0.99
0
0.1
0.2
0.3
0.4
0.5
0.6
0.7
0.8
0.9
1

4

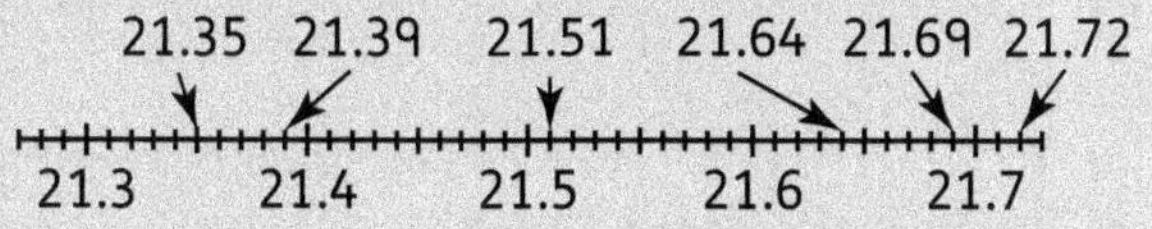
21.35
21.39
21.51
21.64
21.69
21.72
21.3
21.4
21.5
21.6
21.7

5 **a** 0.3 < 1 **b** 0.05 < 0.5 **c** 2.5 < 2.9
d 0.7 > 0.69 **e** 0.09 < 0.1 **f** 0.33 > 0.3

Round decimals

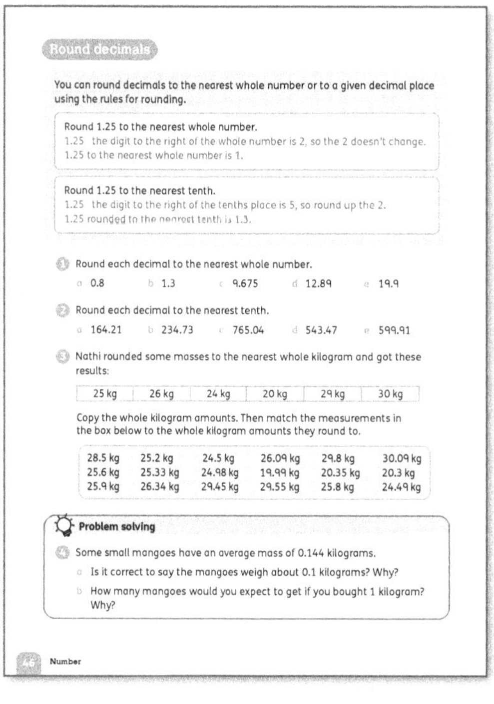

Round decimals

You can round decimals to the nearest whole number or to a given decimal place using the rules for rounding.

Round 1.25 to the nearest whole number.
1.25 the digit to the right of the whole number is 2, so the 2 doesn't change.
1.25 to the nearest whole number is 1.

Round 1.25 to the nearest tenth.
1.25 the digit to the right of the tenths place is 5, so round up the 2.
1.25 rounded to the nearest tenth is 1.3.

1 Round each decimal to the nearest whole number.
a 0.8 b 1.3 c 9.675 d 12.89 e 19.9

2 Round each decimal to the nearest tenth.
a 164.21 b 234.73 c 765.04 d 543.47 e 599.91

3 Nathi rounded some masses to the nearest whole kilogram and got these results:

25 kg	26 kg	24 kg	20 kg	29 kg	30 kg

Copy the whole kilogram amounts. Then match the measurements in the box below to the whole kilogram amounts they round to.

28.5 kg	25.2 kg	24.5 kg	26.09 kg	29.8 kg	30.09 kg
25.6 kg	25.33 kg	24.98 kg	19.99 kg	20.35 kg	20.3 kg
25.9 kg	26.34 kg	29.45 kg	29.55 kg	25.8 kg	24.49 kg

Problem solving

4 Some small mangoes have an average mass of 0.144 kilograms.
a Is it correct to say the mangoes weigh about 0.1 kilograms? Why?
b How many mangoes would you expect to get if you bought 1 kilogram? Why?

46 Number

Materials

0–5 number line, marked in tenths (for display); 1.2–1.5 number line, marked in hundredths (for display); small counters; place-value tables (page 21); highlighters

Warm-up

Use any 'Rounding and estimating' activity using whole numbers from the Activity bank (page 24) as a warm-up for this lesson.

Focus

- Display a 0–5 number line. Mark the tenths between the whole numbers. Remind the children that we often round numbers to give approximate numbers. Ask them how they round whole numbers to the nearest ten and hundred.
- Explain that we are now going to look at rounding decimals in a similar way. Mark 3.7 on the number line. Ask: *Which whole number is this nearest to?*
- Tell the class that 3.7 rounded to the nearest whole number is 4. Repeat for other examples.
- Mark 2.5 on the number line and ask the class what they think they should do. If necessary, remind the children that when we round whole numbers, if they are halfway between two numbers, we round up. Explain that we do the same with decimals, so 2.5 rounded to the nearest whole number is 3.
- Now introduce rounding to the nearest tenth. Use a number line from 1.2 to 1.5, with the hundredths marked. Ask the children to position numbers such as 1.38 and then say which tenth they are nearest to (1.4).
- Emphasise that when we round to the nearest tenth, if the hundredths digit is 5 or more we round up the tenths digit. If the hundredths digit is less than 5 the tenths digit stays the same.
- Ask the children to read through the explanation and examples on **Pupil book 5 page 46** to consolidate what you have just taught.
- The children can complete question 1 and question 2 independently. Check the answers before moving on.
- Discuss with the class how they will keep track of their work in question 3. They could suggest working systematically through the masses in the box and classifying them one by one. Alternatively, they could start by finding all the masses that round to 25 kg and putting a small counter on them to show they have been used, then moving on to the masses that round to 26 kg, and so on.
- Problem solving: Let the children talk through and solve the problem in question 4 orally. They can use a calculator if they need to for part b.

Challenge

Give the children a set of decimals and a set of clues for one of the decimals. Once they have found the decimal, let them write their own clues for three different decimals and exchange them with a partner. For example:

8.062	8.861	8.820	8.723
8.702	8.828	8.559	8.653

(Answer: 8.828)

Support

Encourage the children to write the numbers in a place-value table and use a highlighter to mark or underline the digit in the place they are rounding to. Then have them focus only on the digit to the right. Avoid using the term 'round down' as the children may think they have to reduce the digit by 1 when you say this. It is better to say: *If the digit is 5 or more, round up. If not, round off the number.*

Interesting mistakes

- Some children might round to the nearest ten instead of to the nearest tenth. For example, they may round 36.34 to 40 rather than 36.3. Continue to show numbers on a place-value table or encourage them to write the place value above the digits.
- The children may round up to the nearest tenth correctly but still include the remaining numbers in their answer. For example, they may round 3.467 to 3.567 instead of 3.5.
- Another interesting mistake is rounding in more than one step from the 'end' of the number: rounding to hundredths before rounding to tenths. This will sometimes result in the right answer, but sometimes not. For example, using this method, 3.447 would round to 3.45 and then to 3.5 (which is wrong), but 3.437 would round to 3.44 and then to 3.4 (which is correct, but using an incorrect method).

If any of these errors arise, have a number talk (pages 17–18) about what children did and why. Show the decimals by 'zooming in' on number lines to help the children visualise the position of a number and what it is closest to.

Answers for Pupil book 5 page 46

1   a   1   b   1   c   10
    d   13   e   20

2   a   164.2   b   234.7   c   765.0
    d   543.5   e   599.9

3   20 kg: 19.99 kg, 20.35 kg, 20.3 kg
    24 kg: 24.49 kg
    25 kg: 25.2 kg, 24.5 kg, 25.33 kg, 24.98 kg
    26 kg: 26.09 kg, 25.6 kg, 25.9 kg, 26.34 kg, 25.8 kg
    29 kg: 28.5 kg, 29.45 kg
    30 kg: 29.8 kg, 30.09 kg, 29.55 kg

4   a   Individual answers, for example: Yes, because to the nearest tenth of a kg, 0.144 kg rounds to 0.1 kg.
    b   7

More decimals

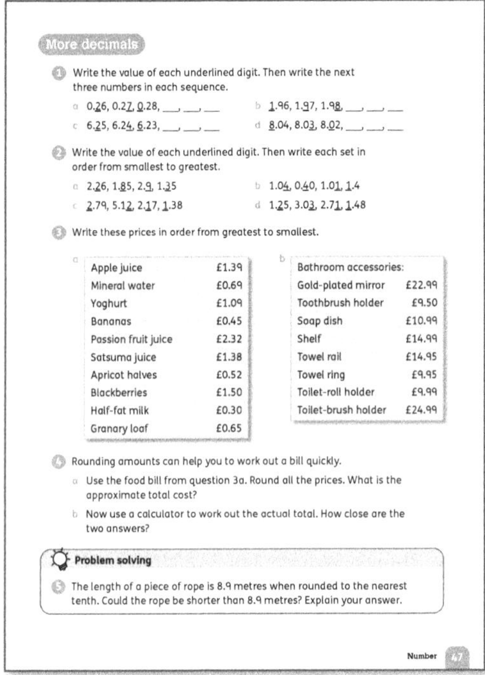

Materials

Blank number lines; place-value tables (page 21)

Warm-up

Do some skip counting activities using numbers with one decimal place as a warm-up for this lesson.

Focus

- Start with a number with two decimal places, for example 5.27. Tell the children this is the first term in a sequence, with term-to-term rule + 0.1. Ask them to generate the first five terms (5.27, 5.37, 5.47, 5.57, 5.67).
- Repeat with a different first term, and term-to-term rule − 0.01.
- Once they are confident counting up and back in tenths and hundredths, give term-to-term rules such as + 0.4, − 0.02
- Use questions 1–4 on **Pupil book 5 page 47** to informally assess whether the children can work with and compare decimals with one and two places. Revise any concepts they seem unsure of.
- Problem solving: Let the children share their ideas for question 5 in groups.

Support

Let the children use number lines to help them order sets of numbers and place-value tables to help them identify the value of each digit.

Answers for Pupil book 5 page 47

1 a 2 tenths, 7 hundredths, 0 ones, 0.29, 0.30, 0.31
  b 1 one, 9 tenths, 8 hundredths, 1.99, 2.00, 2.01
  c 2 tenths, 4 hundredths, 6 ones, 6.22, 6.21, 6.20
  d 8 ones, 3 hundredths, 0 tenths, 8.01, 8.00, 7.99

2 a 2 tenths, 8 tenths, 9 tenths, 3 tenths;
     1.35, 1.85, 2.26, 2.9
  b 4 hundredths, 4 tenths, 1 hundredth, 1 one;
     0.40, 1.01, 1.04, 1.4
  c 2 ones, 2 hundredths, 1 tenth, 1 one;
     1.38, 2.17, 2.79, 5.12
  d 2 tenths, 3 hundredths, 1 hundredth,
     1 one; 1.25, 1.48, 2.71, 3.03

3 a £0.30, £0.45, £0.52, £0.65, £0.69, £1.09, £1.38,
     £1.39, £1.50, £2.32
  b £9.50, £9.95, £9.99, £10.99, £14.95, £14.99,
     £22.99, £24.95

4 a Rounding each amount to the nearest 10p
     gives £10.40.
  b £10.29. The estimate is only 11p from the
     true value.

5 Yes; it could be any length between 8.85 m
   and 8.9 m.

Percentages

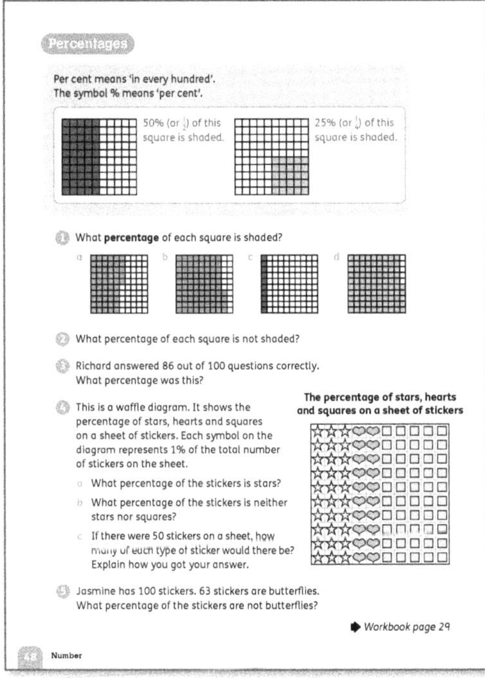

Materials
100 charts (large square divided into 100 small squares, search for this online); advertisements showing percentages; coloured pencils or marker pens

Warm-up
Select any suitable 'Geometry' activity from the Activity bank (pages 28–29) as a mental starter for this lesson.

Focus
- Give each child a blank 100 chart and some coloured pencils or marker pens.
- Explain that you are going to say some multiplications and you want them to shade rectangles on the grid to show each fact as an area model. They should use a different colour for each fact and their rectangles can touch each other but should not overlap. Read out three multiplications (for example: 9×3, 7×6 and 4×3).
- Ask the children how many squares are not coloured on their grid (19). Record this as $\frac{19}{100}$ next to the grid. Write 19% next to the fraction and ask the children what it means (19 per cent). Remind them that per cent means 'out of every hundred'. Say: *We can say that 19 per cent of the squares are not coloured.*
- Let the children say what per cent of the grid they have shaded in each colour.
- Turn to **Pupil book 5 page 48** and look at the two examples with the class. Remind them that we can often write fractions as simpler equivalent fractions (for example: $\frac{50}{100} = \frac{5}{10} = \frac{1}{2}$).
- Let the children work independently to complete questions 1–3. Check the answers before moving on.
- For question 4 and question 5, ask the children to look at the diagram and to explain how it works. A waffle diagram is a proportion chart – most are 100 charts with 10×10 cells. They show the share of data in different categories.
  Let the children answer the questions in pairs.

> It is very important that the children make the connection between *percentages* and fractions of 100 (hundredths) as this will allow them to connect their understanding of decimal fractions to their work on percentages. Colouring grids and making waffle diagrams will help them to make this connection.

Follow-up
Let the children complete **Workbook 5 page 29** independently to consolidate the work on percentages and waffle diagrams.

Challenge
As an additional challenge, give the children 20 counters, in a variety of colours, and ask them to draw a waffle diagram to represent these. (They will need to work out that 5 squares on the waffle diagram represent 1 counter.)

Support
Show the children advertisements involving percentages. Let the children work in groups to create a poster showing percentages in use in everyday life, with each percentage illustrated with a waffle diagram.

Percentages, decimals and fractions

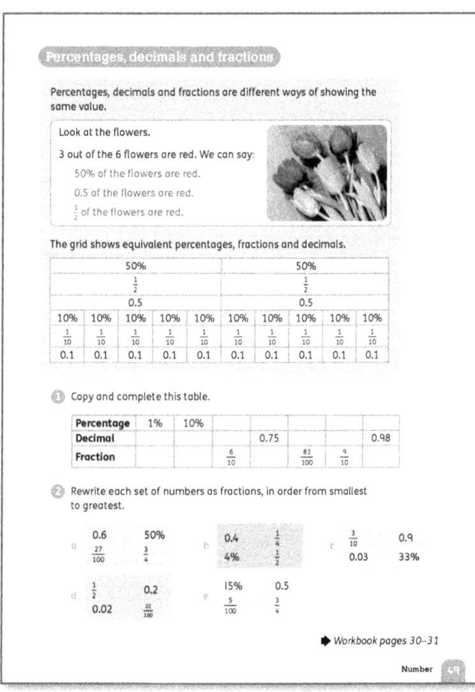

Materials

A set of ten objects in which some objects are the same (for example, a set of 10 marker pens with 3 red, 2 blue and 5 black); 100 charts; sets of flashcards with equivalent fractions, decimals and percentages (one per card); base-ten blocks (pages 20–21); thin card; calculators

Warm-up

- Show the class the set of ten objects. Ask questions to relate the number of items to fractions and decimals. For example:
  - *There are 3 red pens, what fraction of the pens are red?* ($\frac{3}{10}$)
  - *What is this as a decimal?* (0.3)

Focus

- Display a 100 chart with one column of 10 squares shaded. Ask the children what fraction is shaded ($\frac{1}{10}$) and what percentage (10%). Repeat with different numbers until the children are confident that $\frac{1}{10}$ = 10%, $\frac{2}{10}$ = 20%, and so on.

- Look at the pens again.
  - *What percentage of the pens are red?* (30%) *How did you decide?* $\left(\frac{3}{10} = 30\%\right)$
  - *Is it correct to say that 20% of the pens are blue?* (Yes) *How do you know?* $\left(\frac{2}{10} = 20\%\right)$
  - *Half of the pens are black. Is this the same as 50%?* (Yes) *How do you know?* (50% is half of a 100 chart, or $\frac{1}{2} = \frac{5}{10} = 50\%$)
- Show the class a percentage flashcard (for example, 40%) and ask them to write it as a fraction ($\frac{40}{100}$) and a decimal (0.4). Revise how to simplify a fraction, for example: $\frac{40}{100} = \frac{4}{10} = \frac{2}{5}$. Repeat for other percentages, such as 70%, 80%, 25% and 75%.
- Demonstrate how to convert a fraction into a percentage by finding the *equivalent fraction* with a *denominator* of 100.
- Work through the equivalence table on **Pupil book 5 page 49** with the class.
- Ask the children to copy the table in question 1 onto thin card to make a bookmark for reference. Ask the children to look at their tables and discuss in pairs how to convert from a decimal to a percentage. Confirm that you multiply by 100. *How do we convert from a percentage to a decimal?* (Divide by 100.)
- Discuss how to answer questions like the ones in question 2. Make sure the children understand that they can convert the numbers to any forms they like to compare them, but that they must use the original numbers given when they write the answers.

Follow-up

- Ask the children to complete **Workbook 5 page 30** as an informal assessment activity. They should work independently to answer the questions and then check each other's work.
- Then use **Workbook 5 page 31** to consolidate the work on percentages, decimals and fractions in this lesson.

Challenge

- Ask the children to work in groups to create a game of dominoes using equivalent percentages, decimals and fractions. Give them time to discuss and plan their game, make the dominoes and play a game.
- Allow each group to explain their game to the class. Use these games to revise and consolidate the concepts from this lesson.

Support

Use flashcards with equivalent fractions, decimals and percentages to play matching games:

- 'Snap': Each child starts with the same number of cards. The children take turns to place one card at a time face up on a pile. If the card played is equivalent to the one on the top of the pile, the child who says 'Snap!' first takes all the cards in the pile.
- 'Memory game': Spread all the cards face down. The children take turns to turn over two cards. If the cards are equivalent, the child takes the cards. If not, they turn the cards back over and the next child has a turn.

Interesting mistakes

- The children may still confuse tenths and hundredths and say that 0.4 = 4% rather than 0.4 = 40%. They may also forget to use 0 as a place holder and write 6% as 0.6 rather than 0.06.
- Let the children refer to a 100 chart or use base-ten blocks and let them count or colour the number so that they can see what mistake they have made.

Answers for Pupil book 5 page 49

1

1%	10%	60%	75%	83%	90%	98%
0.01	0.1	0.6	0.75	0.83	0.9	0.98
$\frac{1}{100}$	$\frac{1}{10}$	$\frac{6}{10}$	$\frac{75}{100}$	$\frac{83}{100}$	$\frac{9}{100}$	$\frac{98}{100}$

2
a $\frac{27}{100}, \frac{50}{100}, \frac{60}{100}, \frac{75}{100}$
b $\frac{4}{100}, \frac{25}{100}, \frac{40}{100}, \frac{50}{100}$
c $\frac{3}{100}, \frac{30}{100}, \frac{33}{100}, \frac{90}{100}$
d $\frac{2}{100}, \frac{20}{100}, \frac{22}{100}, \frac{50}{100}$
e $\frac{5}{100}, \frac{15}{100}, \frac{50}{100}, \frac{75}{100}$

Answers for Workbook 5 page 30

1
a 10 by 10 square with 15 squares shaded
b 10 by 10 square with 20 squares shaded
c 10 by 10 square with 1 square shaded

2
a 
b 
c 

3
a 50   b 25   c 10   d 60

4
a $5   b $2.50   c $0.10

Answers for Workbook 5 page 31

1

Fraction	Decimal	Percentage
$\frac{1}{2}$	0.5	50%
$\frac{1}{10}$	0.1	10%
$\frac{4}{10}$	0.4	40%
$\frac{7}{10}$	0.7	70%
$\frac{2}{100}$	0.02	2%
$\frac{17}{100}$	0.17	17%
$\frac{27}{100}$	0.27	27%
$\frac{88}{100}$	0.88	88%
$\frac{90}{100}$	0.9	90%
$\frac{45}{100}$	0.45	45%
$\frac{60}{100}$	0.6	60%

2
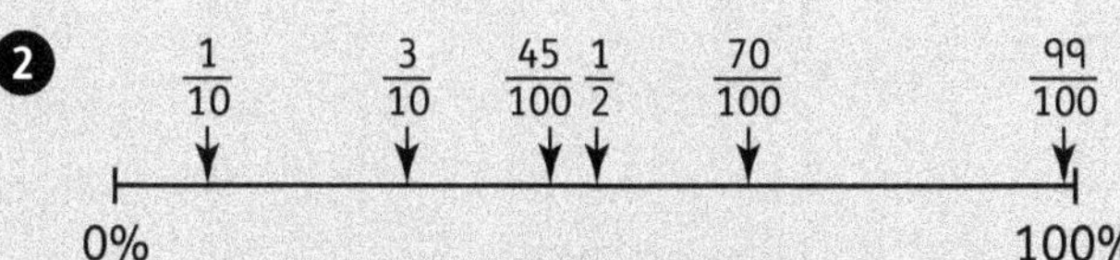

3
a 2.6 kg   b 0.95 kg
c 11.03 kg   d 25.425 kg

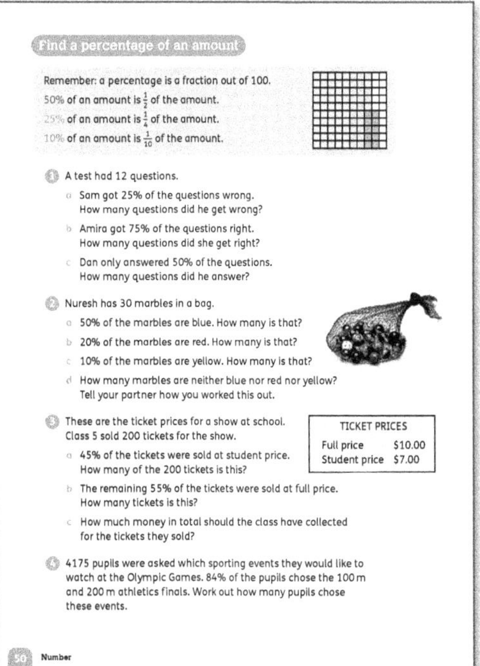

Find a percentage of an amount

Materials
Calculators

Warm-up
Play 'Three or four in a row' from the Activity bank (page 28) to practise finding equivalent fractions, decimals and percentages.

Focus
- Teach the children how to find percentages of quantities by converting the percentage into a fraction, then finding the fraction of the quantity (for example, to find 25% = $\frac{1}{4}$, divide by 4; to find 75% = $\frac{3}{4}$, divide by 4 then multiply by 3).
- Once they are confident with dividing by 10 to find 10%, encourage the children to use mental methods (for example, to find 60%, first divide by 10 to find 10% and then multiply by 6).
- Write a number in a circle in the middle of the board (for example, 320). Draw 'arms' coming from the circle and label each with a percentage (for example, 10%, 25%, 5%, 45%). Invite children in turn to write each percentage of 320.
- Encourage the children to find their answers mentally and discuss the strategies they used. For example, 'I found 10% by dividing by 10 and I knew 5% would be half of that.'
- Turn to **Pupil book 5 page 50**. Let the children discuss and then complete questions 1–4. Then have a number talk (pages 17–18) about how the children worked and what they found easy and challenging.

Support
Ask the children to design a poster explaining how to find percentages of quantities. Discuss the key points they included on their posters to check that they have expressed the process correctly. Display the posters in the classroom for reference.

Answers for Pupil book 5 page 50

1 a 3 b 9 c 6
2 a 15 b 6 c 3
 d 6; add together the marbles you know the colour of and subtract the answer from 30
3 a 90 b 110 c \$1730
4 3507 pupils

Equivalent percentages, fractions and decimals

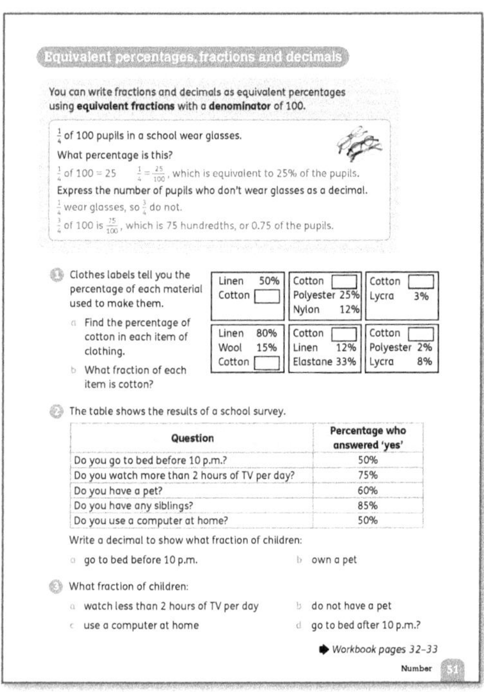

Materials

Calculators; examples of percentages in advertisements or newspaper articles

Warm-up

As a mental starter for this lesson, ask the children to add and subtract pairs of whole numbers that total 100, then decimals with one place that total 1 and, finally, decimals with two places that total 1.

Focus

- Start with **Workbook 5 page 32**. Let the children work on their own or in pairs to complete the activities on equivalent percentages, fractions and decimals. This page gives an example of converting $\frac{1}{4}$ to a decimal by dividing the numerator by the denominator. Children can use calculators for these types of calculation. Use this work to identify any misunderstandings about converting between different forms of the same number.
- Show the children examples of percentages in advertisements or newspaper articles and ask questions based on these. For example, *20% of shoppers prefer one product. What percentage of shoppers do not? 87% of people want a new medical centre. What percentage do not?*

- Work through the examples on **Pupil Book 5 page 51** with the children, then let them work in pairs to complete question 1.
- Discuss the table showing the school survey results with the class, then work through questions 2 and 3 orally.

Follow-up

Use the mixed activities on **Workbook 5 page 33** to informally assess how well the children have mastered the concepts in this unit.

Answers for Pupil book 5 page 51

1 a top line: 50%, 63%, 97%
 bottom line: 5%, 55%, 90%
 b top line: $\frac{1}{2}, \frac{63}{100}, \frac{97}{100}$
 bottom line: $\frac{5}{10}\left(=\frac{1}{20}\right), \frac{55}{100}\left(=\frac{11}{20}\right), \frac{90}{100}\left(=\frac{9}{10}\right)$
2 a 0.5 b 0.6
3 a $\frac{25}{100}\left(=\frac{1}{4}\right)$ b $\frac{40}{100}\left(=\frac{4}{10} \text{ or } \frac{2}{5}\right)$ c $\frac{1}{2}$ d $\frac{1}{2}$

Answers for Workbook 5 page 32

Here are some example of fractions and decimals you could use to complete the equivalent fraction wheels.

1, **2**

	Examples of equivalent fractions	Equivalent decimal
1 a $\frac{1}{4}$	$\frac{2}{8}, \frac{3}{12}, \frac{4}{16}, \frac{5}{20}$	0.25
b $\frac{1}{5}$	$\frac{2}{10}, \frac{3}{15}, \frac{4}{20}, \frac{20}{100}$	0.2
c $\frac{1}{10}$	$\frac{2}{20}, \frac{3}{30}, \frac{4}{40}, \frac{10}{100}$	0.1
d $\frac{3}{4}$	$\frac{6}{8}, \frac{9}{12}, \frac{12}{16}, \frac{75}{100}$	0.75
e $\frac{2}{5}$	$\frac{4}{10}, \frac{6}{15}, \frac{8}{20}, \frac{40}{100}$	0.4
f $\frac{3}{10}$	$\frac{6}{20}, \frac{9}{30}, \frac{12}{40}, \frac{30}{100}$	0.3

2

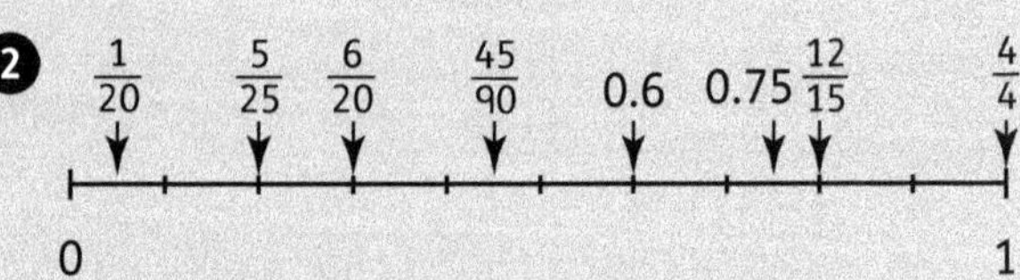

Answers for Workbook 5 page 33

1

Tenths	$\frac{1}{10}$	$\frac{2}{10}$	$\frac{3}{10}$	$\frac{4}{10}$	$\frac{5}{10}$	$\frac{6}{10}$	$\frac{7}{10}$	$\frac{8}{10}$	$\frac{9}{10}$	$\frac{10}{10}$
Decimals	0.1	0.2	0.3	0.4	0.5	0.6	0.7	0.8	0.9	1.0
Percentages	10%	20%	30%	40%	50%	60%	70%	80%	90%	100%

2 A $\frac{2}{10}$, 0.25, 0.3, 50%, 0.7, $\frac{3}{4}$ B $\frac{0}{10}$, 25%, $\frac{2}{5}$, $\frac{1}{2}$, 75%, 0.8
 C $\frac{6}{100}$, 0.4, $\frac{1}{2}$, 60%, 0.66, 0.7
3 A 0.25 and $\frac{3}{4}$ B 75% and 25%
 C 60% and 0.4
4 a 0.75 and $\frac{2}{10}$, difference 0.55; $\frac{0}{10}$ and 0.8, difference 0.8; $\frac{6}{100}$ and 0.70, difference 0.64
 b 0.55 0.64 0.8
 0   1

Ask some or all of these questions to assess how well the children have understood the concepts in this unit.

- *What is the value of the 9 in each number 3.97, 12.09, 4.259* (nine tenths, nine hundredths, nine thousandths)
- *What is the value of 4 in the number 23.184? Give your answer as a decimal.* (4 thousandths, 0.004)
- *How do you say this number?* (Give some different decimals.)
- *What is 0.7 as a fraction?* ($\frac{7}{10}$)
- Provide a set of decimals. Ask the children to order the decimals. Then ask: *When you ordered these decimals what did you look at first? Then what did you do?* (For example: Write the numbers in a column with the decimal points lined up. Compare digits, starting from the left. Stop at the first digit that is different and compare these.)
- Provide a set of decimals. Ask: *Which numbers do you find hardest to order? Why?*
- *A child says '3.009 is greater than 3.1 because 9 is greater than 1'. How can you show the child that this is incorrect?* (For example: Show both numbers in a place-value table or on a number line.)
- *Give me a number that lies between 3.2 and 3.3. How many answers do you think there are to this question?* (For example: 3.21, 3.279. There is an infinite number of answers, as you can always create another number by going to the next decimal place.)

- *I have a rope that measures 25 cm to the nearest centimetre. What is the longest the rope could be if we measure to two decimal places? Explain why.* (25.49 cm. This is the greatest length that will round to 25 cm, as 25.5 cm will round to 26 cm.)
- *How could you explain to someone how to round decimal fractions to the nearest whole number? And to the nearest tenth?* (Nearest whole number: Look at the number in the tenths place; if it is less than 5, round to the previous whole number; if it is 5 or more, round to the next whole number. Nearest tenth: Look at the number in the hundredths place; if it is less than 5, round to the previous tenth; if it is 5 or more, round to the next tenth.)
- Provide a mixed set of fractions, decimals and percentages. Ask: *What is the largest number? What is the smallest number?*
- *What percentages can you easily work out in your head? Give me some examples.* (For example: 10%, 25%, 50%.)
- *What is (give a percentage) as a fraction? And as a decimal?*
- *'To calculate 10% of a quantity you divide by 10, so to find 30% of a quantity you divide by 30.' Is this statement true? Why not?* (The statement is false. 30% is three times as much as 10%, so you divide by 10 and then multiply by 3.)

UNIT 6

Time

Learning objectives

- Understand that there are time zones and compare times in different zones

- Solve problems involving converting between units of time

- Understand time intervals of less than one second

- Find time intervals in seconds, minutes and hours that bridge 60

- Recognise that time intervals can be expressed as decimals or using mixed units

Unit introduction

Materials

Large time zone map; animation/video explaining time zones or a globe (there are many of these online), or a globe and torch

Teaching guidance

- Display a large time zone map for the class. The children can also look at the map on **Pupil book 5 page 52**.
- Ask the children to locate your country and discuss how the time in your country compares with the time in other countries. Try to elicit that different countries in the world right now have times that are different to your time.

- The idea that the time is different in different places may be familiar to the children who have travelled to other countries or who have watched televised events in another part of the world. For example, they may watch a live football or cricket match in the morning that is being played in another country at night.
- Show the class an animation or video to explain time zones or use a globe to demonstrate why different places experience different times of night and day at simultaneously.
  - Use a torch to represent the Sun and shine it onto the globe. The side facing the torch/Sun is lit up at the same time as the places on the other side of the globe are not lit up.
  - Slowly turn the globe, showing the class that the places that were in the dark now move into the light and the day begins. At the same time, the places on the other side of the globe move out of the light into the dark and the night begins.
  - Emphasise that these events take place at the same time to make the point that morning in one part of the world is evening in another. Show how each place moves through the times of day (and night) and that this is repeated each day.
  - Find your own country. Then find countries that have night when you have day.
- Once the children understand the reasons for time differences, it is easier for them to grasp that one place may be one or more hours ahead or behind them in time.
- Let the children work out from the map how many time zones there are (24, one per hour of the day). Explain that clocks are usually set one hour earlier or later than the adjacent time zone but not always. For example: in South Africa, the west coast and east coast are in different time zones, but the country uses a single time (the west coast time). Some countries use daylight saving, so they jump forward an hour in summer, effectively using the next time zone's time.
- Place the children in groups and give each group a different question to investigate. Here are some examples:
  - *Which countries have more than one time zone?*
  - *Which countries cover more than one time zone but use just one standard time?*
  - *How does having different time zones in a country (for example, Russia, India and the USA) affect how businesses operate?*
  - *What is the International Date Line?*
  - *What do the abbreviations GMT, UTC and CAT mean? Can you find any other time zone abbreviations?*
  - *How do you work out what time it is in another country?*
  - *How does daylight saving work? How does this affect time comparisons between countries?*

Once the children have done their research, let the groups feed back to the class.

Time zones

Materials
Calendar

Warm-up
Display the calendar for one month and ask the children to look for patterns. See 'Calendar patterns' in the Activity bank (page 28).

Focus
- Discuss why sports matches and other events that take place in other parts of the world are sometimes broadcast at strange times on local television. The children should use what they have found out about time zones to answer this question.
- Think and share: Let the children work in pairs or small groups to discuss the map on **Pupil book 5 page 52** and to answer the questions. They will need to be able to identify the time zone in which they live. If possible, project the map on the board and mark the position of your country.
- Make sure the children can read and make sense of the map. Then let them work in pairs to complete question 1. Allow time for the children to share the strategies they used to work out the times (question 2).

Support
The children may need support with time zone questions. Encourage them to use the time zone map as a number line.

Answers for Pupil book 5 page 52
1 a 7.15 p.m.   b 12.15 p.m.   c 11.15 a.m.
  d 2.15 a.m. the next day
  e 3.15 a.m. the next day
  f 8.15 p.m.
2 Individual answers.

Convert units of time

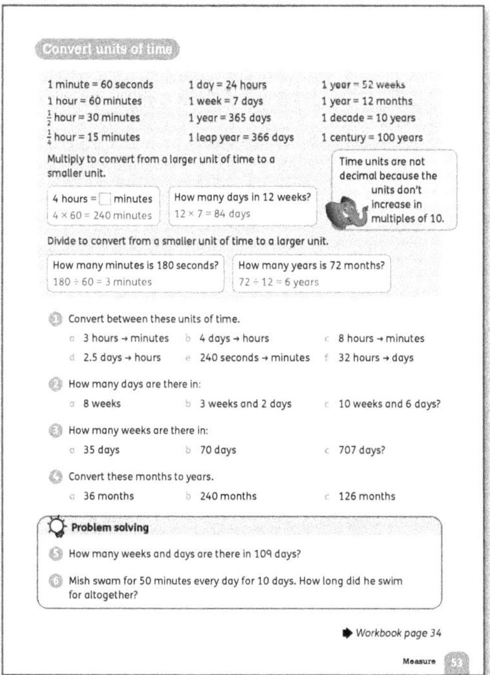

Materials
Sets of cards with the equivalent times from **Pupil book 5 page 53** shown separately (for example, a card with 1 minute and a separate card with 60 seconds, and so on); calculators (if needed)

Warm-up
Choose any suitable multiplication or division activity from the 'Calculation skills' section of the Activity bank (pages 25–28) as a mental starter for this lesson.

Focus
- In order to convert units of time, the children have to know how the units relate to each other. They have already learnt about time units, but it is useful to revise them, as follows.
- Give each group a set of time cards that has been shuffled. Let the groups sort the cards and make equivalent times. Once they have made the pairs, they can check their answers on **Pupil book 5 page 53**.
- Work through the example conversions with the class. These show how to convert between units of time by multiplying or dividing by the appropriate conversion factor.
- Let the children work independently to complete questions 1–4. Check answers as a class.
- Problem solving: Allow the children to use calculators if they need additional support with the calculations in questions 5 and 6, as the focus here is on the correct procedure for converting. Compare answers to question 6. *Has everyone given their answer in the same units? For how many hours did he swim?*

Follow-up
Use **Workbook 5 page 34** as an informal assessment task to check that the children understand the relationships between different units of time, can decompose and recompose times, and can correctly convert between units of time.

Interesting mistakes
The children might find the answers shown on a calculator confusing because they do not understand what to do with the decimal part of the answer. For example, if they use a calculator to convert 109 days to weeks, they will get an answer of 15.5714.

Have a number talk (pages 17–18) about what the calculations mean and how to interpret the remainder.
- For example, ask the children how many whole weeks there are in the answer 15.5714. (There are 15 whole weeks.)
- Then discuss how to use that to work out the remainder as a number of days. (15 × 7 = 105 days, so there are 15 weeks and 4 days in 109 days.)

Answers for Pupil book 5 page 53
1 a 180 minutes   b 96 hours
  c 480 minutes   d 60 hours
  e 4 minutes   f 1 day 8 hours
2 a 56   b 23   c 76
3 a 5 weeks   b 10 weeks   c 101 weeks
4 a 3 years   b 20 years
  c 10 years 6 months
5 15 weeks and 4 days
6 8 hours and 20 minutes

Answers for Workbook 5 page 34
1 a 7   b 56   c 245   d 52
  e 12   f 13   g 25   h 4
  i 32   j 12   k 60   l 240
  m 3   n 7
2 a 525   b 190   c 504
  d 8 hours, 17 minutes
  e 1440   f 60
3 a 40 minutes = 2400 seconds
  b $4\frac{1}{4}$ days = 4 days and 6 hours
  c 130 minutes = 2 hours and 10 minutes
  d $25\frac{3}{4}$ minutes = 1545 seconds

Less than a second

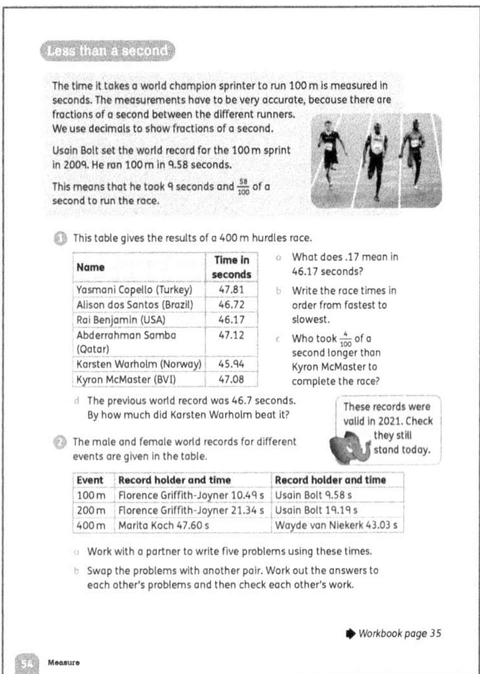

Materials

Electronic stopwatch (either on a smartphone or on the Internet so you can display it to the class)

Warm-up

- Remind the class that a second is a small unit of time by asking: *What can you do in a second?* List their suggestions on the board. Some examples are: blink, breathe, snap your fingers once, say 'one alligator', say 'a thousand and one'.
- Show the class the electronic stopwatch. Tell them that you are going to time 5 seconds. (If the children have smartphones, they can do this themselves.) Ask the children to focus on the 'seconds' to start with, then ask them to look at how quickly the numbers to the right of the seconds change. Explain that these are fractions of a second. Let the children experiment with stopping the timer at different intervals. For example, they could time themselves doing some simple activities.

Focus

- Time two children tying their shoelaces, stacking a set of 10 cups, or carrying out a similar task. Ask the children: *Who was fastest? Who took the longer time?* Choose a time, such as 12.68 seconds. Circle the 12 and tell the class that this means there are 12 whole seconds. Next, point to the .68 and remind the children that this is a decimal and that it means $\frac{68}{100}$ of a second.
- Talk about track and field events such as the 100-metre sprint race. Explain that race times are recorded by sophisticated clocks and are given to decimal fractions (tenths and hundredths) of seconds.
- Point out that, although the units we use to measure time are not decimal, we often use decimals to measure parts of units of time. It is important that the children know the difference between times such as 48.1 seconds (which is 48 seconds and $\frac{1}{10}$ of a second) and 3 minutes and 23 seconds (which is not the same as 3.23 minutes).

- Turn to **Pupil book 5 page 54**. Let the children work in pairs or small groups to read through the information and then complete question 1.
- For question 2, encourage the children to write different kinds of problems and to try to include some multi-step problems. Ask them to produce at least one problem that they think is very challenging.

> The world records given on **Pupil book 5 page 54** were accurate in 2021. You may like to check, or get the children to check, that these are still up to date. The International Amateur Athletics Federation website keeps these records up to date and also gives the regional records for each event.

Follow-up

Let the children work on their own to complete **Workbook 5 page 35**. Check their answers as a class and discuss any problems or interesting mistakes.

Challenge

Give the children these two examples of times that include fractions of seconds:

- In 1972, a Swedish swimmer won a 400-metre race by 0.002 seconds. A day later, the swimming body made a ruling that times would be measured to 0.01 seconds.
- In 2008, Michael Phelps beat Milorad Cavic by 0.01 seconds in a swimming race final even though it looked to the human eye that the swimmers touched the end of the pool at the same time. Cavic disputed the results. The Olympic body reviewed the times to 0.001 seconds and declared Phelps the winner.

Ask the children to explain what each decimal fraction means. Then, ask them to find out what a millisecond is and what is measured in milliseconds today (mostly computer processing speeds).

Answers for Pupil book 5 page 54

1 a one tenth and 7 hundredths of a second
   b 45.94 s, 46.17 s, 46.72 s, 47.08 s, 47.12 s, 47.81 s
   c Abderrahman Samba
   d 0.76 seconds (assuming he would need to run a time of 46.7 seconds to beat the world record).

2 a, b Individual answers.

Answers for Workbook 5 page 35

1 19 seconds and $\frac{4}{10}$ of a second = 19.4 seconds
   19 seconds and $\frac{65}{100}$ of a second = 19.65 seconds
   19 seconds and $\frac{1}{4}$ of a second = 19.25 seconds
   19 seconds and $\frac{3}{4}$ of a second = 19.75 seconds
   19 seconds and $\frac{4}{5}$ of a second = 19.8 seconds
   19 seconds and $\frac{4}{100}$ of a second = 19.04 seconds

2 a 9.58, 9.96, 9.99, 10.00, 10.80
   b 9.96 s; 9.60 s; 10.26 s

3 a It is faster because 21.339 is less than 21.340.
   b 21.24 s; 24.34 s; 21.39 s

Calculate with times

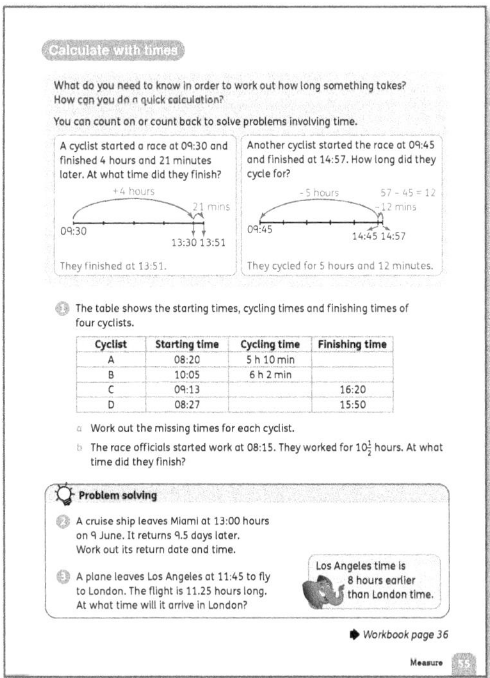

Materials
Calendar

Warm-up
Select any suitable 'Mental problem solving' activity from the Activity bank (pages 24–25) as a starter for this lesson.

Focus
- Turn to **Pupil book 5 page 55**. Discuss the example questions with the class.
- The second example involves counting back to find the duration (difference between start and end times). Ask the children how they could find the duration by counting on. (Count on from the start time, to the end time.)

Set the class some problems, such as the following. Then ask the children what information they are given in each problem and what they need to do to solve it.
- *A school concert started at 5:45 p.m. and finished at 10:18 p.m. How long was the concert?* (The start and end times are given. The children have to work out the difference between these times.)
- *I started a hike at 4:55 p.m. I hiked for 2 hours and 25 minutes. What time did I get home?* (The start time and the duration are given. The children have to add the duration to the start time (count on).)
- *I watched a movie for 2 hours and 16 minutes. The movie finished at 10:05 p.m. What time did it start?* (The end time and the duration are given. The children have to subtract the duration from the end time (count back).)

Discuss how the children can add and subtract times when working with different units and non-decimal numbers. Let them share their strategies, for example:
- *I jumped in half-hour intervals.*
- *I jumped to the next hour, then added the remaining minutes.*
- *I counted back in minutes to subtract 16 minutes, then I jumped back 2 hours.*
- Discuss what the table in question 1 shows. Let the class suggest how to find the missing times in each column before asking the children to work them out.
- <u>Problem solving:</u> For question 2, the children can refer to a calendar if they need to.
- Discuss how the children can deal with the time difference in question 3. Elicit that they can *either* add 8 hours to the start time *or* work out the answer using the given times and subtract 8 hours from the answer.

Follow-up
Use **Workbook 5 page 36** to consolidate calculating durations and drawing a suitable graph to compare durations.

Challenge
The children can create their own time problems. Collect these in to make a class set of problems, then hand them out randomly for the children to complete.

Support
Demonstrate how to draw and use a timeline to help with time calculations.
- For example, start with a timeline marked from 2 p.m. to 5 p.m. with the hours divided into 5-minute intervals. Draw an arrow or place a marker on the timeline to show a time such as 2.15 p.m. Start with simple questions such as: *What is the time 1 hour later? 2 hours later? 3 hours later? What was the time 15 minutes earlier?* And so on.
- Move on to durations that involve hours and minutes and involve crossing or bridging the hour. For example, *What is the time 1 hour and 20 minutes later?*
- Keep reminding the children that there are 60 minutes in one hour.
- When the children are more confident, give them problems such as: *A TV programme started at 3.20 p.m. and ended at 4.40 p.m. How long was the programme?* (1 hour 20 minutes) Let the children demonstrate how to work this out on the timeline.

Interesting mistakes
- Children often try to calculate time differences by first finding the difference between the whole hours. Look at this problem: *I caught the bus at 1.45 p.m. (13:45) and arrived at my location at 3.11 p.m. (15:11). How long was the bus journey?* Some children might think: 1.00 p.m. to 3.00 p.m. is 2 hours, 45 minutes + 11 minutes is 56 minutes so the journey time is 2 hours and 56 minutes. One way to overcome mistakes like this is to ask the children to show the steps they followed on a time line.
- The children may try to add or subtract the times as normal decimal numbers, forgetting that there are 60 seconds in a minute and 60 minutes in an hour, not 100. For example, when asked to work out the time difference between 3.24 p.m. and 4.11 p.m., the children might calculate the answer as 411 – 324 = 87 minutes. Encourage the children to estimate the answers to time problems to help them

to decide whether their answers seem reasonable. If they realise that 3.24 p.m. and 4.11 p.m. are less than 1 hour apart (less than 60 minutes), they can see that 87 minutes cannot be correct. Again, showing the working on a number line will help the children to make sense of the calculations.

Answers for Pupil book 5 page 55

1 a A: 13:30, B: 16:07, C: 7 h 7 min, D: 7 h 23 min
   b 18:45
2 19th June, 01:00
3 07:10 the next day

Answers for Workbook 5 page 36

1 2 hours and 5 minutes; 10:55; 2 hours and 45 minutes; 10:20; 11:45; 14:20
2 Chennai 3 hours, Bangalore 2 hours and 45 minutes, Hyderabad 2 hours and 20 minutes, Kolkata 2 hours and 15 minutes, Mumbai 2 hours and 5 minutes, Jaipur 50 minutes
3 Individual bar chart with Duration of flight labelled on the vertical axis and Destination on the horizontal axis. Gaps between the bars. Chennai 3 hours, Bangalore 2 hours and 45 minutes, Hyderabad 2 hours and 20 minutes, Kolkata 2 hours and 15 minutes, Mumbai 2 hours and 5 minutes, Jaipur 50 minutes

End-of-unit check

Ask some or all of these questions to assess how well the children have understood the concepts in this unit.

- *Why don't people in other countries have lunch at the same time as us?* (Different countries are in different time zones.)
- *Which countries have times that are later than ours?* (For example: Kenya, China.)
- *Which countries have times that are earlier than ours?* (For example: The USA, Canada.)
- *How do you work out what time it will be in another country?* (For example: Look at a time zone map to find out how many hours the country is ahead/ behind the time in the UK (Greenwich Mean Time).)
- *If it is 02:45 in Colombo (Sri Lanka), which is $5\frac{1}{2}$ hours ahead of GMT, what time is it in London?* (21:15)
- *A digital clock shows a time of 05:57. What will the time be 10 minutes later?* (06:07)
- *How do you convert seconds to minutes? Why?* (Divide by 60, because there are 60 seconds in 1 minute.)
- *What do you need to do to convert hours to seconds? Why?* (Multiply by 3600, because there are 60 minutes in 1 hour and 60 seconds in 1 minute. 60 × 60 = 3600)
- *Which is longer: 3.25 hours or 200 minutes?* (200 minutes; 3.25 hours is 195 minutes)
- *Miri says that 3 months is equal to 90 days. Ana says that 3 months is equal to 92 days. How do you think these children worked out the answer? Who is correct? Why?* (They are both correct. They used different

months. For example, Miri could have used January, February and March, and Ana could have used October, November, December.)
- *Do these conversions: 4 years to months, 18 months to years, 6 days to hours, $3\frac{1}{2}$ hours to minutes, 4.5 minutes to seconds, 720 seconds to minutes.* (48 months, 1.5 years, 144 hours, 210 minutes, 270 seconds, 12 minutes)
- *A bus that is due to arrive at 15:20 is delayed and will be half an hour late. At what time will it arrive?* (15:50)
- *A lesson lasts one and a quarter hours. The lesson starts at 13:05. At what time will the lesson finish?* (14:20)
- *What time is 0.1 seconds slower than 23.09 seconds?* (22.99 seconds)
- *What time is 3 hundredths of a second faster than 9.99 seconds?* (10.02 seconds)

Mixed practice 1

Answers to Mixed practice 1 Pupil Book 5 pages 56–57

1 a, b   A 9̲0̲3 798   B 35̲2 049   C 1 2̲5̲0 373
   c 1 250 373, 903 798, 352 049   f A 888 798
   d 904 000   B 337 049
   e 400 000   C 1 235 373
   g 96 202
   h subtract 250 373

2 a   26465   b   25825
     + 5937   + 3271
      32402   29096
   c   3637   d   14325
     − 946   − 9236
      2691   5089

3 A = 2.25, B = 2.75, C = 2.125, D = 2.875

4

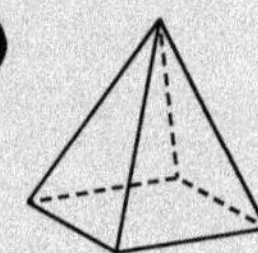

5 a cuboids   b Individual answers.
   c Each net has 6 faces; the sizes of the rectangles are different in each net
6 5.04, 4.55, 4.544, 5.05, 4.54
7 11:35
8 a 20°; acute   b 90°; right
   c 85°; acute   d 145°; obtuse
9 a 24%   b 117
   c 24% travel in private cars. As a fraction, this is $\frac{24}{100}$ which is not equivalent to $\frac{1}{5}$. So the statement is incorrect.
10 $2\frac{1}{2}$ hours = 150 minutes
   7200 seconds = 120 minutes
   $4\frac{1}{2}$ days = 108 hours
   $2\frac{1}{2}$ days = 3600 minutes
   20 minutes = 1200 seconds
11 a = 30°, b = 70°, c = 105°, d = 105°

Multiplication and division 1

Learning objectives

- Build increasing fluency in multiplication table facts and related division facts
- Multiply and divide numbers mentally drawing upon known facts
- Identify multiples and factors, including finding all factor pairs of a number, common factors of two numbers and lowest common multiples
- Express numbers as a product of two factors
- Know and use the vocabulary of prime numbers, prime factors and composite numbers
- Work out whether a number up to 100 is prime and recall prime numbers to 19
- Multiply and divide whole numbers by 10, 100 and 1000
- Recognise and use square numbers and cube numbers and use notation for squared and cubed
- Solve problems involving multiplication and division including using their knowledge of factors and multiples, squares and cubes
- Understand that the four operations follow a particular order
- Use laws of arithmetic to simplify calculations

Key words

multiple   common multiples
lowest common multiple (LCM)   square numbers
squared   cube numbers   factor   common factor
highest common factor (HCF)   composite number
prime number

Unit introduction

Materials

Pictures and real-life examples of arrays; paper and other materials for making a poster (for example, a tray of eggs, a box of chocolates, an eyeshadow compact, a muffin baking tray, a chessboard, lockers in a corridor, windows on an apartment block)

Teaching guidance

- <u>Think and share:</u> Turn to **Pupil book 5 page 58** and use the examples shown to revise how to use arrays. Work with the whole class. Arrangements in rows and columns like these are called arrays.
- Let the children suggest how arrays can help them to work out and remember multiplication and division facts. (For example, from the first one you can work out 1 × 8, 2 × 8, etc up to 6 × 8; you can work out 48 ÷ 2, 48 ÷ 3, etc.) Accept all reasonable suggestions and let individuals show methods that other children might not know.
- Working independently, the children list five multiplication and division facts using the arrays (both of which make 48). For example, 6 × 8 = 48 and 48 ÷ 6 = 8.
- Next, tell the class that they are going to work in groups to make posters of different arrays used in real life. They can either photograph or draw these arrays themselves or find images online to use on their posters.
- Identify some examples of arrays with the class to get them started.
- The groups then find images, plan and make their posters. They should list some of the multiplication and division facts they can derive from each image.
- Let the groups present their posters and ask the rest of the class questions about the multiplication, division, repeated addition and subtraction shown in the arrays.
- Display the posters in the classroom to act as a reference throughout this unit.

What do you already know?

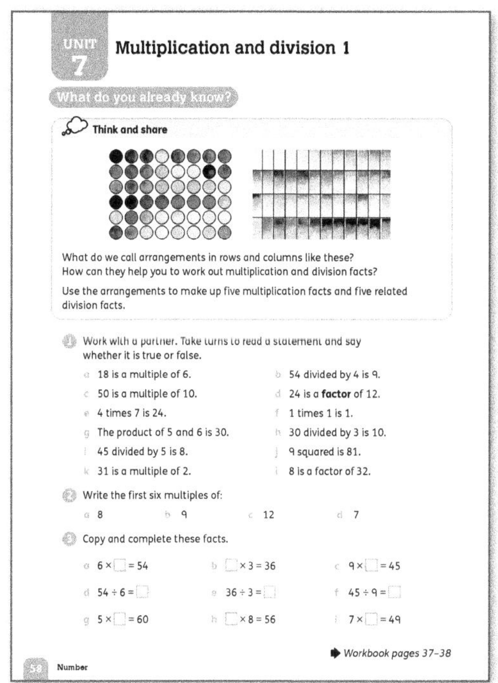

Materials

Arrays on display in the classroom; squared paper

Warm-up

Use question 1 on **Workbook 5 page 37** as a mental warm-up for this lesson. The children should work independently and complete the multiplication grid as accurately as they can. Remind them to focus on accuracy rather than speed.

Focus

- Remind the class that you get a *multiple* of a number when you multiply it by another number. For example, 15 is a multiple of 5 (5 × 3). It is also a multiple of 3 (3 × 5). The first multiple of any number is the number itself. For example, the first multiple of 5 is 5 (5 × 1). Let the children list the multiples of 4 and read them to each other to check them.
- Remind the class that a factor is a number that divides into another number. For example, 3 is a *factor* of 15, because 15 ÷ 3 = 5. *What are the other factors of 15?* (1, 5, 15)
- Let the children complete question 1 on **Pupil book 5 page 58** orally in pairs. Encourage them to refer to the arrays on display in the classroom if they are not sure of an answer.
- The children should work individually on questions 2 and 3. Prompt them to look for related facts in question 3.

Follow-up

- Use question 2 on **Workbook 5 page 37** to consolidate multiplication facts. You can work on the three fact sets at different times during this unit if you wish. Again, remind the children to focus on accuracy and spend some time talking about how children can improve their recall of multiplication facts.
- Use **Workbook 5 page 38** to revise and consolidate division facts. Again, accuracy is more important than speed. Follow up by letting the children share their strategies for learning division facts.

Support

Let the children make their own 12 × 12 multiplication grid on squared paper and use this to check facts as they need to.

Interesting mistakes

- Sometimes children do not apply the commutative property of multiplication to make learning facts easier. For example, they may remember that 9 × 4 = 36 but need support to work out 4 × 9.
- Some children may see multiplication and division as completely separate operations and ignore the fact that they are linked as inverse operations. The children may have fairly good recall of multiplication facts but need additional support with division facts. For example, a child may know that 6 × 7 = 42 but not use this to work out that 42 ÷ 7 = 6.

Continued work with arrays will help the children to visualise the relationships. For example, using the first array on **Pupil book 5 page 58**, you could ask: *How many dots are there altogether?* (48) *What multiplication facts give us 48?* (6 × 8 and 8 × 6) *How many rows is the array divided into?* (6 rows of 8) *Can you give me a division fact for that?* (48 ÷ 6 = 8 dots per row) *What if we divide the array into columns?* (We get 8 columns with 6 dots in each.) *How do we write that as a division fact?* (48 ÷ 8 = 6) *What fact family can we make using the numbers 6, 8 and 48?* (6 × 8 = 48, 8 × 6 = 48, 48 ÷ 6 = 8, 48 ÷ 8 = 6)

Answers for Pupil book 5 page 58

1

a true	b false	c true
d false	e false	f true
g true	h true	i false
j true	k false	l true

2
a 8, 16, 24, 32, 40, 48
b 9, 18, 27, 36, 45, 54
c 12, 24, 36, 48, 60, 72
d 7, 14, 21, 28, 35, 42

3

a 9	b 12	c 5
d 9	e 12	f 5
g 12	h 7	i 7

Answers for Workbook 5 page 37

1

	3	6	4	8	1	10	0	12	9
× 4	12	24	16	32	4	40	0	48	36
× 8	24	48	32	64	8	80	0	96	72
× 7	21	42	28	56	7	70	0	84	63
× 9	27	54	36	72	9	90	0	108	81
× 11	33	66	44	88	11	110	0	132	99

2 Fact check 1: 20, 24, 18, 9, 24, 21, 14, 25, 16, 27, 30, 70, 42, 27, 32, 36, 36, 50, 24, 6

Fact check 2: 30, 56, 16, 18, 70, 12, 72, 15, 36, 35, 28, 64, 49, 81, 48, 60, 63, 54, 40, 63

Fact check 3: 20, 81, 72, 90, 12, 45, 48, 40, 45, 35, 20, 18, 32, 54, 42, 56, 28, 15, 60, 36

Answers for Workbook 5 page 38

1

	6	12	36	28	24	32	48	60	72
÷ 2	3	6	28	14	12	16	24	30	36
÷ 4		3	9	7	6	8	12	15	18
÷ 3	2	4	12		8		16	20	24
÷ 6	1	2	6		4		8	10	12
÷ 12		1	3		2		4	5	6

2 Fact check 1: 8, 10, 3, 9, 6, 9, 5, 6, 8, 10, 8, 6, 5, 5, 8, 7, 9, 8, 9, 7

Fact check 2: 4, 9, 10, 10, 5, 6, 5, 8, 3, 7, 9, 5, 7, 6, 8, 6, 9, 4, 8, 3

Fact check 3: 3, 4, 3, 10, 2, 0, 9, 7, 6, 4, 9, 7, 7, 6, 7, 2, 4, 4, 8, 8

Multiples

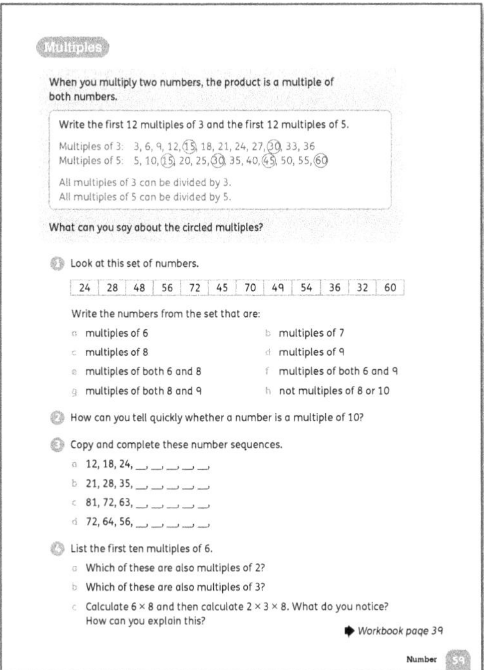

The children have already learnt about the concept of multiples. This year, they are expected to recognise and work with multiples in a higher range. They will also begin to explore using patterns and breaking up numbers to make multiplication easier.

Materials
Arrays from the previous lesson; small counters; coloured marker pens

Warm-up
Select any suitable 'Mental problem solving' activity from the Activity bank (pages 24–25) as a warm-up for this lesson.

Focus
- Work through the example on **Pupil book 5 page 59** The question in the example box asks what the children can say about the circled multiples. (These are multiples of 3 and 5.)
- Let the children work on their own to complete question 1 and then have them compare and check their answers in pairs or small groups.
- Discuss question 2 orally with the class.
- For question 3, remind the class that a number sequence is a set of numbers that increases or decreases according to a rule.
- Let the children complete question 4 independently and then discuss their results as a class.

Follow-up
Use **Workbook 5 page 39** to informally assess how well the children have understood and can apply the concepts covered in this lesson.

Challenge
Let the children work on their own or in pairs to make up their own 'Circle the multiples' question, like question 2 on **Workbook 5 page 39**. They can use a calculator to find multiples and swap and solve each other's questions.

Answers for Pupil book 5 page 59
1 a 24, 48, 72, 54, 36, 60
   b 28, 56, 70, 49
   c 24, 48, 56, 72, 32
   d 72, 45, 54, 36
   e 24, 48, 72
   f 36, 54, 72
   g 72
   h 28, 45, 49, 54, 36
2 It ends in zero.
3 a 12, 18, 24, 30, 36, 42, 48
   b 21, 28, 35, 42, 49, 56, 63
   c 81, 72, 63, 54, 45, 36, 27
   d 72, 64, 56, 48, 40, 32, 24
4 a 6, 12, 18, 24, 30, 36, 42, 48, 54, 60
   b 6, 12, 18, 24, 30, 36, 42, 48, 54, 60
   c You get the same answer; 2 × 3 = 6, so 2 × 3 × 8 is the same as 6 × 8.

Answers for Workbook 5 page 39
1 a 4, 8, 12, 16, 20, 24, 28, 32, 36, 60
   b 7, 14, 21, 28, 35, 42, 49, 56, 63, 70
   c 12, 24, 36, 48, 60, 72, 84, 96, 108. 120
   d 30, 40, 50, 60, 70, 80, 90, 100, 110, 120
   e 12, 15, 18, 21, 24, 27, 30, 33, 36, 39
   f 22, 33, 44, 55, 66, 77, 88, 99, 110, 121
2 Multiples of 5: 45, 50, 60, 75, 80, 85, 100, 125, 140, 185, 190, 200, 305, 310, 15, 150, 180, 455, 500, 505, 515, 625, 675, 800, 940, 965, 1000, 1025
   Multiples of 10: 50, 60, 80, 100, 140, 190, 200, 310, 150, 180, 500, 800, 940, 1000
3 Multiples of 100: 300, 3000, 3500, 3800, 1200, 12 000, 12 500, 15 200, 400, 800, 8400, 4800, 48 800, 84 400
4

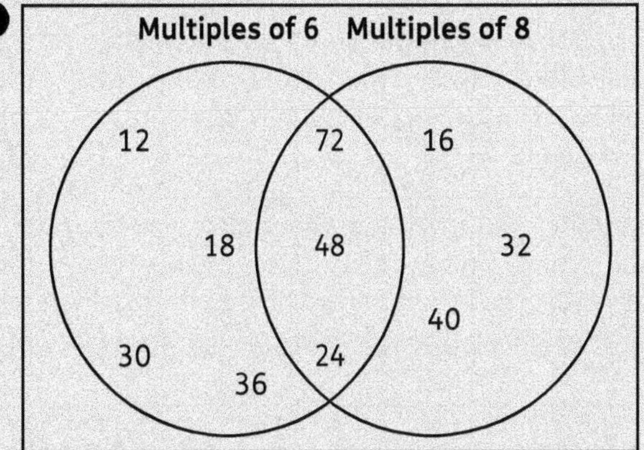

Sequences of multiples

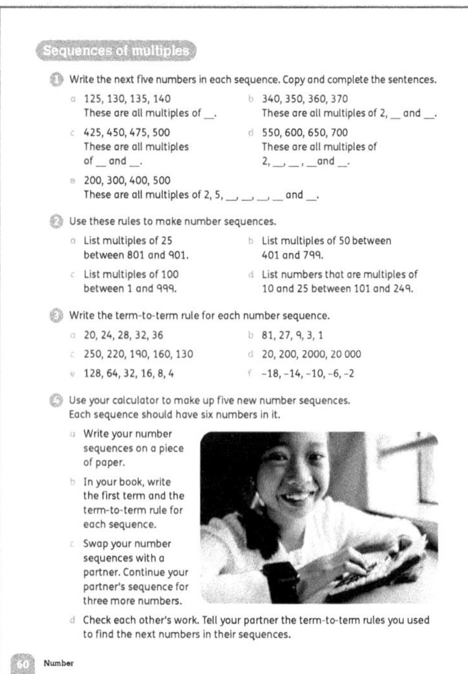

Materials

Calculators; copies of the grid shown in the notes below

Warm-up

Write 100 in a circle on the board. Ask the children to tell you factors of 100, and draw these connected to 100 by lines. Then ask them for factors of these factors. For example, 10 is a factor, and 1, 2, 5, and 10 are factors of 10, so are also factors of 100. Continue until all the factors are shown, remembering to include 1 and 100. (1, 2, 4, 5, 10, 20, 25, 50, 100)

Display the sequence 25, 50, 75, and ask the class for the term-to-term rule (+ 25) and the next five terms. (100, 125, 150, 175, 200) Ask: *Are these all multiples of 5?* (Yes, 25 is a multiple of 5, so all multiples of 25 are multiples of 5.)

Focus

- Let the children complete questions 1–4 on **Pupil book 5 page 60** on their own. Observe them as they work to assist and identify any children who need support.
- For question 4, make sure that the children have calculators and let them work on their own sequences before swapping with a partner.
- Give pairs of children a grid and ask them to find a path through it by following multiples of the same number. Tell them that they can move up, down, left or right. They can use counters to cover the numbers as they move. Compare results from the class, to see if they found all the multiples. Repeat using multiples of a different number.

Start	144	77	70	66	90	400
99	72	120	65	132	1200	60
121	11	48	100	96	130	300
25	33	132	156	108	14	168

Answers for Pupil book 5 page 60

1 **a** 145, 150, 155, 160, 165; multiples of 5
b 380, 390, 400, 410, 420; multiples of 2, 5 and 10
c 525, 550, 575, 600, 625; multiples of 5 and 25
d 750, 800, 850, 900, 950; multiples of 2, 5, 10, 25 and 50
e 600, 700, 800, 900, 1000; multiples of 2, 4, 5, 10, 20, 25, 50 and 100

2 **a** 825, 850, 875, 900
b 450, 500, 550, 600, 650, 700, 750
c 100, 200, 300, 400, 500, 600, 700, 800, 900
d 150, 200

3 **a** Term-to-term rule +4
b Term-to-term rule divide by 3
c Term-to-term rule −30
d Term-to-term rule multiply by 10
e Term-to-term rule divide by 2
f Term-to-term rule + 4

4 Individual answers.

Use multiples to solve problems

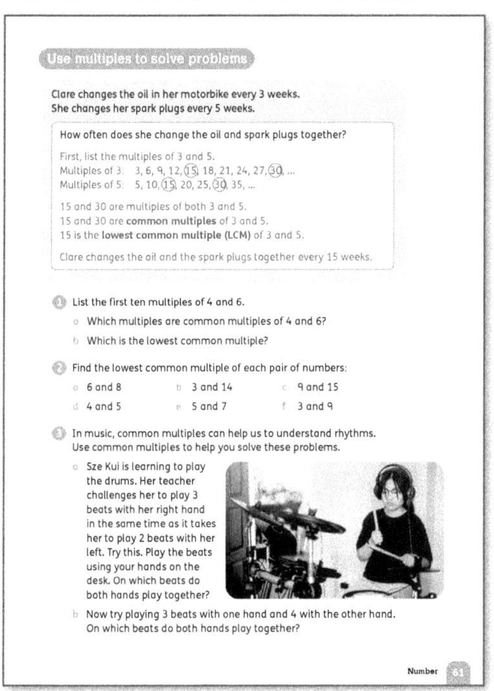

Materials

Recordings of polyrhythmic music, if possible (African traditional music and classical Indian music are both often polyrhythmic.)

Warm-up

Use known multiples to count on and back from different starting numbers.

Focus

- Let the children read the information at the top of **Pupil book 5 page 61**. Ask the children to work in pairs to find the *lowest common multiple* (LCM) of 5 and 9, then 4 and 12. Discuss what they notice, and ask them to find another pair of numbers where the LCM is one of the pair (for example, 5 and 25). Ask them to suggest how they can tell if the lowest common multiple of two numbers will be one of the numbers. (When one is a factor/multiple of the other.)

- The children can work independently on questions 1 and 2.
- For question 3, explain to the class that polyrhythmic music is any piece of music in which two patterns (of beats) are played at the same time. Play some music and let the children listen and try to identify when the rhythms 'overlap' (when a beat from each rhythm is played at the same time). African traditional music often has drums and percussion instruments playing different rhythms at the same time. Indian classical musicians may play complicated rhythms on a pair of small hand drums (tabla), sometimes playing 12 beats with one hand and 11 with the other. When every common multiple of the two numbers occurs, the beat will be played together.

Answers for Pupil book 5 page 61

1 Multiples of 4: 4, 8, 12, 16, 20, 24, 28, 32, 36, 40
Multiples of 6: 6, 12, 18, 24, 30, 36, 42, 48, 54, 60
- **a** 12, 24, 36 **b** 12

2 **a** 24 **b** 42 **c** 45
d 20 **e** 35 **f** 9

3 **a** 6, 12, 18, 24, … **b** 12, 24, 36, 48, …

Square numbers

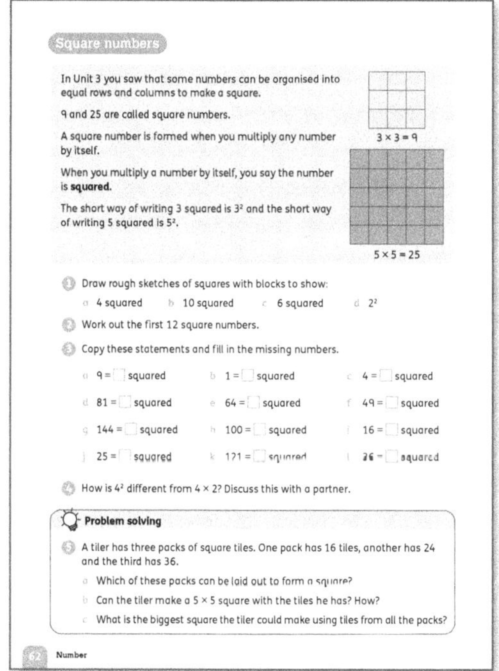

Materials

Squared paper; counters

Warm-up

Select any suitable 'Mental problem solving' activity from the Activity bank (pages 24–25) as a starter for this lesson.

Focus

- Use counters to demonstrate how some numbers (for example, 9 and 4) can be represented in a square array. Give each group some counters and ask them to show some numbers that cannot be represented in a square array (for example, 8 and 20).
- Then hand out squared paper to the class. Ask them to draw a 4×4 square array on the grid. Point out that the product of 4×4 is 16. Tell the class: *A number that can be arranged into a square array, such as 16, is called a square number.* Introduce square notation, $4^2 = 4$ squared $= 16$.
- Turn to **Pupil book 5 page 62** and read through the explanation with the class.
- For question 1, allow the children to draw their diagrams on squared paper. The children can work independently on questions 1–3.
- Have a class discussion about question 4. Make sure the children understand that the notation 4^2 means 4×4 and not 4×2.
- Problem solving: Children can work on question 5 in pairs. Encourage them to use counters or draw diagrams if they need to.

Answers for Pupil book 5 page 62

1 **a** 4 by 4 square **b** 10 by 10 square
c 6 by 6 square **d** 2 by 2 square

2 1, 4, 9, 25, 36, 49, 64, 81, 100, 121, 144

3 **a** 3 **b** 1 **c** 2 **d** 9
e 8 **f** 7 **g** 12 **h** 10
i 4 **j** 5 **k** 11 **l** 6

4 $4^2 = 4 \times 4$; you multiply 4 by itself
4×2 means multiply 4 by 2

5 **a** The pack of 16 and the pack of 36
b Yes; use a total of 25 tiles from the three packs; use tiles from the pack of 36, or use the pack of 24 and one tile from one of the other packs
c He has 76 tiles. The biggest square he could make is 8×8, which needs 64 tiles.

Cube numbers

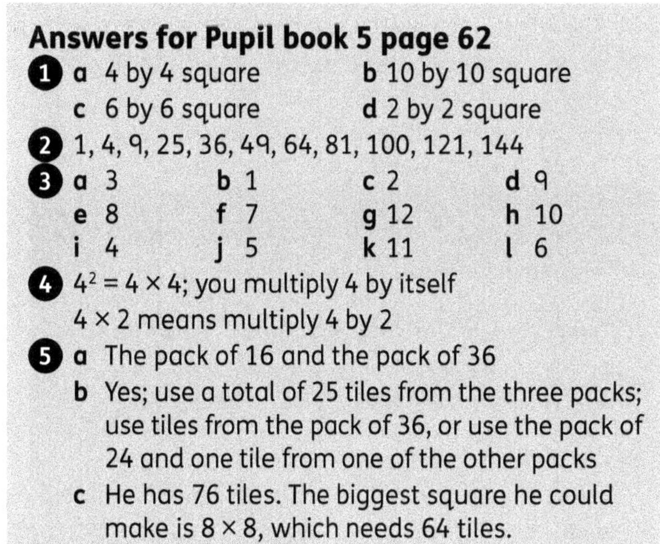

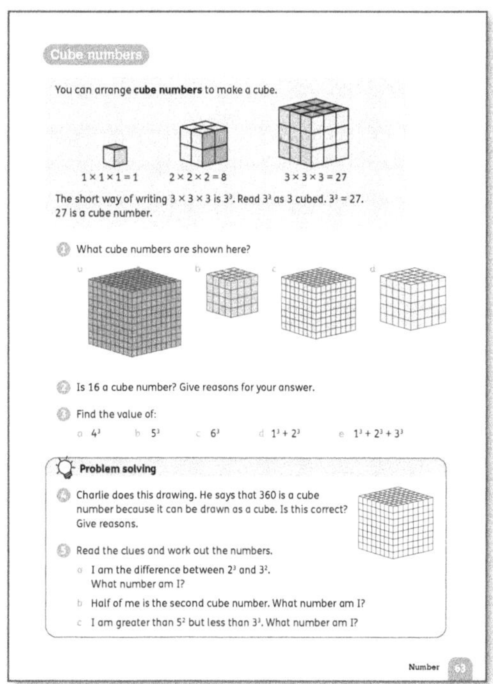

Materials

Rubik's cube (optional); ten frames; base-ten blocks:
100-flats, 10-rods and 1-cubes (pages 20–21)

Warm-up

Select any suitable 'Mental problem solving' from the
Activity bank (pages 24–25) as a starter for this lesson.

Focus

- Show the class a Rubik's cube if you have one. Point
  out that there are the same number of blocks in the
  length, width and height of the cube. Then show
  them a 100-flat. Point out that you can use 10 of
  these to build a cube that is 10 long, 10 high and
  10 wide.
- Remind the class that a cube is a 3D shape with
  square faces, so all its edges (length, width and
  height) are the same number of units.
- Turn to **Pupil book 5 page 63**. Read through the
  information and introduce cube notation,
  $4^3 = 4$ cubed.
- Point out that 4^2 means 4×4 and 4^3 means $4 \times 4 \times 4$.
  The power tells you how many times to multiply the
  number by itself.
- Ask the children how they could work out question
  1 without trying to count all the small cubes (for
  example, multiply length by width by height).
- For question 2, the children can try to build a cube
  using 16 1-cubes.
- Problem solving: The children can work in pairs on
  questions 4 and 5, and you can then discuss them
  as a class. For question 5, explain that the 1st cube
  number is 1^3, the second is 2^3 and so on.

Interesting mistakes

- Children (even up to high school level) sometimes
  interpret a number like 4^3 as 4×3. This is particularly
  problematic if they visualise it as having 3 rows of
  4 blocks.
- Address the superscript notation with the class and
  show them a $4 \times 4 \times 4$ cube as well as three sets of
  4 cubes so they can see the difference. Say 4^3 as 4
  cubed and ask the children to think of this as a cube
  with edges 4 units long.

Answers for Pupil book 5 page 63

1 a 1000 (which is 10^3)   b 64 (which is 4^3)
   c 512 (which is 8^3)   d 125 (which is 5^3)

2 No. $2^3 = 8$ and $3^3 = 27$. There is no cube number in
   between.

3 a 64   b 125   c 216
   d 9   e 36

4 No. All the sides of a cube must have equal
   length. $7^3 = 343$ and $8^3 = 512$. There is no
   cube number in between, so 360 is not a cube
   number.1

5 a 1   b 16   c 26

Factors

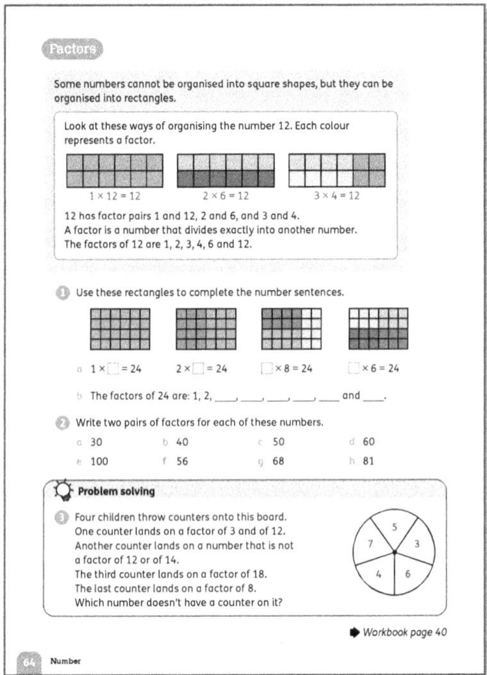

Materials

Geoboard and elastic bands (optional); counters

Warm-up

Revise some division facts for tables to 12×12 as a
mental starter for this lesson.

Focus

- Remind the children of the meaning of factor. Write
  24 on the board.
- Ask the children to say factors of 24 and record them.
  Ask the children how they can be sure that they have
  found all the factors of 24.
- Talk about working up systematically from 1 and
  pairing each number with the number it is multiplied
  by to make 24, until all factor pairs have been recorded.
- Illustrate factor pairs by drawing arrays for 24 or
  making them on a geoboard.
- Repeat for another number. Include some arrays for
  odd numbers (for example, 45) so that the children
  do not think that only even numbers can have several
  factor pairs.
- Turn to **Pupil book 5 page 64**. Ask the children to
  read through the text and examples and then ask
  them questions to make sure they understand
  and can explain what the colour shading shows on
  the diagrams. You may need to adapt this activity for
  children who cannot distinguish the colours.
- Children can work independently on questions 1 and 2.
- Problem solving: For question 3, the children can use
  counters to model the throws. They may assume
  that each counter lands on a different number, but
  the third counter could land on 3 or 6. Allow them
  to discuss this if it arises and to make their own
  decisions based on the fact that a problem might
  have more than one solution.

Challenge

Ask the children whether they think bigger numbers have more factors than smaller numbers. Ask them to give examples to show when this is true and when it is not true.

Support

Use **Workbook 5 page 40** to provide additional practice and to help the children develop methods of working with pairs of factors.

Interesting mistakes

There is a lot of terminology in this unit that the children need to become familiar with and that they may find confusing. For example, they may confuse the terms *factors* and *multiples*. To overcome this, provide short regular activities that require the children to use the correct terminology to describe numbers and their properties.

Answers for Pupil book 5 page 64

1 a 24; 12; 3; 4

b 1, 2, 3, 4, 6, 8, 12, 24

2 a any two pairs from 1 and 30, 2 and 15, 3 and 10, 5 and 6

b any two pairs from 1 and 40, 2 and 20, 4 and 10, 5 and 8

c any two pairs from 1 and 50, 2 and 25, 5 and 10

d any two pairs from 1 and 60, 2 and 30, 3 and 20, 4 and 15, 5 and 12, 6 and 10

e any two pairs from 1 and 100, 2 and 50, 4 and 25, 5 and 20, 10 and 10

f any two pairs from 1 and 56, 2 and 28, 4 and 14, 7 and 8

g any two pairs from 1 and 68, 2 and 34, 4 and 17

h any two pairs from 1 and 81, 3 and 27, 9 and 9

3 7

Answers for Workbook 5 page 40

1 18: 1 and 18, 2 and 9, 3 and 6;
56: 1 and 56, 2 and 28, 4 and 14, 7 and 8;
28: 1 and 28, 2 and 14, 4 and 7;
48: 1 and 48, 2 and 24, 3 and 16, 4 and 12, 6 and 8; 42: 1 and 42, 2 and 21, 3 and 14, 6 and 7;
60: 1 and 60, 2 and 30, 3 and 20, 4 and 15, 5 and 12, 6 and 10

2 12, 2, 6, 8, 6, 10, 10, 9; 8, 9, 8, 9, 7, 7, 8, 9

3 a 1, 2, 3, 4, 6, 9, 12, 18, 36

b 1, 2, 5, 10, 25, 50

c 1, 2, 4, 8, 16, 32, 64

4 They are square numbers. 6 is its own partner for 36, and 8 is its own partner for 64.

Find factors of a number

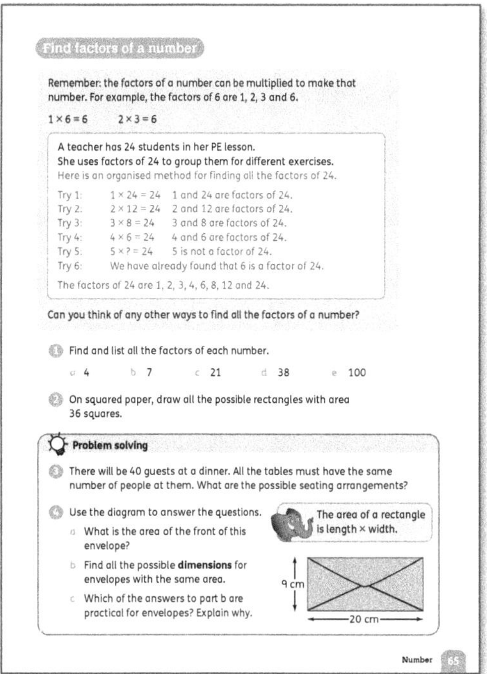

Find factors of a number

Remember: the factors of a number can be multiplied to make that number. For example, the factors of 6 are 1, 2, 3 and 6.

$1 \times 6 = 6$ $2 \times 3 = 6$

A teacher has 24 students in her PE lesson. She uses factors of 24 to group them for different exercises. Here is an organised method for finding all the factors of 24.

Try 1: $1 \times 24 = 24$ 1 and 24 are factors of 24.
Try 2: $2 \times 12 = 24$ 2 and 12 are factors of 24.
Try 3: $3 \times 8 = 24$ 3 and 8 are factors of 24.
Try 4: $4 \times 6 = 24$ 4 and 6 are factors of 24.
Try 5: $5 \times ? = 24$ 5 is not a factor of 24.
Try 6:   We have already found that 6 is a factor of 24.

The factors of 24 are 1, 2, 3, 4, 6, 8, 12 and 24.

Can you think of any other ways to find all the factors of a number?

1 Find and list all the factors of each number.

a 4   b 7   c 21   d 38   e 100

2 On squared paper, draw all the possible rectangles with area 36 squares.

Problem solving

3 There will be 40 guests at a dinner. All the tables must have the same number of people at them. What are the possible seating arrangements?

4 Use the diagram to answer the questions.

a What is the area of the front of this envelope?

b Find all the possible **dimensions** for envelopes with the same area.

c Which of the answers to part b are practical for envelopes? Explain why.

Number 65

Materials

Squared paper; counters

Warm-up

Revise multiplication facts up to 12×12 as a mental starter for this lesson.

Focus

- Work through the example on **Pupil book 5 page 65** with the class. Make sure the children understand that once they reach a number that is already a factor, they can stop looking. The question in the example box asks the children for some other ways of finding all the factors of a number. (You could divide by all the numbers less than the number. If the answer is a whole number, then the number you are dividing by is a factor. Or you could write all the multiplication calculations that give the number as the answer. The numbers being multiplied are factors of the number.)

- Spend some time talking about the children's strategies and ideas for checking that they have all the factors of a number.

- Let the children work on their own to complete questions 1–4.

- Problem solving: In question 3, check that the children understand 'seating arrangements', and that they can use as many as or few tables as they like, as long as they have equal numbers of people at them. In question 4b, explain that the dimensions are the length and width. For 4c part c, children may find it useful to draw possible envelopes accurately. Then go through the answers as a class. For question 2, ask: *Did you use the factors of 36 to help you to draw the rectangles? How?* (Factor pairs give the side lengths.)

Challenge

- Children work in pairs. Ask them to find a number less than 50 that has ten different factors and to list all the factors. How many numbers can they find? (There is only one, 48.)

Answers for Pupil book 5 page 65

1 a 1, 2, 4   b 1, 7
  c 1, 3, 7, 21   d 1, 2, 19, 38
  e 1, 2, 4, 5, 10, 20, 25, 50, 100

2 1 square by 36 squares; 2 squares by 18 squares;
4 squares by 9 squares; 3 squares by 12 squares

3 1 table of 40 people; 2 tables of 20 people;
4 tables of 10 people; 5 tables of 8 people;
8 tables of 5 people; 10 tables of 4 people;
20 tables of 2 people; 40 tables of 1 person

4 a 180 cm²
  b 1 cm by 180 cm, 2 cm by 90 cm, 3 cm by 60 cm,
4 cm by 45 cm, 5 cm by 36 cm, 6 cm by 30 cm,
9 cm by 20 cm, 10 cm by 18 cm, 12 cm by 15 cm
  c 9 cm by 20 cm, 10 cm by 18 cm, 12 cm by
15 cm; Individual answers.

Common factors and multiples

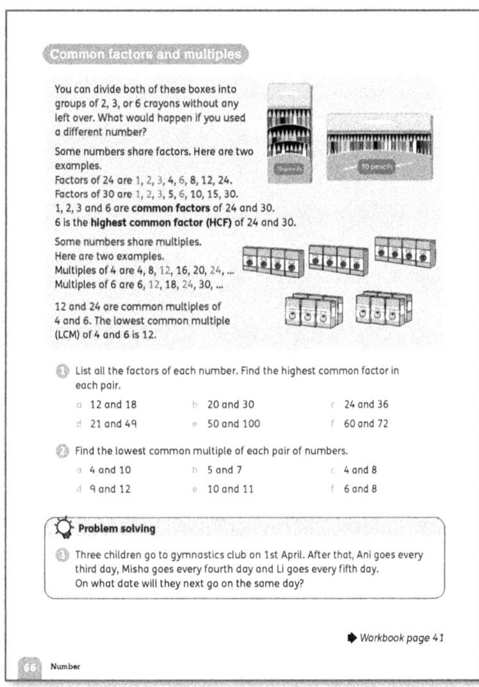

Materials
Calendars showing month to a page

Warm-up
As a mental starter for this lesson, prepare a set of mixed multiplication and division facts. Place the children in teams and ask different teams in turn to give the answer. Allow them to check with others in their team before answering so that this does not become a pressurised activity.

Focus
- Read through the examples at the top of **Pupil book 5 page 66** with the class. Use the pictures as arrays to talk about the factors and multiples.
- The children can work independently on questions 1 and 2.

- Problem solving: For question 3, give the children a calendar or have a large one available for them to check the date. They can circle the days that each person attends and use that to find the LCM.

Follow-up
Use **Workbook 5 page 41** to consolidate work on common factors and multiples. Let the children complete the activities independently and then check the answers as a class. Make time to discuss any disagreements and interesting mistakes.

Answers for Pupil book 5 page 66

1 a Factors of 12: 1, 2, 3, 4, 6, 12
    Factors of 18: 1, 2, 3, 6, 9, 18
    HCF = 6
  b Factors of 20: 1, 2, 4, 5, 10, 20
    Factors of 30: 1, 2, 3, 5, 6, 10, 15, 30
    HCF = 10
  c Factors of 24: 1, 2, 3, 4, 6, 8, 12, 24
    Factors of 30: 1, 2, 3, 5, 6, 10, 15, 30
    HCF = 6
  d Factors of 21: 1, 3, 7, 21
    Factors of 49: 1, 7, 49
    HCF = 7
  e Factors of 50: 1, 2, 5, 10, 25, 50
    factors of 100: 1, 2, 4, 5, 10, 20, 25, 50, 100
    HCF = 50
  f Factors of 60: 1, 2, 3, 4, 5, 6, 10, 12, 15, 20, 30 and 60
    Factors of 72: 1, 2, 3, 4, 6, 8, 9, 12, 18, 24, 36 and 72
    HCF = 12

2 a 20   b 35   c 8
  d 36   e 110   f 24

3 60 days after 1st April; 31st May

Answers for Workbook 5 page 41

1 a 1   b 25, 50, 75, 100   c 12
  d 10   e 20, 40, 60, 80

2 24: 1, 2, 3, 4, 6, 8, 12, 24
36: 1, 2, 3, 4, 6, 9, 12, 18, 36
HCF is 12
32: 1, 2, 4, 8, 16, 32
48: 1, 2, 3, 4, 6, 8, 12, 16, 24, 48
HCF is 16
42: 1, 2, 3, 6, 7, 14, 21, 42
72: 1, 2, 3, 4, 6, 8, 9, 12, 18, 24, 36, 72
HCF is 6

3 a 3: 3, 6, 9, 12, 15, 18 , 21
    7: 7, 14, 21, 28, 35
    LCM is 21
  b 10: 10, 20, 30, 40, 50, 60
    12: 12, 24, 36, 48, 60
    LCM is 60
  c 6: 6, 12,18, 24, 30, 36
    9: 9, 18, 27, 36
    LCM is 24

Prime numbers

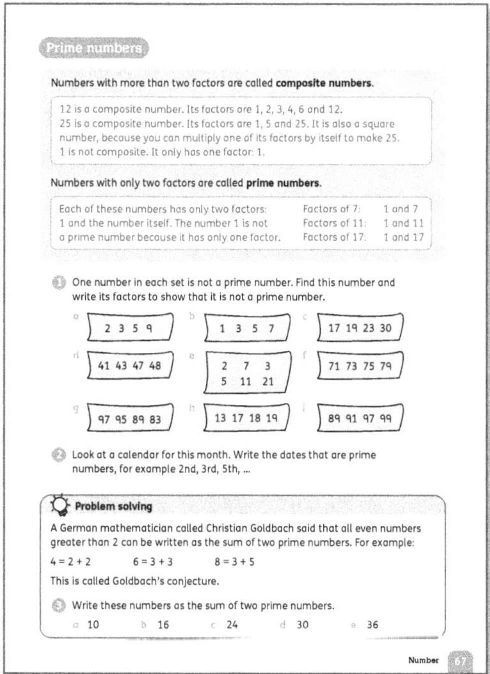

Materials
A 100 chart with the numbers 1–100 for each child (search online); calendar for this month

Warm-up
Select any suitable 'Mental problem solving' activity from the Activity bank (pages 24–25) as a warm-up for this lesson.

Focus
- Give each child a 100 chart. Explain that they are going to try to discover which numbers have no other factors than 1 and themselves.
- Start by crossing out every multiple of 2 except 2.
- Then ask the children to cross out all the multiples of 3 except 3, and so on.
- Ask the children why they did not need to cross out all the multiples of 4 or 6 and why they can stop once they have crossed out the multiples of 7.
- Explain that we can classify numbers by looking at their factors. Turn to **Pupil book 5 page 67** and read through the explanations of *prime* and *composite numbers* with the class.
- Ask the children to list the prime numbers to 100 based on their completed grids.
- For question 1, talk about strategies that the children might use to decide whether a number is not prime. (For example, unless it is 2, it cannot be an even number. If it has 5 in the ones place, it cannot be prime as it is a multiple of 5, and so on.)
- Provide calendars or planners if needed for question 2, but children can also refer back to their 100 charts.
- <u>Problem solving:</u> Give the children some time to think about question 3 and to work out the sums. Share the answers as a class and talk about the strategies the children used to find the two prime numbers that give each total. Note that they may choose different prime numbers.

Challenge
- Write a 2-digit number on a piece of paper and keep it hidden. The children must ask questions to which you can reply *yes* or *no* in order to try to guess the number.
- Encourage them to ask questions about number properties (for example, 'Is it a multiple of 6?') as well as about the size of the number.
- Discuss with the children what your answer to each question tells them. Help them to work out which numbers they can eliminate after each answer to narrow down the options.
- When the children guess the correct number, show the paper to confirm this. Invite a child to write a number down and respond to other children's questions.

Support
Display a 2- or 3-digit number and ask the children to describe its properties (for example, composite or prime number, the factors of the number, some multiples of the number and so on). The children could also show one or two visual representations of the number to revise composing and decomposing numbers.

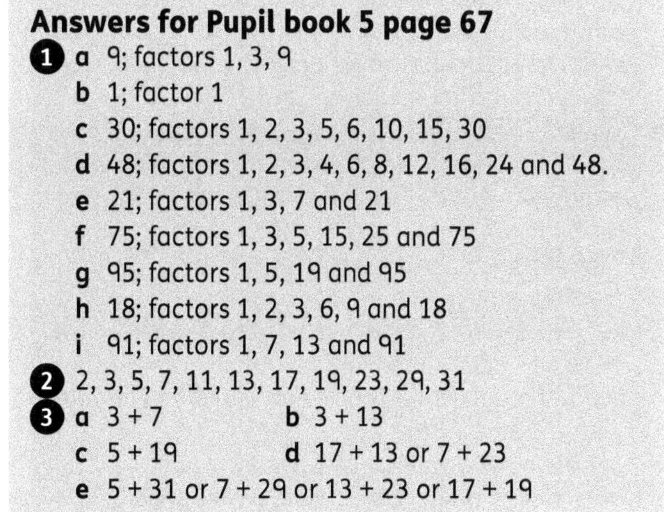

Answers for Pupil book 5 page 67

1. a 9; factors 1, 3, 9
    b 1; factor 1
    c 30; factors 1, 2, 3, 5, 6, 10, 15, 30
    d 48; factors 1, 2, 3, 4, 6, 8, 12, 16, 24 and 48.
    e 21; factors 1, 3, 7 and 21
    f 75; factors 1, 3, 5, 15, 25 and 75
    g 95; factors 1, 5, 19 and 95
    h 18; factors 1, 2, 3, 6, 9 and 18
    i 91; factors 1, 7, 13 and 91
2. 2, 3, 5, 7, 11, 13, 17, 19, 23, 29, 31
3. a 3 + 7   b 3 + 13
    c 5 + 19   d 17 + 13 or 7 + 23
    e 5 + 31 or 7 + 29 or 13 + 23 or 17 + 19

Multiply by 10, 100 or 1000

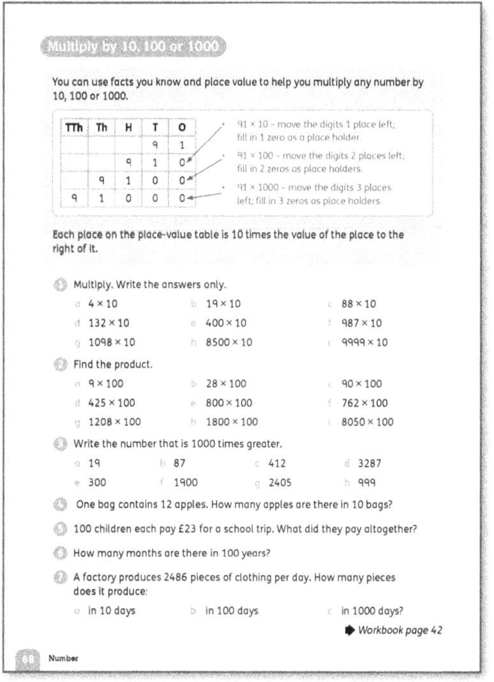

Materials

Place-value table (page 21); place-value cards (page 21) calculators

Warm-up

As a starter for this lesson, have a number talk (pages 17–18) about the relationship between powers of 10. For example, *10 is 10 times the size of 1, and 100 is 10 times the size of 10*. Focus on language such as *10 times more, 10 times greater* and so on.

Focus

- Ask the children to relate the warm-up discussion to different numbers. For example, ask: *What number is 100 times greater than 3?* (300) *What number is 10 times the size of 15?* (150)
- Revise multiplication by 10 using place-value tables and calculators. Record the results and make sure the children understand that to multiply by 10 they move the digits one place to the left.
- Next, ask: *What happens when you multiply by 10 and then multiply by 10 again?*
- Discuss the results. Explain that this is the equivalent of multiplying by 100. Work through some examples with the class to establish that to multiply by 100 you move the digits two places to the left.
- Write 13 × 1000 = 13 000. Ask: *Can you explain why the digits move three places to the left when you multiply by 1000?* (Because 1000 = 10 × 10 × 10.)
- Have a class discussion about why it is mathematically incorrect to say that you add 0 to a number when you multiply it by 10. Use an example such as 3 + 0 = 3 but 3 × 10 = 30 to make sure the children understand.
- Work through the examples on **Pupil book 5 page 68** with the class.
- Let the children work independently to complete and check the mental calculations in questions 1–3, then ask them to solve the word problems in questions 4–7.

Follow-up

Let the children complete **Workbook 5 page 42** to consolidate their understanding of multiplying by 10, 100 and 1000.

Interesting mistakes

- The rules for multiplying (and dividing) by 10 and 100 may confuse children who need support with understanding place value. Use place-value cards and tables to model calculations.
- The children could also use a calculator to do several calculations involving 10 and 100 to help them to remember the patterns.

Divide by 10, 100 or 1000

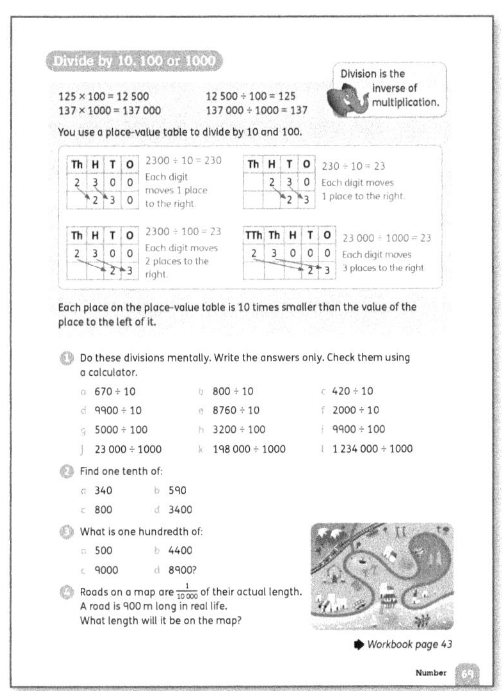

➜ Workbook page 43

Materials

Place-value table (page 21); place-value cards (page 21); calculators

Warm-up

As a starter for this lesson, have a number talk (pages 17–18) about the patterns involved when dividing numbers by 10. For example, say: *1 is $\frac{1}{10}$ of 10 and 10 is $\frac{1}{10}$ of 100*. Use phrases such as *10 times smaller, $\frac{1}{10}$ of the size* and so on.

Focus

- Ask the children to relate the warm-up discussion to different numbers. For example, ask: *What number is 10 times smaller than 300?* (30) *What number is $\frac{1}{10}$ of 150?* (15)
- Remind the children that division is the inverse of multiplication. Establish the inverse rules by adapting the activities from the previous lesson for division. Make sure the children understand that to divide by 10 they move the digits one place to the right, to divide by 100 they move the digits two places to the right and so on.
- Make sure the children understand that to find one tenth they have to divide by 10, to find one hundredth they have to divide by 100 and so on.
- Work through questions 1–3 on **Pupil book 5 page 69** with the class to teach and consolidate division by 10, 100 and 1000. In question 4, they need to divide by 10 000, moving the digits four places to the right.

Follow-up

Once the class has completed and checked the Pupil book questions, let the children work in pairs to complete **Workbook 5 page 43**.

Challenge

- Ask the children to choose any two digits from 1–9 and use them to make a set of equivalent statements using × 10, × 100, × 1000, ÷ 10, ÷ 100, ÷ 1000, including zeros and decimal points as they need them. For example:
  $23 \times 10 = 2300 \div 10$
  $23 \times 100 = 2.3 \times 1000$
  $230 \div 10 = 2300 \div 100$
- Encourage them to try to make equivalent statements with a × and a ÷ sign as well as statements with two × signs and two ÷ signs.
- Let them use a calculator to check their work.

Answers for Pupil book 5 page 69

1
a 67	b 80	c 42
d 990	e 876	f 200
g 50	h 32	i 99
j 23	k 198	l 1234

2
a 34	b 59
c 80	d 340

3
a 5	b 44
c 90	d 89

4 0.09 m or 9 cm

Answers for Workbook 5 page 43

1 140 000, 14 000, 14, 1400, 140 000, 140, 14
2 1, 24, 134, 99, 1300, 75
3 Individual answers.

Make multiplying simpler

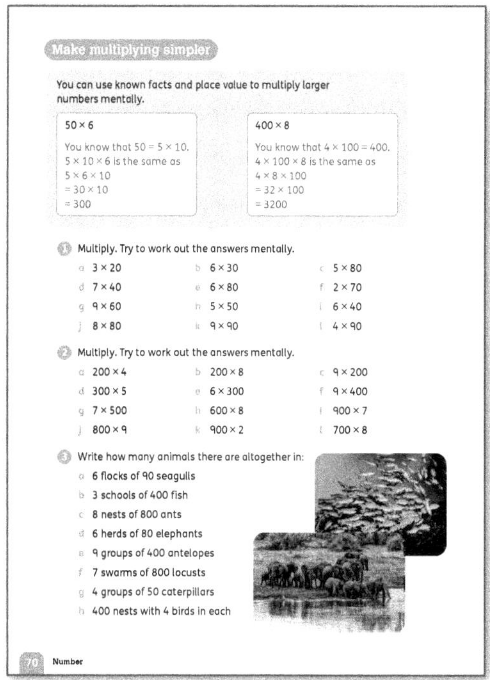

Make multiplying simpler

You can use known facts and place value to multiply larger numbers mentally.

50×6

You know that $50 = 5 \times 10$.
$5 \times 10 \times 6$ is the same as
$5 \times 6 \times 10$
$= 30 \times 10$
$= 300$

400×8

You know that $4 \times 100 = 400$.
$4 \times 100 \times 8$ is the same as
$4 \times 8 \times 100$
$= 32 \times 100$
$= 3200$

1 Multiply. Try to work out the answers mentally.

a 3×20	b 6×30	c 5×80
d 7×40	e 6×80	f 2×70
g 9×60	h 5×50	i 6×40
j 8×80	k 9×90	l 4×90

2 Multiply. Try to work out the answers mentally.

a 200×4	b 200×8	c 9×200
d 300×5	e 6×300	f 9×400
g 7×500	h 600×8	i 900×7
j 800×9	k 900×2	l 700×8

3 Write how many animals there are altogether in:

a 6 flocks of 90 seagulls
b 3 schools of 400 fish
c 8 nests of 800 ants
d 6 herds of 80 elephants
e 9 groups of 400 antelopes
f 7 swarms of 800 locusts
g 4 groups of 50 caterpillars
h 400 nests with 4 birds in each

70 Number

Materials

Sticky notes

Warm-up

Have a quick class quiz to revise multiplication and division facts up to 12 × 12 as a mental warm-up for this lesson. Let the children write the answers only and then check and rate their accuracy.

Focus

- Ask the children whether 4 × 2 is equal to 2 × 4. Ask them to give other examples that show that multiplying the numbers in any order gives the same product.
- Set a problem that requires multiplication by multiples of 10. For example: *A goat has 4 legs. How many legs do 30 goats have?* You can use cards or sticky notes to model this for the class:

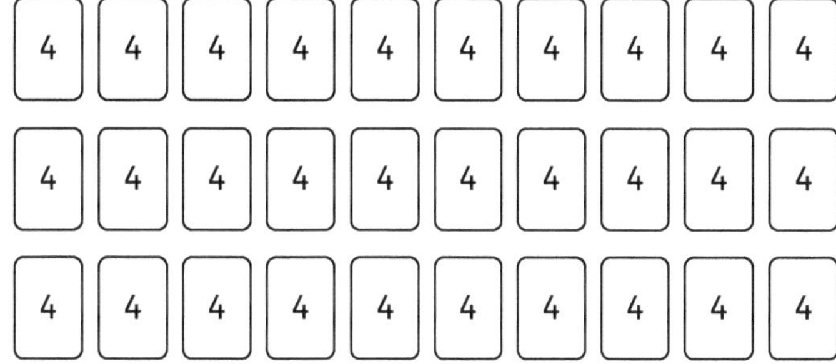

- This model shows that 30 goats are the same as 3 × 10 goats. So 30 × 4 = 3 × 10 × 4. We can multiply numbers in any order, so rearrange to 3 × 4 × 10 = 12 × 10 = 120.
- Ask the class to suggest how they could find the number of legs on 60 goats (double the answer) and 90 goats (multiply the answer by 3) or 15 goats (halve the answer).

- Discuss with the class how looking at the numbers in a calculation can help us to choose a strategy for working out the solution. In these examples, the number we are multiplying by is a multiple of 10. Let the children share their own methods and ideas.
- Work through the examples on **Pupil book 5 page 70** with the class. Then ask them to complete questions 1–3.
- Observe the children as they work to see what strategies they use and suggest alternatives where necessary.

Support

If any children have difficulty doing calculations mentally, encourage them to use jottings to break a number into factors. They do not need to show all the steps as in the worked examples. Instead, they could write something like this:

$$4 \times 90$$
$$4 \times 9 \times 10$$
$$360$$

Answers for Pupil book 5 page 70

1 a 60 b 180 c 400
   d 280 e 480 f 140
   g 540 h 250 i 240
   j 640 k 810 l 360
2 a 800 b 1600 c 1800
   d 1500 e 1800 f 3600
   g 3500 h 4800 i 6300
   j 7200 k 1800 l 5600
3 a 540 b 1200 c 6400
   d 480 e 3600 f 5600
   g 200 h 1600

Write a number as a product of its factors

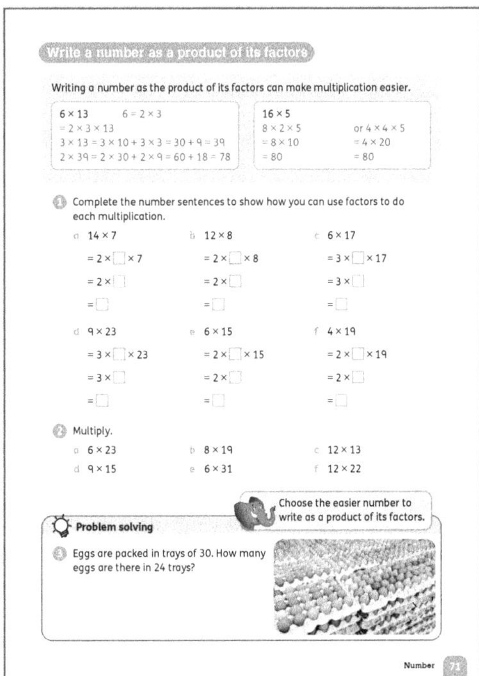

Materials

Cubes or other small objects

Warm-up

As a mental starter, do some activities that involve finding factor pairs of numbers. This is a fundamental concept for this lesson and you need to make sure that the children understand it before moving on to the work in the Pupil book.

Focus

- Write 9×16 on the board. Underneath it write:
  $3 \times 3 \times 16$ $9 \times 2 \times 8$
- Ask the children: *Can you explain why both of these represent the same calculation as 9 × 16?* (Because 3×3 is the same as 9 and 2×8 is the same as 16.)
- Let the children suggest how writing one or both of the numbers as a product of its factors can help them to multiply.
- Let them share their ideas before asking them which of these calculations they find easier to do. Ask them to explain their choices.
- Look at $9 \times 2 \times 8$. Tell the class that you are going to work this out. Say: *9 × 8 is 72. Double 72 is 144.*
- Ask the class to explain what you did. Make sure they mention that you can multiply numbers in any order.

Turn to **Pupil book 5 page 71**. Let the children read through the examples on their own before completing the questions.

- The layout of question 1 forces the children to think about the factors and how to combine them. The focus is more on the strategy than the answers.
- No method is prescribed for question 2. The children may use factors, but they might find it easier to use other methods that they know.
- <u>Problem solving:</u> Talk about the hint box for question 3 and explain that different people might break up the numbers differently. Emphasise that this is acceptable and that the children should use the method they find easiest to do calculations.

Support

If necessary, spend some time revising doubling numbers using place value. This will help the children to see that writing an even number as 2 × another factor can help them with multiplication. Models like these can help the children to understand this process:

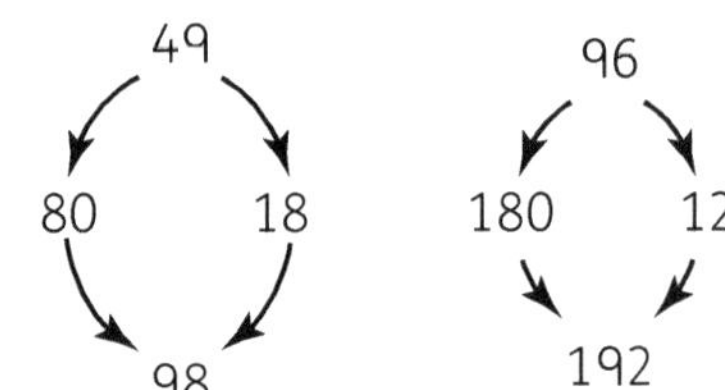

Answers for Pupil book 5 page 71

1 a $14 \times 7 = 2 \times 7 \times 7 = 2 \times 49 = 98$

b $12 \times 8 = 2 \times 6 \times 8 = 2 \times 48 = 96$

c $6 \times 17 = 2 \times 3 \times 17 = 3 \times 34 = 102$

d $9 \times 23 = 3 \times 3 \times 23 = 3 \times 69 = 207$

e $6 \times 15 = 2 \times 3 \times 15 = 2 \times 45 = 90$

f $4 \times 19 = 2 \times 2 \times 19 = 2 \times 38 = 76$

2 a 138   b 152   c 156

d 135   e 186   f 264

3 720

Order of operations

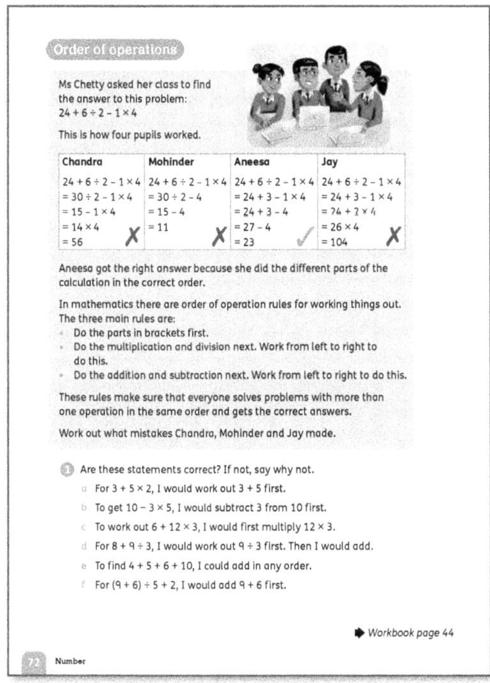

Materials

Calculators (Make sure that the calculators follow the order of operations.)

Warm-up

Select any suitable 'Mental problem solving' activity from the Activity bank (pages 24–25) as a starter for this lesson.

Focus

- Write this calculation on the board and ask the children to try to solve it:
  $25 \div 5 + 3 \times 6 − 2$
- Ask the children to share their answers. Ask: *Did you all get the same answer?*
- Ask them to decide who got the answer correct. Do not confirm the correct answer at this stage.

Turn to **Pupil book 5 page 72**. Let the children work in pairs or small groups to read through the incorrect and correct worked examples and the order of operations rules. Let them share their ideas with the class when they have finished. (Chandra added 24 and 6 and then divided by 3, but should have divided 6 by 2 and then added the result to 24. Then they subtracted 1 from 15 and multiplied the result by 4 instead of multiplying 1 by 4 and subtracting the result. Mohinder has added 24 and 6 and then divided by 3, but should have divided 6 by 2 and then added the result to 24. Jay correctly worked out 6 divided by 2 first to get 3, but then subtracted 1 from 3 and multiplied the result by 4 instead of multiplying 1 by 4 and subtracting the result from 24 + 3.)

- Explain to the class that we have to do certain operations before others. Use a real-life example to illustrate this. (For example, *8 children come to a party. Each child gets 1 chocolate and 2 lollipops in their party pack. How many sweets do I use altogether to make the packs?*)
- Draw the 8 party packs on the board, and add up how many sweets would be used altogether (24). Then ask the children what number sentence we could use to work this out quickly. Elicit the number sentence: $(2 + 1) \times 8$. Point out that the brackets tell us to do that part of the calculation first.
- Use this and similar examples to illustrate the order of operations rules.
- Point out that we can use brackets to 'promote' operations that would otherwise come later. For example, if you need to work out the addition part of a sum first (as in the party example above), it should appear in brackets.
- Return to the problem you wrote on the board at the start of the lesson: $25 \div 5 + 3 \times 6 − 2$. Ask them to work out the correct answer (21). Discuss any mistakes that the children made.
- Do question 1 orally with the class. Give the children time to read the statements and decide whether they are correct or not. Then ask for their answers and reasons for their decisions.

Follow-up

Use **Workbook 5 page 44** as a written consolidation task to check that the children understand and can follow the order of operations rules. Give the children time to make up their own challenging calculations. Let them complete and check each other's work.

Interesting mistakes

- The children may need additional support with the idea that they will not always work from left to right in a maths problem. Encourage them to make a flow chart with words like *first, second, next, then …* to consolidate the idea that the rules tell them when to do different operations.
- If you teach the rule as brackets first, then orders (indices), then multiplication and division, and lastly addition and subtraction, the children may assume that they always have to multiply before they divide and always add before they subtract. This is not the case – they need to work from left to right through both pairs of operations. When you discuss the rules and the order, try to always say *multiply and divide from left to right, add and subtract from left to right*.
- Lastly, the children may not realise that any operations within the brackets still follow the rules, so if you have $(2 + 3 \times 2)$, they still need to multiply before they add. Give the class some examples and model how to work them out.

Answers for Workbook 5 page 44

1 **a** $(10 - 2) - 3 = 8 - 3 = 5$

b $8 + 5 - 3 = 13 - 3 = 10$

c $(12 - 9) \times 24 - 6 = 3 \times 24 - 6 = 72 - 6 = 66$

d $144 - 21 \times 2 = 144 - 42 = 102$

e $10 + 7 - 2 \times 6 = 10 + 7 - 12 = 17 - 12 = 5$

f $36 \div 6 - 2 = 6 - 2 = 4$

g $5 - 10 \div 2 = 5 - 5 = 0$

h $15 \div 3 - (3 + 2) = 15 \div 3 - 5 = 5 - 5 = 0$

i $3 \times 4 - 2 \times 6 = 12 - 12 = 0$

j $7 - 24 \div 6 = 7 - 4 = 3$

k $6 \times 3 - 17 = 18 - 17 = 1$

l $3 \times 3 - 3 = 9 - 3 = 6$

m $4 \times 2 - 16 \div 2 = 8 - 8 = 0$

n $2 + 2 + 2 \times 2 = 2 + 2 + 4 = 8$

o $3 \times 9 - 5 \times 5 = 27 - 25 = 2$

p $(8 + 3) \times (20 \div 2) \div 11 = 11 \times 10 \div 11 = 110 \div 11 = 10$

q $2 \times (6 - 3) + 5 = 2 \times 3 + 5 = 6 + 5 = 11$

r $(12 + 6) \div (5 - 2) = 18 \div 3 = 6$

s $(99 + 22) \div 11 = 121 \div 11 = 11$

t $8 \div (16 - 14) - 1 = 8 \div 2 - 1 = 4 - 1 = 3$

2 **a–f** Individual answers.

3 **a–f** Individual answers.

Work in order

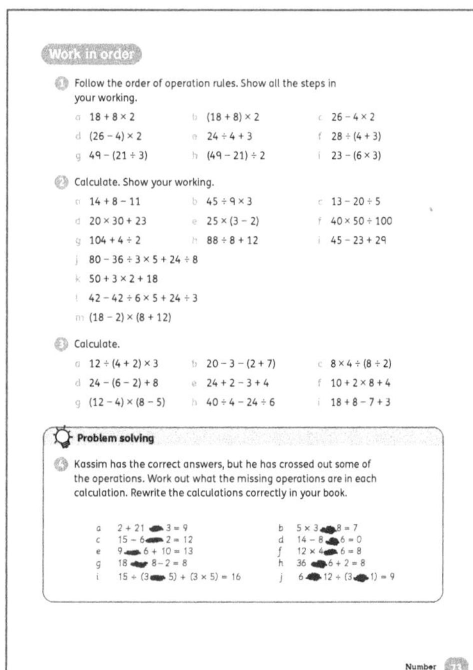

Warm-up

Do any 'Calculation skills' activity (pages 25–28) as a warm-up for this lesson.

Focus

- The children can work through questions 1–3 on **Pupil book 5 page 73** independently to consolidate their understanding of the order of operations rules. You can set some of the work for homework if it is too much for a class lesson.

- <u>Problem solving</u>: The children can work on their own to solve the problems in question 4. Make sure they realise that they need to find operation signs $(+, -, \times, \div)$ and not numbers.

Answers for Pupil book 5 page 73

1 **a** $18 + 8 \times 2 = 18 + 16 = 34$

b $(18 + 8) \times 2 = 26 \times 2 = 52$

c $26 - 4 \times 2 = 26 - 8 = 18$

d $(26 - 4) \times 2 = 22 \times 2 = 44$

e $24 \div 4 + 3 = 6 + 3 = 9$

f $28 \div (4 + 3) = 28 \div 7 = 4$

g $49 - (21 \div 3) = 49 - 7 = 42$

h $(49 - 21) \div 2 = 28 \div 2 = 14$

i $23 - (6 \times 3) = 23 - 18 = 5$

2 **a** $14 + 8 - 11 = 22 - 11 = 11$

b $45 \div 9 \times 3 = 5 \times 3 = 15$

c $13 - 20 \div 5 = 13 - 4 = 9$

d $20 \times 30 + 23 = 600 + 23 = 623$

e $25 \times (3 - 2) = 25 \times 1 = 25$

f $40 \times 50 \div 100 = 2000 \div 100 = 20$

g $104 + 4 \div 2 = 104 + 2 = 106$

h $88 \div 8 + 12 = 11 + 12 = 23$

i $45 - 23 + 29 = 22 + 29 = 51$

j $80 - 36 \div 3 \times 5 + 24 \div 8 = 80 - 12 \times 5 + 3 = 80 - 60 + 3 = 20 + 3 = 23$

k $50 + 3 \times 2 + 18 = 50 + 6 + 18 = 56 + 18 = 74$

l $42 - 42 \div 6 \times 5 + 24 \div 3 = 42 - 7 \times 5 + 8 = 42 - 35 + 8 = 7 + 8 = 15$

m $(18 - 2) \times (8 + 12) = 16 \times 20 = 320$

3 **a** 6   **b** 8   **c** 8

d 28   **e** 27   **f** 30

g 24   **h** 6   **i** 22

4 **a** $\div$ **b** $-$ **c** $\div$

d $-$ **e** $-$ **f** $\div$

g $-$ **h** $\div$ **i** $\times$

j $+, -$

End-of-unit check

Ask some or all of these questions to assess how well the children have understood the concepts in this unit:

- *What is 6×3?* (18) (or any other tables facts)
- *Start on 21 on a number line and make three jumps of 5. What numbers do you land on?* (26, 31, 36)
- *How many groups of 8 do we need to make 40?* (5)
- *Write another multiplication sentence that has the same answer as 6×5.* (1×30, 2×15, 3×10, 5×6, 10×3, 15×2, 30×1)
- *What is the square of 5?* (25) (or use any other number)
- *What are all the factors of 12?* (1, 2, 3, 4, 6, 12) (or use any other number)

- *Is 9 a multiple of 3? Is 10?* (9 is a multiple of 3. 10 is not a multiple of 3.) (Give a range of examples.)
- *What are the factors of* (give a number)?
- *Give me three multiples of* (give a number).
- *Which of these numbers are prime numbers?* (Give a list of numbers.) *How do you know?* (A prime number has exactly two factors: 1 and the number itself.)
- *What is* (give a number) *multiplied by 10/100/1000?*
- *What is* (give a number) *divided by 10/100/1000?*
- *I divided a number by 100 and got an answer of* (give a number). *What was the original number?*
- *What is one tenth/one hundredth of* (give a number)?
- *What is an easy way to multiply by 25?* (For example: Multiply by 5 and then by 5 again.)
- *Is multiplying by 3 and then multiplying by 10 the same as multiplying by 30? Why or why not?* (It is the same. 3 × 10 = 30)

- *What are the factors of* (give a number)?
- *How can you use factors to make multiplication easier?* (Multiply the number by the first factor and then multiply the answer by the second factor.)
- *Which of these mean the same?* (Give some multiplications using numbers and factors. For example: 16 × 8; 8 × 2 × 8, 4 × 4 × 4 × 2, and so on.)
- *Mindy says 6 + 2 × 3 = 24 and Mandy says 6 + 2 × 3 = 12. Who is correct? Why?* (Mandy is correct. You need to do the multiplication before the addition. 2 × 3 = 6 and 6 + 6 = 12) (and other similar questions)

Unit 8 Measures and money

Learning objectives

- Work with lengths, mass and capacity in different contexts

- Convert between different units of metric measure

- Understand and use approximate equivalences between metric units and common imperial units

- Use all four operations to solve problems related to decimal measure, including scaling

- Divide 1 into 2, 4, 5 and 10 equal parts and read scales/number lines marked in units of 1, 2, 4, 5 and 10 equal parts

- Read and interpret measuring scales in context

- Revise basic money concepts and solve problems involving money amounts

Key words

tonne   capacity   measuring scale   metric   imperial

Unit introduction

Teaching guidance
- Start this unit by having a class discussion about when and where the children use measurements in their daily lives. Let groups discuss this and make lists of examples.

- Take feedback from the class. When the children give examples, they should provide some detail. For example, 'When I help my grandmother bake, I measure the ingredients in grams and millilitres.' 'We have a rain gauge at home that measures the amount of rain that falls in millimetres.' 'I helped my mother measure the lounge in metres so that we could buy a new rug.'
- Next display some jobs that people do in your community. Select jobs that the children are likely to be familiar with (for example, nurse, engineer, farmer, architect, baker, tailor, hairdresser, traffic officer and so on).
- Ask the children to choose two different jobs and to discuss and record all the ways that a person might use measurement in that job. For example, for a nurse they might produce a diagram like this:

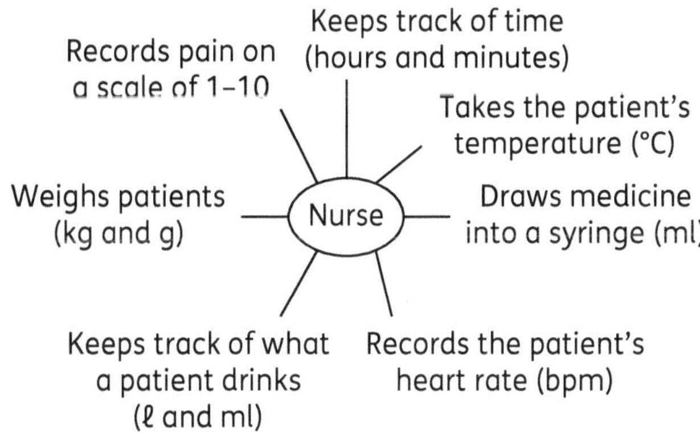

- Let groups show their completed diagrams to the class and ask the other children if they have anything to add.
- Finish the activity by talking about how important measurement is in our daily lives and how often we use measurements and convert between measurements without really thinking about it.

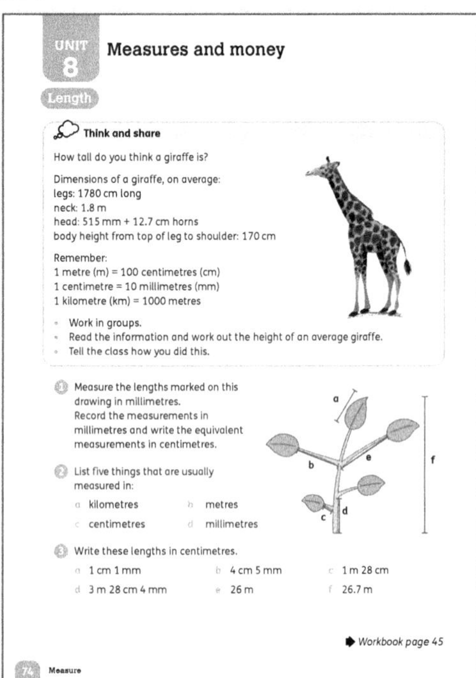

Materials

Rulers marked in millimetres

Warm-up

Have a class quiz with questions that involve multiplying and dividing by 10, 100 and 1000 as a mental warm-up for this lesson.

Focus

Try some of the following activities to teach the children about units of length and how to convert between them.

- Ask the children to draw a line that is 1 centimetre long. Then ask them whether they know of a smaller unit. Ask the children to look at their rulers and notice that a centimetre is divided into 10 smaller units. Revise or introduce the term *millimetre*.
- Ask the children to suggest things that we need to measure in millimetres. Encourage them to think of not only small things but also situations where larger things need to be measured very accurately (for example, in building work).
- Write 5 cm 6 mm on the board. Ask the children: *How many millimetres is 5 cm 6 mm in total?* Encourage the children to explain how they worked it out. (For example, 'There are 50 mm in 5 cm so it is 50 + 6, which is 56 mm.')
- Ask the children how many centimetres there are in a metre. Then ask: *We know that there are 10 mm in every centimetre, so how many millimetres must there be in a metre?* (1000). Then ask: *What fraction of a metre must a millimetre be?* ($\frac{1}{1000}$)
- Draw a place-value table with columns for metres, centimetres and millimetres on the board. Use this to practise converting measurements between the units.
- On the board write kilometres, metres, centimetres and millimetres. Ask the children to think of things that would be best measured in each unit in turn. Record their suggestions and encourage the children to explain how they made their decisions.

- Ask the children how many metres are in a kilometre. Having established the answer, ask the children what fraction of a kilometre a metre must be.
- Draw a place-value table on the board. Use this to practise converting metres to kilometres and vice versa by dividing/multiplying by 1000.
- Write some lengths and distances on the board, such as *the length of a pencil, the distance between two towns* and so on. Ask the children to give a sensible unit for measuring each distance.
- <u>Think and share:</u> Read through the giraffe problem on **Pupil book 5 page 74**. Let the children work in groups to convert the measurements and answer the question about the giraffe. (The height of an average giraffe is 592.2 cm or 5.922 m. It is found by adding the dimensions of the legs, neck, head and horns, and body.)
- Then ask the children to complete the conversions on **Workbook 5 page 45**. Check that they can convert from bigger to smaller units as well as from smaller to bigger units.
- The children can then complete **Pupil book 5 page 74**. For question 1, make sure that they have rulers marked in millimetres and that they understand how to measure the different lengths. Note that the centimetre lengths may be decimal numbers.
- For question 2, the children can refer back to the discussion about units earlier in the lesson.
- Make sure the children know that they must write all their answers to question 3 in centimetres, so they will need to use decimals for some of the questions.

Support

- Let the children draw their own diagrams to show the conversion factors and allow them to use these for reference when they have to convert between units. A diagram for length may look this:

$$\begin{array}{cccc} \times 1000 & \times 100 & \times 10 \\ \Rightarrow & \Rightarrow & \Rightarrow \\ \text{km} \quad \text{m} & \text{cm} & \text{mm} \\ \Leftarrow & \Leftarrow & \Leftarrow \\ \div 1000 & \div 100 & \div 10 \end{array}$$

- The children can add diagrams for converting between units for measuring mass and capacity when you get to those topics.

Interesting mistakes

Converting between millimetres, centimetres, metres and kilometres requires the children to work with a wide range of place values, which can result in frequent mistakes. Remind the children to work from one unit to the next unit up or down to avoid multiplications becoming too unwieldy. For example, they can convert from millimetres to centimetres before converting to metres.

Answers for Pupil book 5 page 74

1 a 1.4 cm b 1.8 cm c 0.4 cm
    d 1.4 cm e 2.3 cm f 5.4 cm

2 a For example, distances between towns
    b For example, the length of a garden
    c For example, the length of a pencil
    d For example, the length of a beetle

3 a 1.1 cm b 4.5 cm c 128 cm
    d 328.4 cm e 2600 cm f 2670 cm

Answers for Workbook 5 page 45

1 a 23 km = 23 000 m = 2 300 000 cm
134 km = 134 000 m = 13 400 000 cm
2.345 km = 2345 m = 234 500 cm
90.5 km = 90 500 m = 9 050 000 cm
0.2 km = 200 m = 20 000 cm
1.25 km = 1250 m = 125 000 cm

b 4 m = 400 cm = 4000 mm
9.6 m = 960 cm = 9600 mm
43 m = 4300 cm = 43 000 mm
123 m = 12 300 cm = 123 000 mm
9 m = 900 cm = 9000 mm
0.85 m = 85 cm = 850 mm

2 African elephant: height 327 cm; ear length 60.9 cm; trunk 140 cm; length 400 cm
Bengal tiger: shoulder height 110 cm; length 196.5 cm; tail 107 cm
Crocodile: tooth; 9 cm; length 335 cm; tail 213.3 cm

Mass

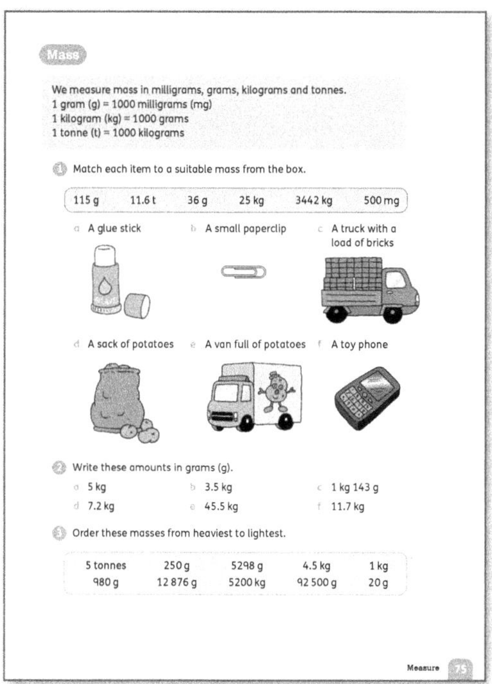

Warm-up

As most conversions between units of mass involve units that are 1000 times greater or smaller than each other (such as *tonnes*), practise multiplying and dividing by 1000, including where the answer is a decimal, as a mental warm-up for this lesson.

Focus

- Have a number talk (pages 17–18). Ask the class: *How can we use what we know about converting between units of length to help us to convert between units of mass?*
- The children can read the units of mass at the top of **Pupil book 5 page 75**. Let the children share their ideas and, if necessary, show the conversions on place-value tables as you did with units of length.

- Once the children understand that the conversions work in the same way, let them complete questions 1–3 on **Pupil book 5 page 75** independently.
- For question 3, tell the children that they can convert between units to compare the masses but that they should write their answers using the units given in the question.

Challenge

Write these two questions on the board. Let the children discuss them and prepare answers to share with the rest of the class.

- Why is it important to convert measurements to the same unit before you compare and order them or calculate with them?
- How do you decide which unit to convert to?

Support

Ask the children to draw a diagram to show how to convert between units of mass. Alternatively, they can add these conversions to the diagrams they drew for units of length in the previous lesson.

Answers for Pupil book 5 page 75

1 a 36 g   **b** 500 mg   **c** 11.6 t
d 25 kg   **e** 3442 kg   **f** 115 g
2 a 5000 g   **b** 3500 g   **c** 1143 g
d 7200 g   **e** 45 500 g   **f** 11 700 g
3 5200 kg, 5 tonnes, 92 500 g, 12 876 g, 5298 g, 4.5 kg, 1 kg, 980 g, 250 g, 20 g

Capacity

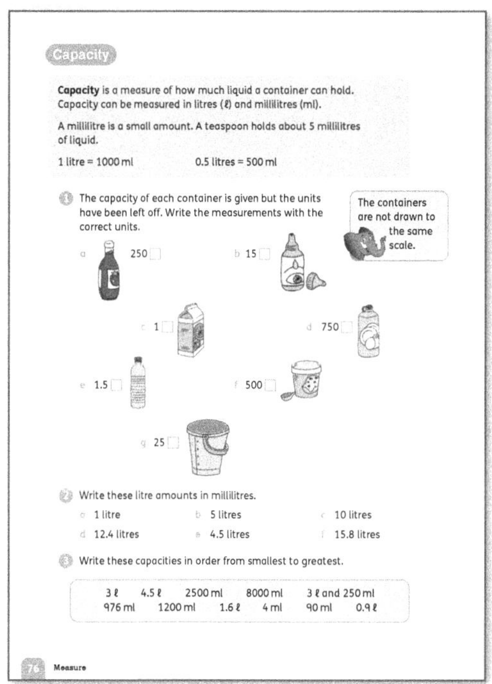

Materials

Containers of different sizes with different capacities marked (try to include litres as well as millilitres; encourage the children to bring some containers with the capacities marked to school so that you have a good range); flashcards with decimal numbers up to three decimal places

Warm-up

Give children flashcards with decimal numbers up to three decimal places, and ask them to put them in order as a warm-up for this lesson.

Focus

- Use the different-sized containers to revise basic units of *capacity* and equivalences between them.
- Spend some time doing practical measuring tasks using water and/or sand. For example, ask: *How many of these small containers do we need to fill this larger container? What does 250 ml look like in this half-litre bottle? What does 250 ml look like in this 2-litre bottle?* and so on.
- Discuss the units of capacity and how they relate to each other. Let the children suggest how to use what they already know to convert between these. You can use place-value tables with units to show the conversions if necessary.
- Let the children work independently through questions 1–3 on **Pupil book 5 page 76.**

Support

Ask the children to draw another diagram to show how to convert between units of capacity or add these conversions to the diagrams they have already drawn.

Interesting mistakes

- Research into errors in converting between units of measure shows that children need additional support with directions of conversion (bigger to smaller and smaller to bigger).
- Interestingly, children also find it easier to convert between units of length than those of other measures, including capacity. This may be due to the fact that lengths are more familiar and they can visualise them.
- It is therefore important to keep referring to the *metric system* when working with all the units of measure, so the children see that there is a pattern and a method that works for all measures, not just lengths. It is also important to spend equal amounts of time on different measures to make sure that the children understand the relationships and can work with them.

Answers for Pupil book 5 page 76

1   a ml   b ml   c ℓ
    d ml   e ℓ   f ml
    g ℓ
2   a 1000 ml   b 5000 ml   c 10 000 ml
    d 12 400 ml   e 4500 ml   f 15 800 ml
3 4 ml, 90 ml, 0.9 ℓ, 976 ml, 1200 ml, 1.6 ℓ, 2500 ml, 3 ℓ, 3 ℓ and 250 ml, 4.5 ℓ, 8000 ml

Measuring scales

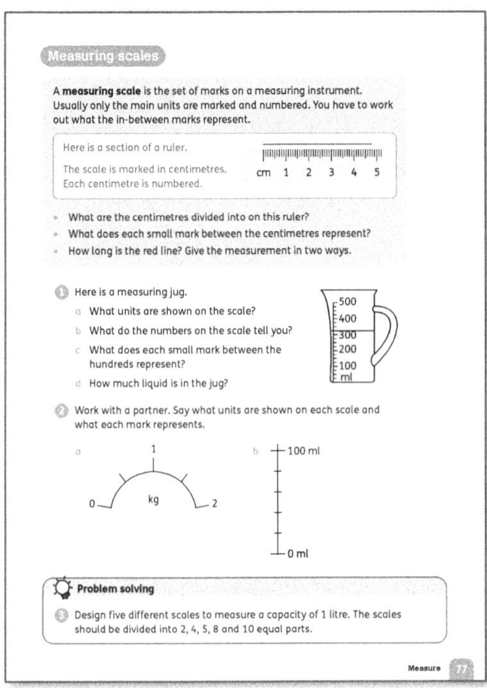

Materials

Measuring instruments with marked scales (collect as many different measuring instruments as possible; you can also download good-quality images from the Internet and display them for the class); rulers and pencils; blank number lines; squared or graph paper

Warm-up

Use 'Round decimals' from the 'Rounding and estimating' section of the Activity bank (page 24) as a starter for this lesson.

Focus

Spend as much time as necessary teaching the children how to read and make sense of different *measuring scales*. Show the class weighing scales, measuring jugs, tape measures and other marked scales.

- Let the children work in pairs to discuss the explanation on **Pupil book 5 page 77**, then share the answers as a class. The centimetres are divided into ten equal parts. 1 millimetre, where 10 mm = 1 cm. The line is 4.8 cm long or 3 cm 8 mm.
- Ask the children to complete questions 1 and 2, then ask them to explain how they made their decisions about each scale. The children work in pairs on question 2.
- <u>Problem solving:</u> Let the children work in pairs to design and draw their scales in question 3.

Challenge

- Ask the children to investigate different and unusual measuring scales. They can photograph or print images of the scales they find and make a poster showing the different scales. They should label each scale with: what the scale measures; what units are shown on the scale; how the scale is divided up. They can include units that they haven't encountered yet in class.

- Note that if the children find some *imperial* measuring scales, the divisions will not be decimals, which could lead to some interesting discussion in class.

Support

If the children need support to design and make their own measuring scales, give them blank number lines to use as a starting point. Working on squared or graph paper will also help them divide the line into equal intervals.

Answers for Pupil book 5 page 77

1 a millilitres (ml)
  b how much liquid there is in the jug
  c 25 ml
  d 325 ml
2 a kg, 0.5 kg
  b ml, 20 ml
3 Individual answers.

Read measuring scales

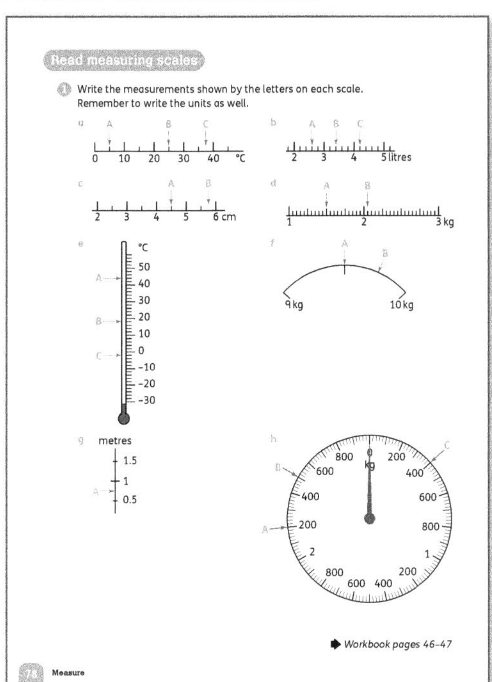

Warm-up

Select any suitable 'Place value and number sense' activity from the Activity bank (pages 22–23) as a starter for this lesson.

Focus

Let the children work on their own to complete the measuring scales questions in question 1 on **Pupil book 5 page 78**. Check the answers orally as a class.

Follow-up

- Use **Workbook 5 page 46** to assess that the children can read and interpret measuring scales.
- Then the children can work on their own to mark the measurements on the scales on **Workbook 5 page 47**. Let the children mark their work in pairs, then spend some time discussing as a class any problems the children experienced.

Interesting mistakes

- The children may need additional support to read and make sense of scales, particularly when the intervals are not labelled. Continue to read and make sense of real measuring scales to show the children how to interpret them.
- Help the children to work out what the intervals are by asking questions such as: *What do these marks on the scale tell you? Why are some marks longer/thicker than others? How many marks are there between the marked units? How can this help you read the scale?*

Answers for Pupil book 5 page 78

1 a A = 5 °C, B = 25 °C, C = 37.5 °C
  b A = 2.6 ℓ, B = 3.4 ℓ, C = 4.2 ℓ
  c A = 4.5 cm, B = 5.75 cm
  d A = 1.5 kg, B = 2.05 kg
  e A = 44 °C, B = 18 °C, C = −2 °C
  f A = 9.5 kg, B = 9.75 kg
  g A = 0.75 m
  h A = 2200 g, B = 2500 g, C = 400 g

Answers for Workbook 5 page 46

1 Possible answers:
  a 20 g, **b** 80 g, **c** 0.5 ℓ, **d** 0.25 ℓ, **e** 1 km,
  f 2.5 km, **g** 3.5 km

2 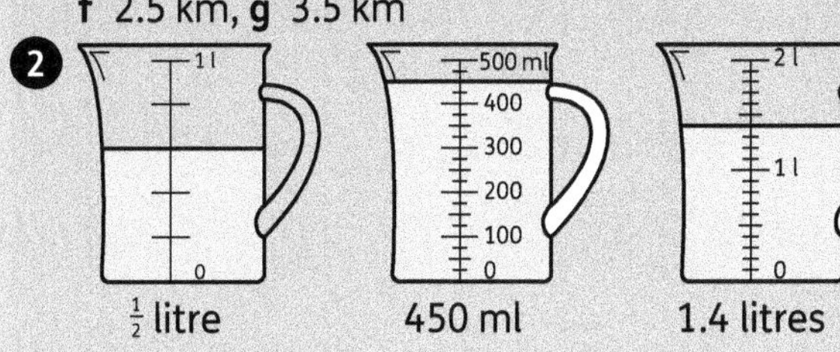

3 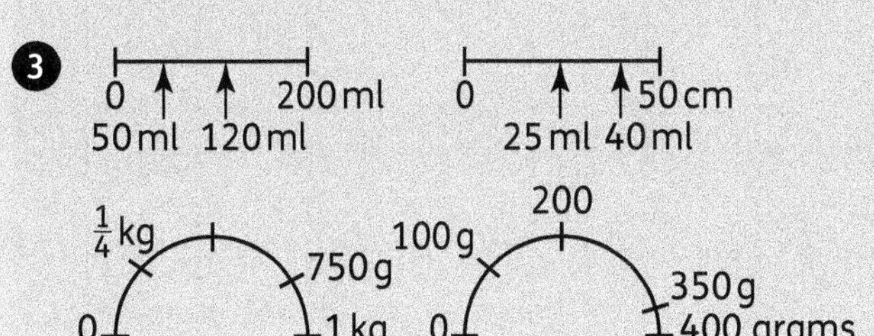

Answers for Workbook 5 page 47

1 a 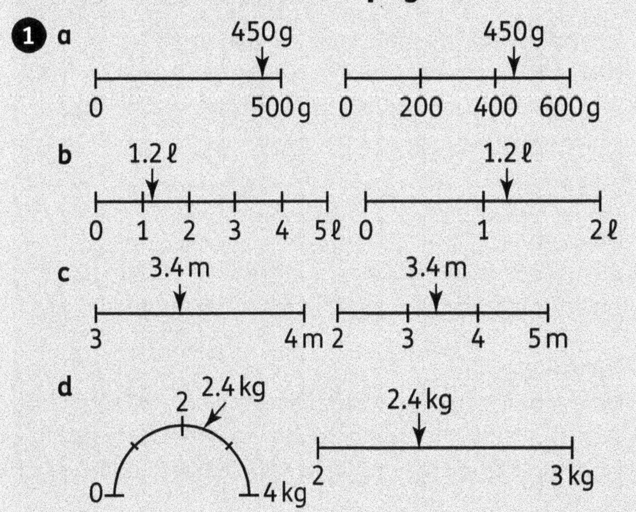

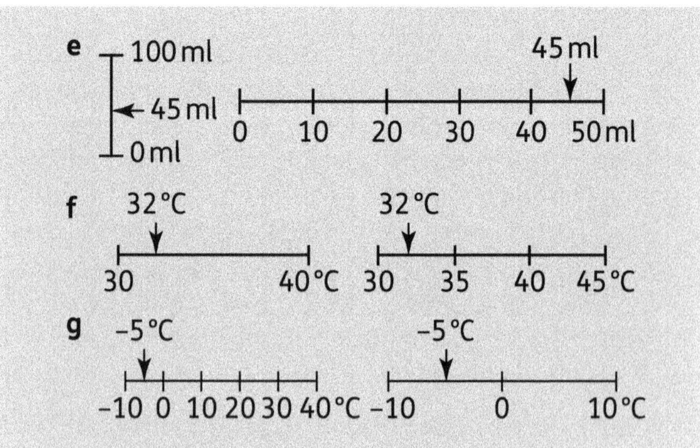

Metric and imperial equivalents

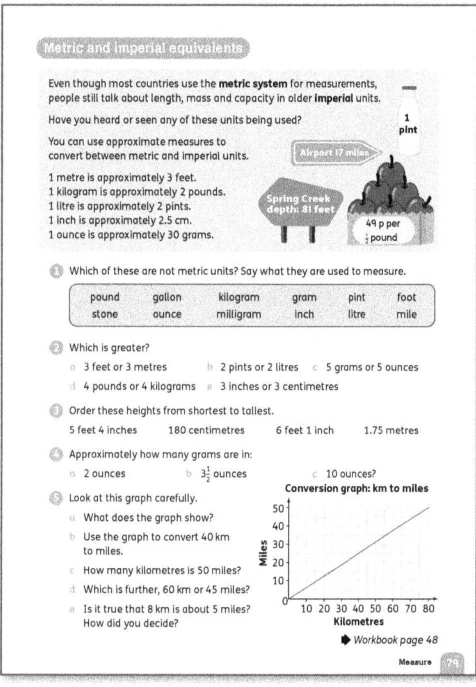

There are only three countries in the world that have not officially adopted the metric system of measures: USA, Myanmar and Liberia. However, even the USA (which has its own version of imperial measures, called 'customary units') recognises and uses the international system of units (sometimes alongside the older imperial units).

People in many countries still use imperial measures in everyday situations, although as the world becomes more digital, people will be more likely to use metric units.

Materials
Tape measures and/or rulers marked in centimetres and inches; access to conversion programs or apps

Warm-up
Give the children sets of whole numbers, decimals and mixed numbers including halves and quarters, for them to compare and sort into descending order as a mental starter for this lesson.

Focus
- Read the information at the top of **Pupil book 5 page 79** and ask the children whether they have heard of any of the imperial units mentioned. Let them give details.
- Explain that a 'rule of thumb' conversion is an approximate conversion between systems. For example, *As a rule of thumb, we say that 1 metre is about 3 feet. In reality, it is 3 feet and just over $3\frac{1}{3}$ inches. For most purposes, the approximate conversions are close enough.*
- You could ask the class to think of situations where accurate conversions between systems are important. Examples might include: if you know your weight in pounds but your doctor needs it in kilograms to decide how much medicine to give you; in engineering, where a millimetre too little or too much might make a structure unsafe.
- Some costly and dangerous problems have resulted from conversion errors. The most famous was the $125 million loss of a space probe in 1999 because one team used metric measures and the other used imperial. The children can find out about this and other mistakes by searching for 'unit conversion disasters' online.
- Investigate conversion apps and programs with the class. This is a fun way for the children to explore the different units and their equivalents. In reality, most people use an online converter to convert between units of measure.
- Children can work through questions 1–5 on **Pupil book 5 page 79** independently. They can use a dictionary or search engine to check whether the measures in question 1 are imperial or metric.
- For questions 2 and 3, remind the children that they need to convert measures to the same units to compare them.
- Before the children do question 5, you might need to revise how to read and interpret line graphs. Also show them how to read a conversion graph.

Follow-up
Use **Workbook 5 page 48** to assess whether the children can use conversion graphs.

Challenge
Let the children research metric units and the International System of Units (*Système international d'unités*). They can share what they find out in groups.

Support
Children who need help with reading conversion graphs can draw lines on the graph or place a ruler at a measurement and read across or up to find the equivalent measurement.

Interesting mistakes
The children sometimes get the impression that imperial or customary measures are less accurate than metric measures. Show the children measuring instruments (such as tape measures) that have scales for both types of units. Demonstrate that you can measure just as accurately using both scales.

Answers for Pupil book 5 page 79

1. pounds – mass, gallon – capacity, pint – capacity, feet – length, stone – mass, ounce – mass, inch – length, mile – length
2. **a** 3 metres **b** 2 litres **c** 5 ounces **d** 4 kilograms **e** 3 inches
3. 5 foot 4 inches, 1.75 metres, 180 cm, 6 feet 1 inch
4. **a** 60 g **b** 105 g **c** 300 g
5. **a** conversion of kilometres to miles
   b 25 miles **c** 80 km **d** 45 miles
   e yes; I used the graph; I drew a vertical line from 8 km on the horizontal axis up to the red line. I then drew a horizontal line from this point across to the vertical axis. The line met the vertical axis at 5 miles.

Answers for Workbook 5 page 48

1. **a** A graph showing the conversion of inches-centimetres
   b 10 cm = 4 inches; 17.5 cm = 7 inches; 20 cm = 8 inches; 5 cm = 2 inches; 22.5 cm = 9 inches
   c From the graph, 10 inches is 25 cm, so multiply this by 3: 30 inches is 75 cm
2. **a** Less; the graph shows that 1 litre is about 1.75 pints
   b 4 litres **c** 7 pints
   d From the graph, 3 litres is 5.2 pints. Doubling gives 6 litres is 10.4 pints. So you can pour 10 pints of water into a 6-litre container.
   e 4.2 pints **f** 4.6 litres

Solve measurement problems

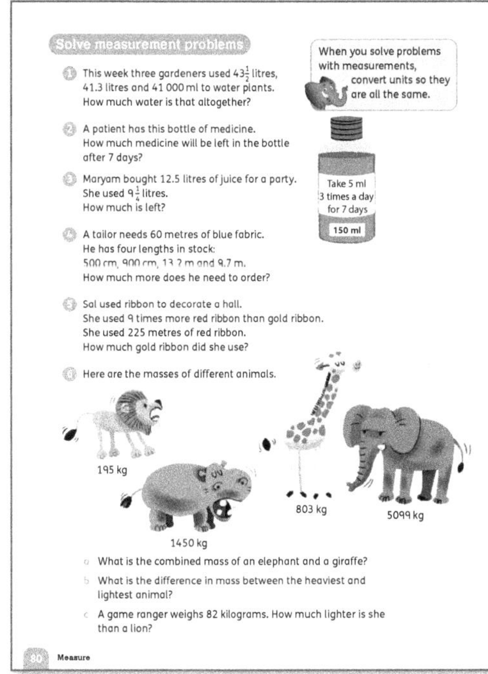

Materials
Calculators

Warm-up
Select any 'Units of measurement' activity from the Activity bank (page 29) as a warm-up for this lesson.

Focus
- Ask the children some simple scaling problems. For example: *I drink 125 ml of orange juice a day. How much will I drink in a week?* (875 ml) *I cut a 2-metre plank of wood into 8 equal pieces. How long is each piece?* (25 cm) *1 brick weighs 1800 g, how much do 3 bricks weigh?* (5400 g or 5.4 kg) Read through the problems on **Pupil book 5 page 80** with the class. Let the children ask questions if they need to and encourage them to describe what they will do to solve each problem.
- Then let the children work on their own or in pairs to answer questions 1–6. Let them check their answers using a calculator and remind them that they need to include the units with all measurements.

Answers for Pupil book 5 page 80
1. 125.8 litres **2** 45 ml **3** 3.25 litres
4. 23.1 m **5** 25 m
6. **a** 5902 kg **b** 4904 kg **c** 113 kg

Work with money

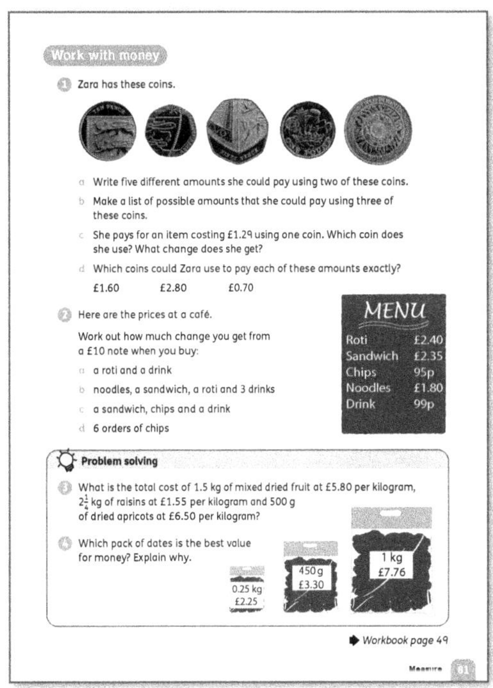

The activities on **Pupil book 5 page 81** use pounds and pence. You can adapt these to use your local currency if you prefer.

Materials
Play money; set of cards with money amounts printed/written on them (The set of cards should be made up of pairs that together make £10 – such as £1.50 and £8.50, £3.25 and £6.75 and so on.); blank card; calculators

Warm-up
Have a 'Make 100' quiz. Give children a number between 1 and 100 and ask what they need to add to make 100.

Focus

- Play a game involving money amounts. Let the children work with play coins to make amounts less than £1 (for example, 27 pence). Have them write this as a decimal (£0.27) and then say what amount they need to add to make a pound.
- Write several money amounts (with two decimal places) less than £10 on cards. Hand these out to the children at random. Tell them that this is the price of an item and ask them to write on a card the amount they would get as change from £10. Discuss how they worked this out.
- Ask the children to complete questions 1 and 2 on **Pupil book 5 page 81**.
- There are many possible answers for the coin combinations in question 1. The children may find different answers. Encourage them to use tables or organised lists to answer this type of question.
- <u>Problem solving:</u> Discuss with the class how they would solve the problems in questions 3 and 4. In question 3, they will have to round up the price for $1\frac{1}{4}$ kg of raisins to the nearest penny. Question 4 is a rate problem and the children need to work out the cost per gram or the cost per kilogram for each pack in order to compare them. The children can use calculators for these questions.

Follow-up

Use **Workbook 5 page 49** to check that the children can add money amounts mentally using what they know about decimals. Let them discuss question 2 in groups.

Challenge

Use money cards to play a 'make £10' memory game in which the cards are shuffled and laid out face down. The children take turns to turn over two cards. If the two amounts make £10, the child can take the cards and score a point.

Support

Play a variation of the 'make £10' game. This time the cards are laid out face up. The children take turns to select two amounts that make £10. They can check the total using a calculator if necessary.

Interesting mistakes

The most common errors the children are likely to make are to put the decimal point in the wrong position or omit it altogether. Remind them to estimate before they calculate and to think about what the quantities mean. For example, say: *If you have two amounts that add up to £10, your answer cannot be more than £10.*

Answers for Pupil book 5 page 81

1 a Individual answers.
  b 80p, £2.30, £1.30, £2.70, £1.70, £3.50, £2.60, £1.60, £3.20, £3.10
  c £2; £0.71 or 71 p
  d £1.60: £1, 50p and 10p
    £2.80: £2, 50p, 20p and 10p
    £0.70: 50p and 20p

2 a £6.61　　**b** £0.48　　**c** £5.71　　**d** £4.30

3 £15.44

4 The 1 kg bag costs £7.76 per kg; the 450 g bag costs £7.33 per kg and the 0.25 kg bag costs £9.00 per kg. The 450 g bag is the best value for money

Answers for Workbook 5 page 49

1 Paris = £15.75; New Delhi = £20; Colombo = £18.75; La Paz = £13.50; Manila = £12.50; Lagos = £13.50; Beirut = £18.75; Jakarta = £15.50; Addis Ababa = £11; Azerbaijan = £21.25

2 No; the letters could be of lower value

3 Individual answers.

4 Individual answers.

End-of-unit check

Ask some or all of these questions to assess how well the children have understood the concepts in this unit.

- *What units would you use to measure this length/ mass/capacity? Why?*
- *Give me an example of when we need to measure lengths accurately using millimetres. (For example: sizes of building bricks, lengths of leaves.)*
- *How can you convert this measurement which is in metres to centimetres? (Multiply by 100.)*
- *How do you read this scale? (Show a range of scales.)*
- *What is this measurement? (Indicate the measurement on a scale.)*
- *This scale shows measurement in litres. Estimate the readings at A and B.*

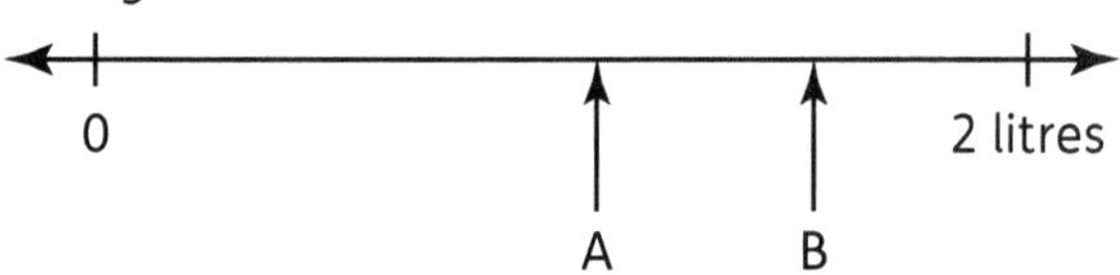

(For example: A = 1.1 litres, B = 1.5 litres.)

- *There was half a litre of water in this jug to start with. How much water was poured out?*

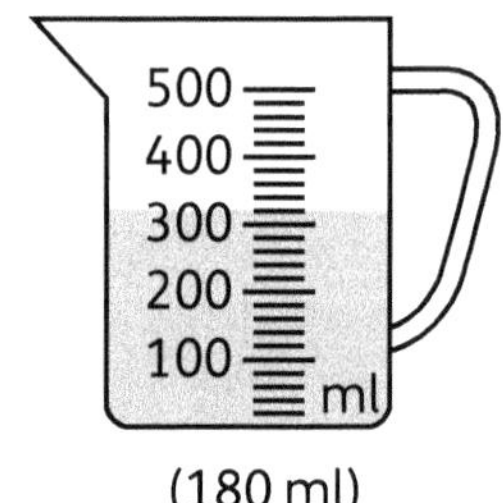

(180 ml)

- *How many mm/cm/ml/g are there in (give a number) m/l/kg? (Provide several conversions.)*
- *A scale reads 780 ml. What is that in litres? (0.78 litres)*
- *Which is heavier, 12 345 g or 123 kg? (123 kg) (Give similar questions using different units.)*
- *How do you convert from imperial to metric measures? (Metres to feet: multiply by 3. Kilograms to pounds: multiply by 2. Litres to pints: multiply by 2. Inches to centimetres: multiply by 2.5. Ounces to grams: multiply by 30.)*
- *What do I need to add to (give a money amount) to make £1?*
- *I have (give a money amount). How much more money do I need to make £5/£10?*
- *How much change will I get from £10 if I buy something that costs (give a money amount)?*

Unit 9 — Perimeter and area

Learning objectives

- Measure and calculate the perimeter of composite rectilinear shapes in centimetres and metres

- Calculate and compare the area of rectangles using standard units of measure

- Use the properties of rectangles to deduce related facts and find missing lengths and angles

- Estimate the area of irregular shapes

- Understand that shapes with the same perimeter can have different areas and vice versa

Key words

perimeter   area   square unit   formula
composite shape

Unit introduction

Materials

Squared paper; masking tape or painter's tape

Teaching guidance

Do some practical drawing and measuring activities to revise the concepts of perimeter and area.

- Hand out squared paper to the children. Challenge them to use block letters (only right angles allowed) to write their initials. They should use a solid colour for the outline and shade the blocks inside the initials. For example, the initials ZM could be written like this:

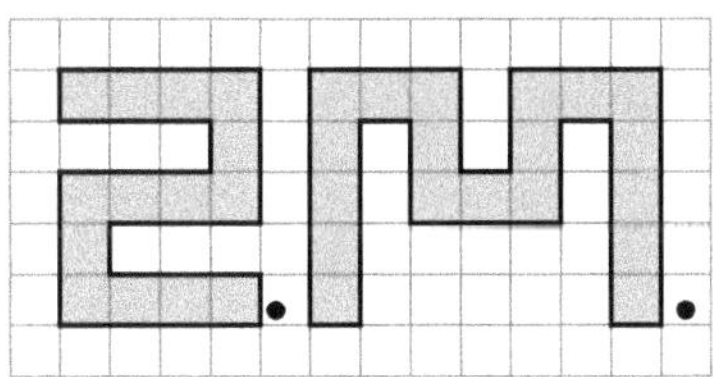

- Once the children have done this, ask them what the term *perimeter* means.
- When you have a definition, get the children to work out the perimeter of each of their initials. Discuss whether they can do this without counting along every side length.
- Then ask the children to work out how many squares are coloured. Again, ask whether there is a way to do this without counting. Ask the children what we call the coloured part of the initials (the *area*) and ask them to explain why we give this measure in squares.

- If you have a tiled floor in the classroom or elsewhere in the school, mark out some *composite shapes* with masking tape or painter's tape (as these both peel off easily).
- Make shapes that cover complete tiles or half tiles, for example:

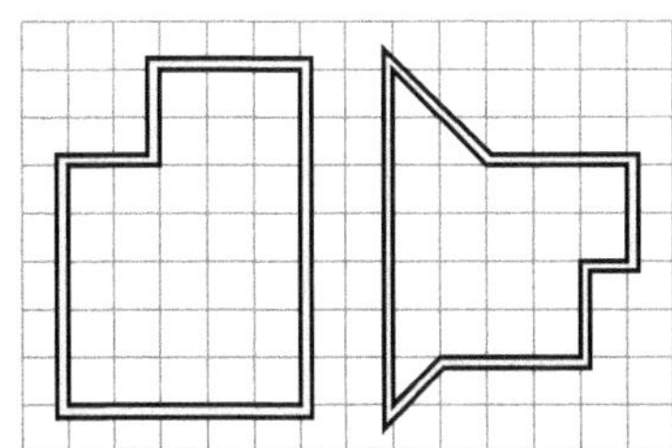

- Let the children measure the perimeter of each shape and work out its area in tiles. If you have time, give each group of children an area (for example, eight tiles) and ask them to use tape to mark out three different shapes with the same area but different perimeters.
- Finish with a number talk (pages 17–18) about perimeter and area. Ask the children: *What is similar and what is different about perimeter and area? How do you work out the perimeter/area of a shape? In which everyday situations is it useful to know either the perimeter or area or both?*

Perimeter

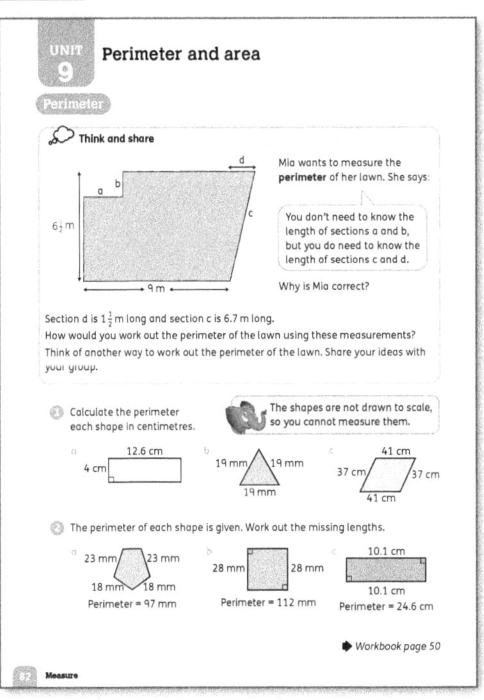

Materials

Small plastic or cardboard shapes (If these are not available, prepare a sheet with a number of shapes drawn on it to copy and distribute to the class.)

Warm-up

Practise mentally doubling numbers as this will help the children find the perimeter of rectangles.

Focus

- Revise measuring accurately in centimetres and millimetres by doing some practical activities using lines and shapes.
- Arrange the class into groups. Hand out a number of small shapes to each group.
- Let the children draw around each shape to show its perimeter on paper. Then have them measure the perimeter in centimetres and/or millimetres. If you let the children measure in centimetres, you can ask them to convert the measurements to millimetres to revise converting from a larger to a smaller unit of measure.
- Think and share: Turn to **Pupil book 5 page 82**. Discuss Mia's statement with the class and let them explain why it is correct. Then the children can work in pairs to answer the questions. Let different children share their ideas.
- The children can then work independently to complete question 1 and question 2.

Follow-up

Use **Workbook 5 page 50** as an informal assessment task to find out how well the children can apply what they have learnt about perimeter and combine this knowledge with what they know about the properties of shapes.

Interesting mistakes

When the children are given the perimeter of a rectangle and have to work backwards to find the length of two opposite sides, they may forget that the perimeter is $2 \times (l + w)$ and not halve their answer. Ask the children to draw a rough sketch and to label each side with its length to double check that the side lengths add up to the correct perimeter.

Answers for Pupil book 5 page 82

1   a   33.2 cm   b   57 mm   c   156 cm
2   a   15 mm   b   28 mm   c   2.2 cm

Answers for Workbook 5 page 50

1   Individual drawings.
2   Individual drawings.
3   Individual drawings.

Composite shapes

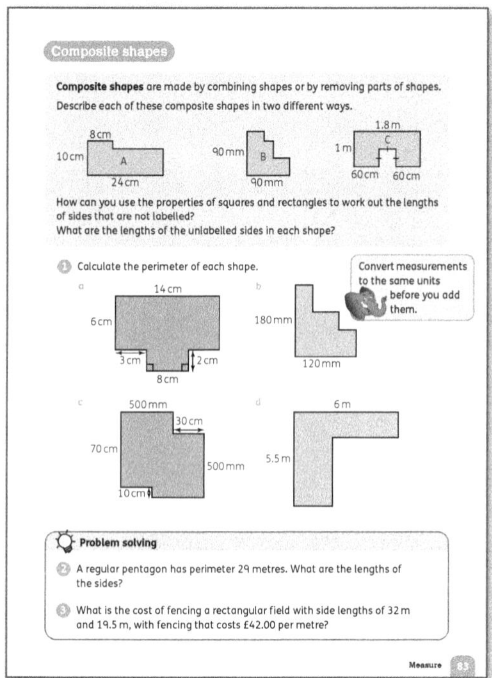

Materials

Rods or drinking straws of different lengths, with some cut into two parts; squared paper

Warm-up

Use the 'Doubling' and 'Halving' activities from the Activity bank (page 26).

Focus

Display a *composite shape* like this one for the class:

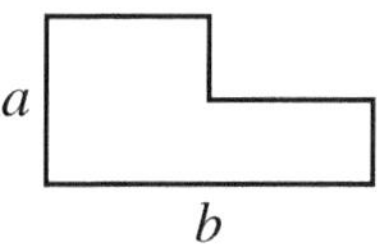

- Discuss how we can work out the perimeter of this shape if we only know the lengths of a and b.
- To do this, allocate lengths to a and b. For example, say: *I know that a is 10 cm and b is 15 cm. What is the perimeter of this shape?*
- Let the children talk about this in groups for a while and then take their responses.
- Use rods or drinking straws to show why the perimeter is $2 \times (a + b)$ even though the shape is not a complete rectangle.
- Ask the children to model a similar shape using rods or straws. Let them move the rods/straws so they can see that the lengths of the two sides of the cut-out section equal the missing parts of the rectangle.
- Model this for the class once they have experimented with their own shapes. For example:

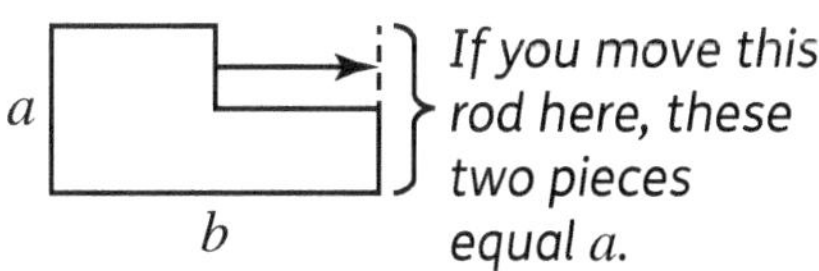

- You can also demonstrate this using a computer program, but physically moving the pieces helps children to make connections. However, it allows the children to experience the concept in a different way.
- Next, ask the children to build a shape that has 'extra lengths', for example,

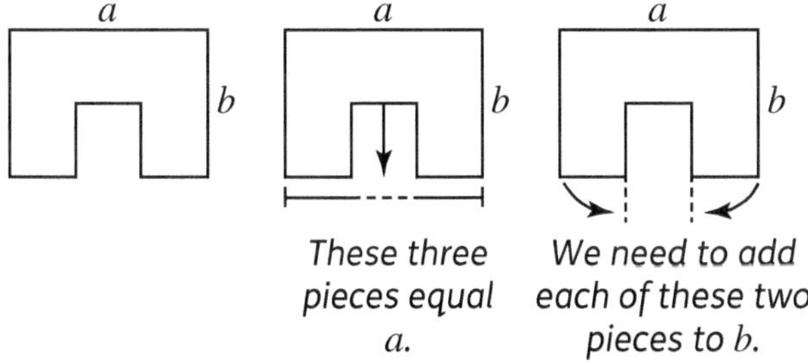

- Explain that, in this case, we have to know the lengths of the shorter vertical sides to find the perimeter.

Ask the children to describe each shape at the top of **Pupil book 5 page 83** in two ways. This encourages them to think of the shapes as *both* two or more rectangles 'added together' *and* as a large rectangle with smaller rectangles 'taken away'. The question in the example box asks the children to explain how to find the unlabelled sides and then to work them out. (To find an unlabelled horizontal length, you can add the horizontal lengths you are given and subtract from the total horizontal length, and, similarly, to find an unlabelled vertical length, add the vertical lengths you are given and subtract from the total vertical length. The unlabelled sides are: Diagram A: horizontal side = 16 cm. The unlabelled vertical sides can be any lengths that add to 10 cm. Diagram B: Both the unlabelled horizontal and vertical sides can be any three lengths that add to 90 mm. Diagram C: The longest vertical side is 1 m. The two shorter vertical sides are equal and can be any length less than 1 m. The unlabelled horizontal side is 60 cm or 0.6 m.) Let them talk about the questions and model the shapes to decide whether or not they need to work out all the missing lengths to find the perimeter.

- Take feedback and let the children share what they have worked out and how they have thought about the shapes.
- Be aware that some children will feel more comfortable working out all the missing lengths and adding them, while others will be confident that they do not need to work out all the missing lengths to find the perimeter in many cases.

- Have a number talk about when we need to include extra lengths and when we can ignore them. The children may say things such as:
  - 'When there are small squares or rectangles cut off the corners of shapes, you can ignore them.'
  - 'When a piece cuts into the shape and it isn't at a corner, you have to include extra lengths.'
  - 'When a piece is added to a shape along a side, you have to include the extra lengths.'
- Allow for discussion and disagreements and ask the children to demonstrate what they mean using diagrams or models.
- For question 1, let the children either draw the shapes and colour sides if they need to or model the shapes using straws or rods so that they can physically move sides.
- <u>Problem solving:</u> For question 2, check that the children remember that a regular pentagon has 5 equal sides. The children can draw a diagram to help them with question 3 if they need to.

Challenge
- Give the children perimeter problems involving composite shapes that include semicircles or triangles. For example, give them the following problem and diagrams:
Shape A has a perimeter of 55 m. What is the perimeter of shape B?

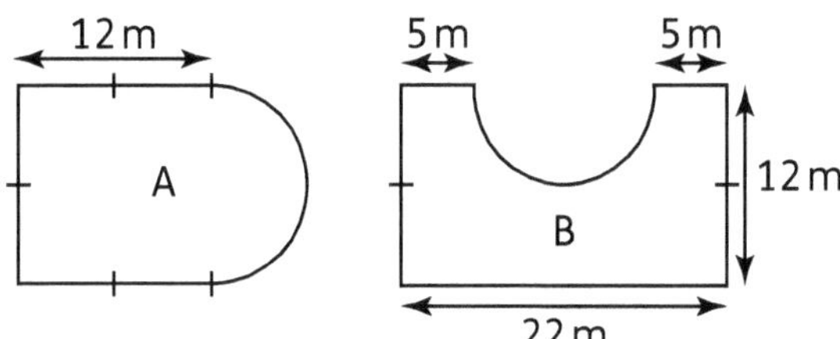

- In this problem, the children have to work out that the semicircles have the same diameter and are therefore the same size. This means that part of each perimeter (the half circumference) is equal. The children can work out that the length of the half circumference is 55 – 36 = 19 m. Therefore, the perimeter of shape B is 75 m.
- Ask the children whether they always need to know all the lengths of a shape to find its perimeter. Encourage them to draw some shapes as part of their explanation. Compare answers as a class.

Support
- Draw some fairly simple composite shapes on squared paper and ask the children to redraw each shape as a rectangle with the same perimeter.
- You could extend this by asking the children to draw a different composite shape with the same perimeter or with a perimeter that is twice as long.

Area

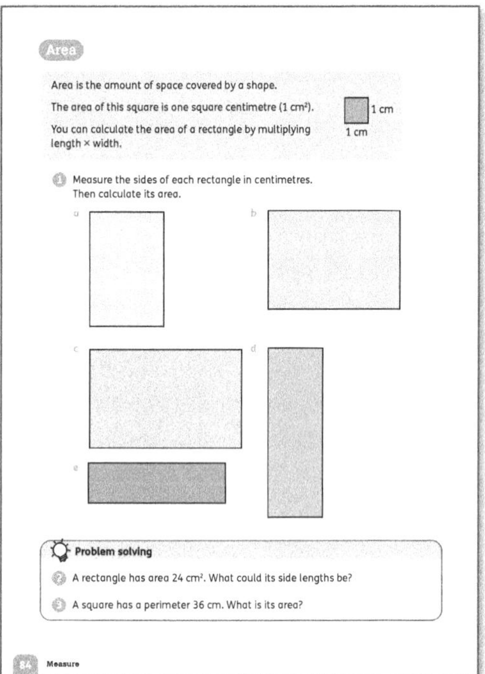

Materials
Squared paper; grids of square centimetres printed onto acetate sheets; set of shapes drawn on paper – squares, rectangles and irregular polygons; rulers; masking tape or painter's tape; cuboid boxes or containers; stickers or marker pens

Warm-up
Revise finding area of shapes and using *square units* using a grid of square centimetres. If possible, prepare an acetate sheet (overhead transparency) or slide with a grid on it for each group. Let them place this over a sheet of shapes and count squares to find the area in square centimetres. For irregular shapes, demonstrate how to estimate the area by counting the whole squares, and squares that are more than half, inside the shape. Ignore squares that are less than half inside the shape.

Focus
- Draw rectangles in different orientations on the board. Label some of the dimensions, for example, some or all of the side lengths. Ask the children: *Can you work out the area of any of these rectangles using the information shown here? Which rectangles need to include more information?*
- Ask the children to find the area of the classroom and of objects within the room (for example, the top of a desk, the cover of a book).
- If you have a tiled floor, mark out shapes using masking tape or painter's tape. Ask the children to measure the sides and find the area of each shape. You can include right-angled triangles so that the children realise they can halve the area of a square or rectangle to work out the area of a triangle.
- Once you have revised how to calculate area, let the children measure the sides and multiply the side lengths to find the areas of the rectangles in question 1 on **Pupil book 5 page 84.**

- Problem solving: Encourage the children to make sketches and label the side lengths to help with questions 2 and 3 if they need to.

Challenge
Give the children a cuboid box or container and ask them to work out the surface area of the (outside) faces. Let them decide how to do this.

Support
Give each group a cuboid box or container. Ask them to choose the face that they think has the largest (or smallest) area. They can place a sticker on this face or mark it with a marker pen. The children can then measure the length and width of each face and calculate its area to check their answers.

Interesting mistakes
The children may forget to include the units when they calculate area. Continue to remind them to do this and refer back to the square grid to reinforce the idea of a square unit.

Answers for Pupil book 5 page 84
1 a $3 \times 4.5 = 13.5 \text{ cm}^2$ b $5.4 \times 3.9 = 21.06 \text{ cm}^2$
    c $6.2 \times 3.9 = 24.18 \text{ cm}^2$ d $2.3 \times 6.7 = 15.41 \text{ cm}^2$
    e $5.6 \times 1.6 = 9.0 = 50.4 \text{ cm}^2$
2 Possible answers: 1 cm and 24 cm; 2 cm and 12 cm; 3 cm and 8 cm; 4 cm and 6 cm; OR non-whole number side lengths which multiply to give 24 cm^2
3 81 cm^2

Use a formula to calculate area

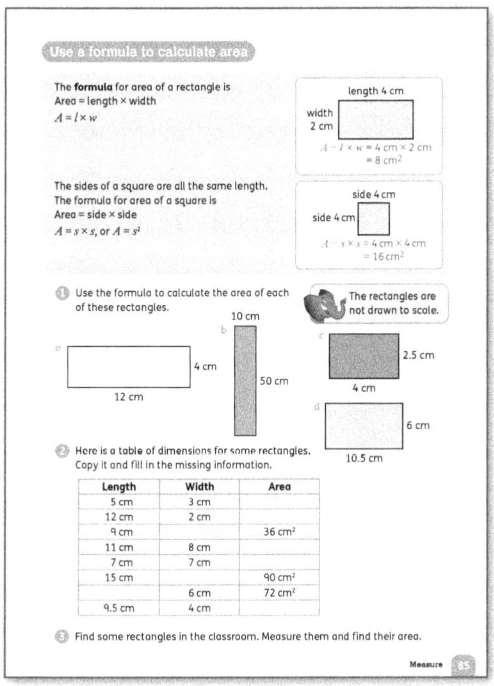

Materials
Measuring instruments: rulers, tape measures, metre rule

Warm-up

As a mental starter for this lesson, revise multiplication facts and square numbers as the children will use these to calculate areas of shapes.

Focus

Ask the class what the word *formula* means in maths. Elicit that this is a 'rule' for a calculation that always works. Here, we are working with the formula for the area of a rectangle. The formula works whatever the dimensions of the shape are.

- Work through the example on **Pupil book 5 page 85** and then do a few more similar examples to make sure that the children understand and can apply the formula.
- Discuss the fact that a square is a rectangle, so the same formula works – we just use the same number for the length and width.
- The children can work independently on question 1.
- Discuss with the class how they can find the missing values for each column of the table in question 2. Ask: *If you know the length and width, what do you do? If you know the area and the length, what do you do?*
- You may need to organise question 3 so that the children work in groups and in different parts of the classroom. The children can check each other's work.

Answers for Pupil book 5 page 85

1   a   48 cm²   b   500 cm²   c   10 cm²   d   63 cm²

2   15 cm², 24 cm², 4 cm, 88 cm², 49 cm², 6 cm, 12 cm, 38 cm²

3   Individual answers.

More area calculations

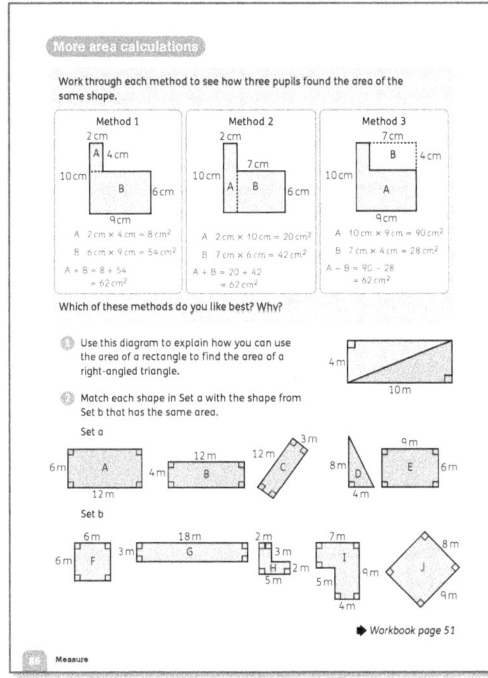

Materials

Squared paper; stickers or marker pens

Warm-up

Do the 'Multiplication grid' activity (page 28) as a starter for this lesson.

Focus

This lesson is a good opportunity for the children to talk through problems, make decisions and ask questions about how they can use what they already know to work out the area of more complicated shapes. Encourage the children to explain which method they prefer. The method they choose is likely to depend on the numbers given in the calculation.

- Give the class time to talk through the different methods of finding the area of a composite shape at the top of **Pupil book 5 page 86**. Have a class discussion afterwards, focusing on what the children found challenging and what they learnt.
- Use **Workbook 5 page 51** to consolidate finding the area of simple composite shapes before moving back to questions 1 and 2 on **Pupil book 5 page 86**.
- In question 1, it is important for the children to realise that a diagonal divides a rectangle into two halves. Each half has half the area of the whole and is a right-angled triangle. The children need to realise that they can view any right-angled triangle as half of a rectangle. Let the children discuss this and explain their thinking in groups before they try to answer question 2.

Challenge

- Let the children investigate overlapping areas and how to solve problems involving overlaps. They can use overlapping stickers to investigate this or they can draw overlapping shapes on squared paper and use the diagrams to investigate different methods.
- Let them focus on what information they need to work out each area. You may need to show them some examples and ask: *What is the total area covered by these two overlapping stickers? How can you work this out? What do you need to know to work it out?*

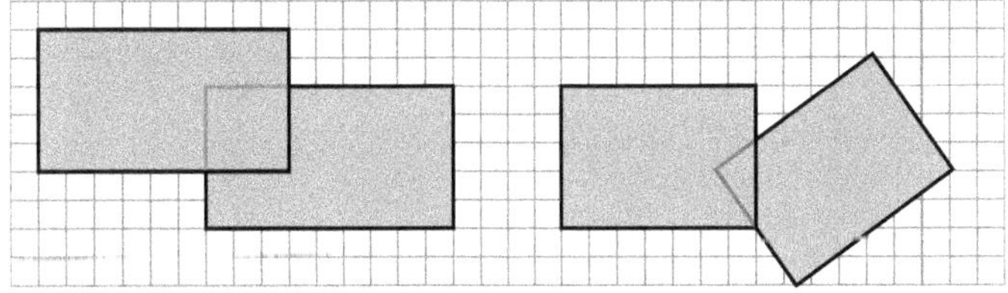

- Let the children discuss when they might find overlapping shapes in real life. Examples might include: two rooms with a shared storage space; two towels, one covering part of the other; a rug covering part of a larger carpet.
- Give the children time to share what they find out. Encourage them to say how these problems are similar to the composite shape problems they solved earlier.
- *Jo marks out a rectangle with an area of 40 m² in the school grounds. The rectangle is 8 m long. Her teacher says that they need the rectangle to have an area of 70 m². Jo increases the length of the rectangle by 2 m. What will be the width of the 70 m² area? (7 m)*

Support

Let the children work in pairs or groups to prepare a chart to teach other children how to decompose shapes to work out the area in parts. They can create diagrams drawn on squared paper to show examples. This work will help them to consolidate and express their understanding.

Interesting mistakes

- When calculating the area of shapes that involve a right angle, the children may use the slant height rather than the correct perpendicular height, particularly when the lengths of all three sides are given and the triangle is in an orientation that makes visualising the rectangle more difficult.
- For example, when given a triangle like this one, the children might work out $17 \times 8 \div 2$, or $17 \times 15 \div 2$. Encourage them to draw the missing sides of the rectangle so they can be sure about which numbers to use.

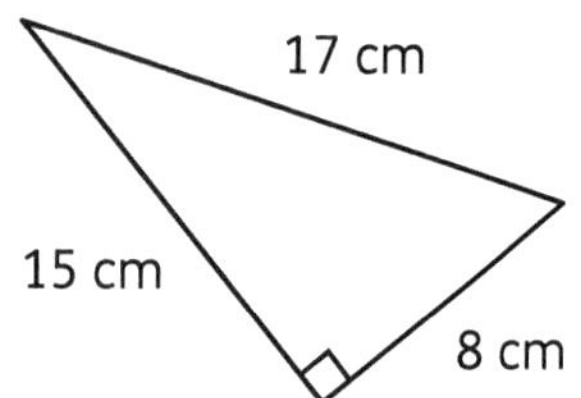

Answers for Pupil book 5 page 86

1. It is half the area of the rectangle
2. A and J; B and I; C and F; D and H; E and G

Answers for Workbook 5 page 51

1. a 513 m² b 3526 m²
   c 1649 m² d 3864 cm²
   e 5255 cm²

Solve perimeter and area problems

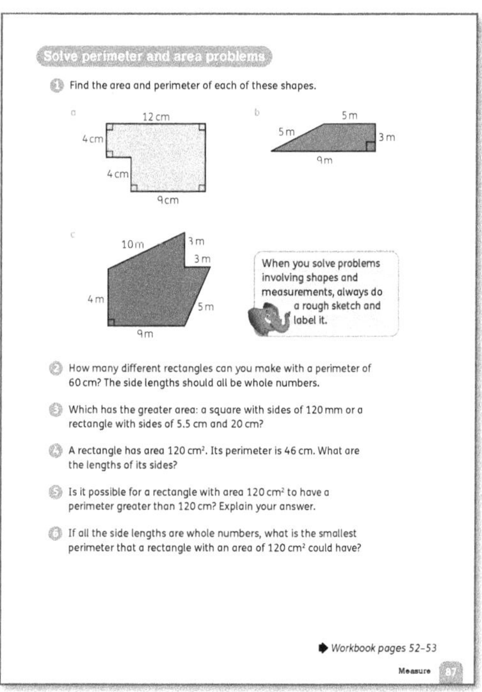

Materials

Squared paper

Warm-up

Start the lesson by asking the children to complete **Workbook 5 page 52**. Check the children's work to make sure they can solve simple perimeter and area problems.

Focus

The children can work through questions 1–6 on **Pupil book 5 page 87** in pairs or in groups. Alternatively, you could give different groups different problems to work on. If you do this, the children can explain to the class how they solved each problem so that they are exposed to different methods of working.

Follow-up

Use **Workbook 5 page 53** as an informal assessment task to check that the children can apply what they have learnt about area and perimeter. The children can work either on their own as a test, or in pairs so you can observe them and identify any misunderstandings.

Support

Let the children draw the shapes on squared paper to help them answer the questions in the Pupil book. Allow them to model the problem situations using real objects.

Answers for Pupil book 5 page 87

1. a area = 84 cm², perimeter = 40 cm
   b area = 21 cm², perimeter = 22 m
   c area = 55.5 cm², perimeter = 34 m
2. 15
3. the square
4. 8 cm by 15 cm
5. Yes, for example if the side lengths are 1 cm, 1 cm, 120 cm and 120 cm, the perimeter is 242 cm.
6. side lengths need to be 10 cm, 10 cm, 12 cm, 12 cm; perimeter is 44 cm

Answers for Workbook 5 page 52

1. a C and D b A and E
   c 7 m² d A and E have the same area
2. a 13 cm b 91 cm²
3. a 25 cm² b 256 cm²

Answers for Workbook 5 page 53

1. a, b Individual answers.
2. Individual answers.

End-of-unit check

Ask some or all of these questions to assess how well the children have understood the concepts in this unit.

- *What information do you need to calculate the area of a rectangle?* (length and width)
- *What information do you need to calculate the area of a square?* (length of one side)

- *What is the area of a 3 cm × 5 cm rectangle?* (15 cm²)
  What is the area of a 30 m × 50 m rectangle? (1500 m²)
- *What pattern do you notice in your two answers?*
  Explain the pattern. (For example: The unit is the square
  of the unit of the lengths. The value for the area of the
  larger rectangle is 100 times the value for the smaller
  rectangle, because the values for the length and width
  are both 10 times larger and 10 × 10 = 100.)
- *What is the formula for calculating the area of a
  rectangle?* (area = length × width)
- *If I draw a rectangle on squared paper, what method
  can I use to calculate the area other than using the
  formula?* (Count the squares.)
- *How could you divide this shape up to find its area?*
  (Show a composite shape.) (Split it into rectangles.)
- Give the children a piece of squared paper and ask
  them to draw a triangle with an area of 16 square
  units. Let them explain how they did this. (The area
  of the triangle is half the area of a rectangle with the
  same length and width, so the height and base of the
  triangle need to multiply together to make 32 cm².)
- *A regular hexagon has sides of 5.5 cm. What is its
  perimeter?* (33 cm)

- Prepare some tables like the one below and ask
  the children to fill in the missing values (shown
  in brackets). Make sure that the children need to
  carry out a range of calculations to find the missing
  perimeter, area and/or one side length. Say: *Complete
  this table. All the shapes are rectangles.*

Length	Width	Perimeter	Area
12 cm	9 cm	(42 cm)	(84 cm²)
(7 cm)	10 cm	34 cm	(70 cm²)
8 cm	(9 cm)	(34 cm)	72 cm²

- Prepare diagrams like those below and ask the
  children to find the unknown values:

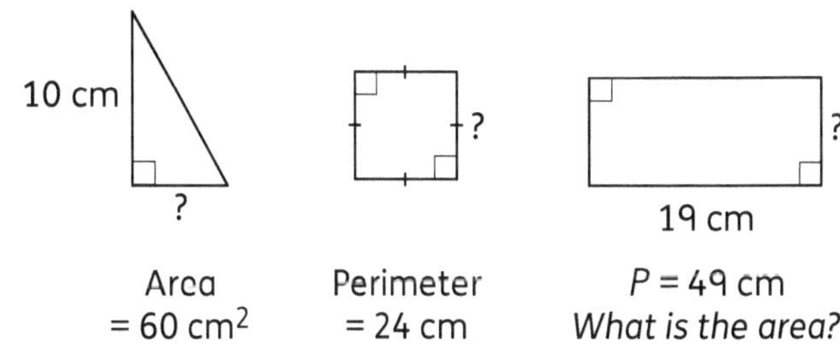

(12 cm, 6 cm, the width is 5.5 cm, so the area is
104.5 cm²)

Unit 10 Statistics

Learning objectives

- Plan and carry out a statistical investigation to
  answer a set of related statistical questions

- Record, organise and represent data using
  appropriate tables, diagrams and graphs

- Solve comparison, sum and difference
  problems using information presented in a line
  graph

- Find the mode, median and range of a set
  of data

- Complete, read and interpret information in
  tables, including timetables

Key words

frequency table   bar chart   bar-line chart
dot plot   Carroll diagram   Venn diagram
line graph   axis, axes   mode   median   range

Unit introduction

Materials

Cup of hot water; graph paper or squared paper;
thermometer; stopwatch

Teaching guidance

Do the following activity with the class to review these
statistics skills: investigating a question; measuring
to collect data; recording data; drawing a graph to
present data.

- Display this question: *How long does it take a cup of
  hot water to cool down to room temperature?*
- Ask the children to estimate and record some of their
  estimates as grouped data. For example, ask: *Who
  thinks it will take less than 5 minutes? Who thinks
  it will take between 5 and 7 minutes?* And so on.
  Draw tallies to record the number of children whose
  estimate falls in each time interval.
- Discuss how the children could answer this question
  scientifically. The children should say that they need
  to measure the room temperature and to measure
  the temperature of the hot water at regular time
  intervals to find out how long it takes for the water
  temperature to equal the room temperature.
- Set up the experiment, either as a class
  demonstration or as a group task.
- Let the children decide how often they will measure the
  temperature and draw up a table to record the data.
- Once the children have recorded the data from the
  experiment, ask them how they could present this data
  as a graph. Line graphs are best for showing how a
  quantity (temperature) changes over time. Discuss and
  review how to draw line graphs.

- Let the children work in groups to draw suitable graphs to show the data.
- Display some of the completed graphs and use them to talk about the features of graphs. Ask the children: *What information do you need to include when you draw a graph?* (a heading, labels, suitable intervals on the scale and so on)

Use data to answer questions

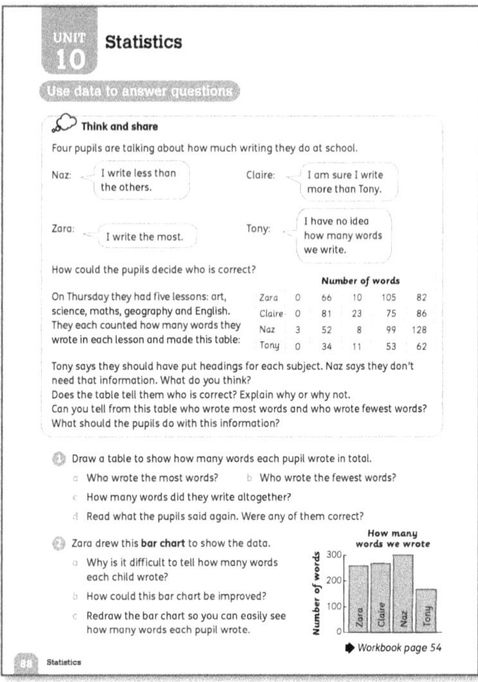

Materials
Graph paper or squared paper; cubes or number rods

Warm-up
Choose a 'Mental problem solving' activity from the Activity bank (pages 24–25) as a starter for this lesson.

Focus
- <u>Think and share:</u> Turn to **Pupil book 5 page 88**. Let the children work in groups to read through the information and to answer the questions about collecting and interpreting data.
- Discuss the children's answers as a class before asking them to complete question 1. (They may suggest adding another column to the table in which to write the total number of words for each child. That will help them to decide who is correct. If they want to know who wrote the most words in each subject, they could add headings for each subject, and if they wanted to know which subject had the highest total number of words, they could add a row in which to write the subject totals.)
- Let the children work in pairs to discuss the graph in question 2 and how it could be improved. Then they can use graph paper or squared paper to draw their own graphs to show the data. You may need to remind the class that the bars in a *bar chart* must be the same width and equally spaced. Let the children compare their charts with others in the class and comment on each other's work.

Follow-up
Let the children work in pairs to compare, discuss and complete the graphs on **Workbook 5 page 54**. Use the answers to question 2 to check that the children understand how the choice of scale affects the appearance of a graph.

Support
Help the children to understand how the choice of scale affects the appearance of a graph by doing some activities with concrete apparatus.
- The children could devise and carry out a survey of ten members of the class. For example, they could ask each child to choose their favourite drink from a choice of four.
- Working in pairs, one child could present their results with cubes or number rods using a simple key of one cube representing one child. The other child could use one cube to represent two children, and so on.
- Working in pairs, one child could make up the results of a survey and display them using different scales. The other child must interpret the results.

Interesting mistakes
- The children may still draw bar charts with bars of unequal widths and with uneven spaces between the bars. Help them to work more accurately by providing squared or graph paper and making sure that the children use a ruler to draw lines.
- The children may not see the importance of spacing the bars, but the spaces are necessary in a bar chart. A chart with no spacing between the bars is a histogram, which the children will learn about in later years. If they ask why it is important, you could explain that the categories in the data are separate, so the bars need to be separate.

Answers for Pupil book 5 page 88

1

Pupil	Total words
Zara	263
Claire	265
Naz	290
Tony	160

a Naz   b Tony   c 978

d Claire is correct – she did write more words than Tony.
Tony *might* be correct – without counting, it is difficult to know how many words you write!

2 a The scale on the vertical axis is very small and there are not very many markers, so it is difficult to tell what each bar reaches to.

b Possible answers: Increase the scale on the vertical axis; add more markers; make the vertical axis start at 200 and finish at 300.

c Individual answers.

Answers for Workbook 5 page 54

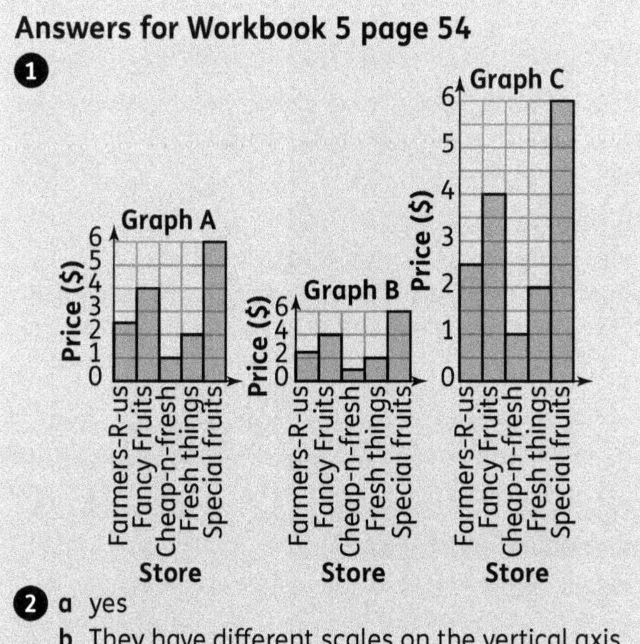

2 a yes

  b They have different scales on the vertical axis.

  c Possible answer: Graph C because it uses the biggest scale.

Tables and graphs

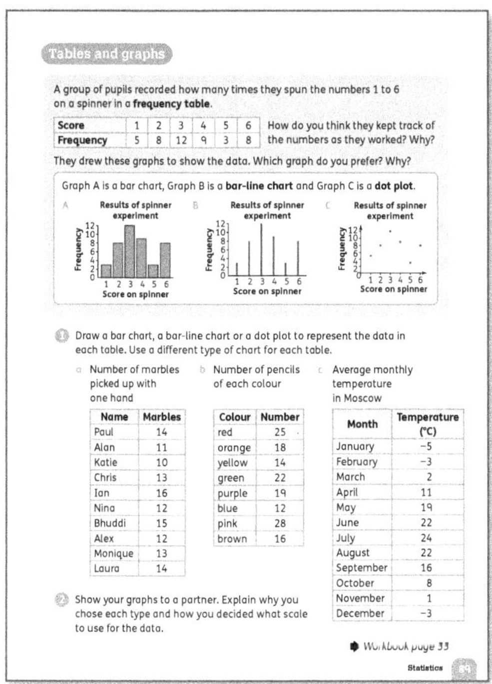

Materials
Selection of different graphs to show the class (try to include pictograms, bar charts, bar-line charts and line graphs); squared or graph paper; rulers and pencils

Warm-up
Select any suitable 'Mental problem solving' activity (pages 24–25) as a starter for this lesson.

Focus
- Show the class a range of different graphs and ask them to say what data each graph shows and how they think the graph was drawn. Compare the graphs in terms of what they look like and how to read them.

- Revise the concept of a *frequency table* and make sure the children can interpret and create them. Use the example at the top of **Pupil book 5 page 89** to check that they can read a frequency table.
- Discuss the three example graphs on **Pupil book 5 page 89** with the class. Make sure they can see that the bar chart, bar-line chart and *dot plot* show the same data and that we read them in the same way. Spend some time discussing the questions with the class.
- Give the children time to talk about the data sets in question 1 and to discuss what type of graph they think would show each set most clearly. There is no correct choice here – all the data is discrete so could be represented as a bar chart, bar-line chart or dot plot. The children may prefer to draw bar charts because they can colour them in, but remind them that they need to leave gaps between the bars. They may find the bar-line charts or dot plots easier to draw. The children can draw their graphs independently and then have the discussion in question 2 with a partner.

Follow-up
Use **Workbook 5 page 55** to provide additional practice in drawing charts and graphs and to reinforce the idea that the same data can be shown in different ways.

Interesting mistakes
Some children may not see the importance of labelling their graphs or simply forget to do so. Present them with some unlabelled graphs and ask them to answer questions about them. Once they realise that this is not possible without the labels, they will understand the need to add labels to their own graphs.

Answers for Pupil book 5 page 89
1 a Bar chart, bar-line chart or dot plot with vertical axis labelled Number of marbles and horizontal axis labelled Name. Title: Number of marbles picked up with one hand. Gaps between the bars.
  Heights of bars: Paul 14; Alan 11; Katie 10; Chris 13; Ian 16; Nina 12; Bhuddi 15; Alex 12; Monique 13; Laura 14

  b Bar chart, bar-line chart or dot plot with vertical axis labelled Number of pencils and horizontal axis labelled Colour. Title: Number of pencils of each colour in stock. Gaps between the bars.
  Heights of bars: red 25; orange 18; yellow 14; green 22; purple 19; blue 12; pink 28; brown 16

  c Bar chart, bar-line chart or dot plot with vertical axis labelled Temperature (°C) and horizontal axis labelled Month. Title: Average monthly temperature in Moscow. Gaps between the bars.
  Heights of bars: January −5; February −3; March 2; April 11; May 19; June 22; July 24; August 22; September 16; October 8; November 1; December −3

2 Individual answers.

Answers for Workbook 5 page 55

1 **a** Bar chart and dot plot with vertical axis labelled Number of newspapers and horizontal axis labelled Day; gaps between the bars of the bar chart.
Heights of bars: Sunday 43; Monday 35; Tuesday 35; Wednesday 36; Thursday 37; Friday 35; Saturday 37

b Bar chart and dot plot with vertical axis labelled Number in class and horizontal axis labelled Class; gaps between the bars of the bar chart.
Heights of bars:1a: 30; 2a 27; 3b 28; 4a 29; 5c 30; 6a 30; 6b 29

Tables and diagrams

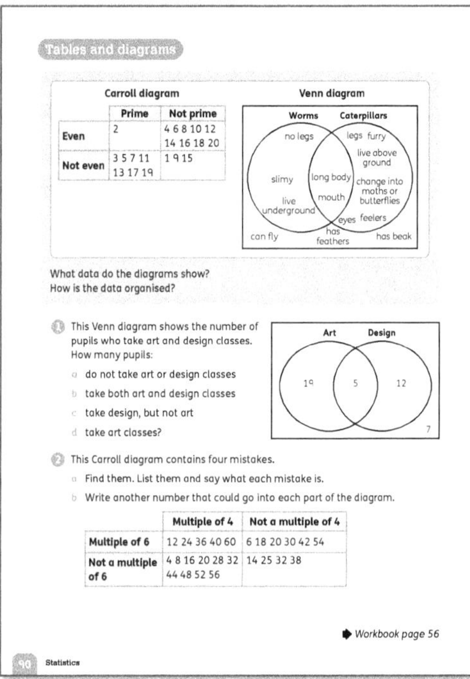

Tables and diagrams

Carroll diagram

	Prime	Not prime
Even	2	4 6 8 10 12 14 16 18 20
Not even	3 5 7 11 13 17 19	1 9 15

Venn diagram

What data do the diagrams show?
How is the data organised?

1 This Venn diagram shows the number of pupils who take art and design classes. How many pupils:

a do not take art or design classes

b take both art and design classes

c take design, but not art

d take art classes?

2 This Carroll diagram contains four mistakes.

a Find them. List them and say what each mistake is.

b Write another number that could go into each part of the diagram.

	Multiple of 4	Not a multiple of 4
Multiple of 6	12 24 36 40 60	6 18 20 30 42 54
Not a multiple of 6	4 8 16 20 28 32 44 48 52 56	14 25 32 38

◆ *Workbook page 56*

90 Statistics

Warm-up

Practise finding factors and multiples of numbers, and deciding whether or not a number is prime, as a mental starter for this lesson.

Focus

• The children have used *Carroll diagrams* and *Venn diagrams* in previous years. Work through the questions and examples at the top of **Pupil book 5 page 90** orally with the class. Check that what children remember is accurate and address any misconceptions. (The Carroll diagram contains the whole numbers up to 20. It shows that 2 is the only even prime numbers and it shows which of these numbers are even and non-prime, odd and prime and odd and non-prime. The Venn diagram shows the features of worms and caterpillars, the features that they have in common and features which neither of them have.)

• Then let the children work on their own to complete question 1 and question 2. Let the children check each other's work and then compare answers as a class.

Follow-up

Use **Workbook 5 page 56** to check that the children can apply what they have learnt to make and complete Venn diagrams and Carroll diagrams.

Challenge

Let the children do some fun activities using Venn diagrams with three intersecting circles. For example, they could draw one of these two Venn diagrams:

• sorting superheroes, using these categories: 'has a super power', 'has great wealth', 'is a genius'

• sorting animals, using these categories: 'will bite you', 'is poisonous', 'lays eggs' (or other categories that the children think of).

Interesting mistakes

The children may find it confusing to use the negative 'not' when they use Carroll diagrams so, for example, they may want to use *odd* and *even* as categories rather than *odd* and *not odd*. Even at this level, it makes sense to work correctly.

• Teach the children that we use a Carroll diagram to sort items using yes/no answers, so it makes more sense to say one thing is even (yes) while the other is not even (no).

• You could also explain that if we say *odd* and *not odd*, we can use the diagram to classify fractions and other numbers, too. If we say *odd* and *even*, we can only sort whole numbers.

Answers for Pupil book 5 page 90

1 **a** 7 **b** 5 **c** 12 **d** 24

2 **a** 40 is not a multiple of 6
20 is a multiple of 4 and not a multiple of 6
48 is a multiple of both 4 and 6
32 is a multiple of 4

b Possible answers:

	Multiple of 4	Not a multiple of 4
Multiple of 6	48	66
Not a multiple of 6	64	7

Answers for Workbook page 56

1 Possible answers:

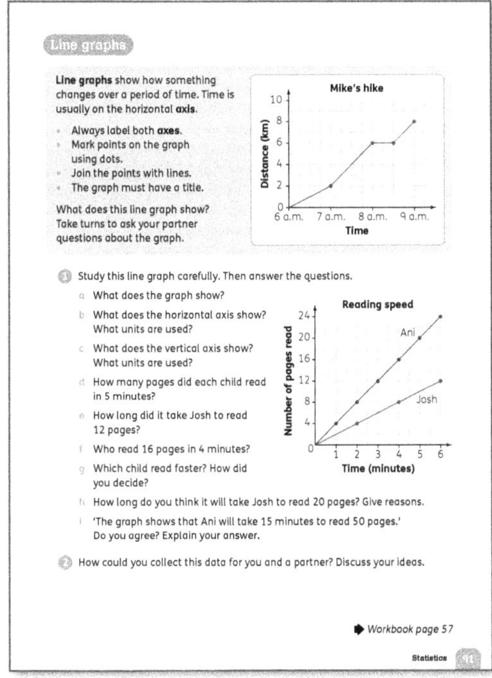

2

	Likes swimming	**Doesn't like swimming**
Has a pet	Ani, Don, Faz, Ike	Bill, Gail, Hans
Doesn't have a pet	Eric	Cal

Line graphs

Materials
Classroom temperatures recorded at the same time each day for a week; rulers

Warm-up
Use any suitable mental calculation or skip counting activities as a starter for this lesson.

Focus
- Draw a line graph for the classroom temperature data to remind the class how this is done.
- Display these vocabulary words: *axis*, *horizontal*, *vertical*, *scale*, *interval*, *label* and *heading*. Ask the children to explain what each term means in the context of *line graphs*.

- Use the example at the top of **Pupil book 5 page 91** to consolidate the key elements of a line graph and to check that the children can interpret graphs. (The graph shows Mike's distance from his starting point at different times during a morning.)
- Let them share some of the questions they asked each other with the class.
- The children can work in pairs to complete questions 1 and 2.

Follow-up
Use **Workbook 5 page 57** to revise drawing line graphs and to check that the children can draw a graph to present data from a table and label their graph correctly.

Ask the children questions that involve finding sums and differences from their graphs, such as: *How many more cans were collected in June than in May?* (100) *How many cans were collected in the November, December and January, in total?* (400)

Answers for Pupil book 5 page 91
1 **a** The reading speed of Ani and Josh.
  b Time, minutes
  c Number of pages read, pages.
  d Josh read 10 pages; Ani read 20 pages.
  e 6 minutes
  f Ani
  g Ani because she read more pages than Josh in the same amount of time.
  h Josh read 10 pages in 5 minutes, so he is likely to read 20 pages in 10 minutes.
  i Ani read 10 pages in 2.5 minutes. Multiplying both by 5 suggests that she will read 50 pages in 12.5 minutes, so the statement is incorrect.
2 Individual answers, for example, time how long each person takes to read 24 pages of the same book. You could round the times to the nearest half minute.

Answers for Workbook 5 page 57

1
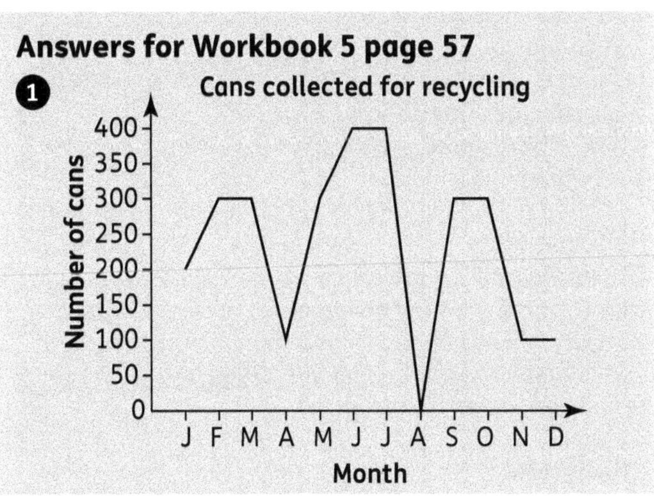

More line graphs

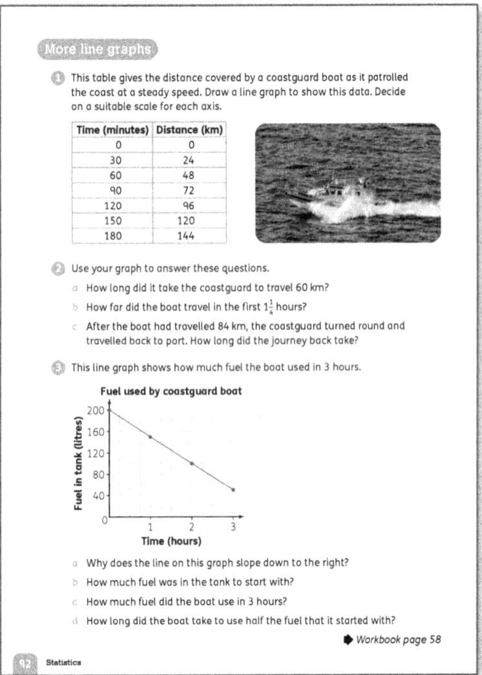

Materials

Graph paper; rulers

Warm-up

Use any suitable mental calculation or skip counting activities as a starter for this lesson.

Focus

- Turn to **Pupil book 5 page 92**. For question 1, the children will need graph paper. Discuss the scales they have chosen to make sure they are suitable before they draw their graphs.
- The table uses 30-minute intervals for the time so the children should do the same; or they could use hours and half hours. For the distance, they could use a scale of 1 cm to 10 km, which means that they need a graph 15 cm high. When they have drawn their graphs, ask: *Did the boat travel the same distance in every 30 minute interval? How do you know?* (Yes, it travelled 24 km every 30 minutes/the graph goes up the same amount in each 30 minutes.)
- The children can do question 2 and question 3 on their own.

Follow-up

Use **Workbook 5 page 58** as an informal assessment task to check that the children can organise data, draw distance–time graphs to show the data and interpret their graphs. Check the children's answers as a class or collect their work and mark it.

Challenge

Let the children investigate using computer programs to draw line graphs. They can use familiar contexts such as

temperature changes to do this. Encourage them to find tables of temperature data online for their own region and then explore how to show this on a graph.

Support

To help the children who have not grasped the basics of drawing line graphs, let them watch step-by-step videos or slide presentations showing how to draw a line graph. There are many different options online, so make sure you select videos that are suited to your class and the contexts in which you work.

Interesting mistakes

- The children may need additional support to allocate data to the correct axis and to develop scales that are suitable and easy to read. You can address this by showing the class impractical and incorrect representations and discussing why they are incorrect and why they do not work.
- The children may also see time and distance graphs as 'pictures' of the journey and ignore the mathematical aspect. For example, they may say: 'He walked up a hill, then walked along a flat road and then walked down a less steep slope.' The correct interpretation of the graph would be that the person travelled at a steady speed, then stopped for a while and then returned at a slower speed. Number talks (pages 17–18) and discussions about the graphs and the data will help to address this misconception.

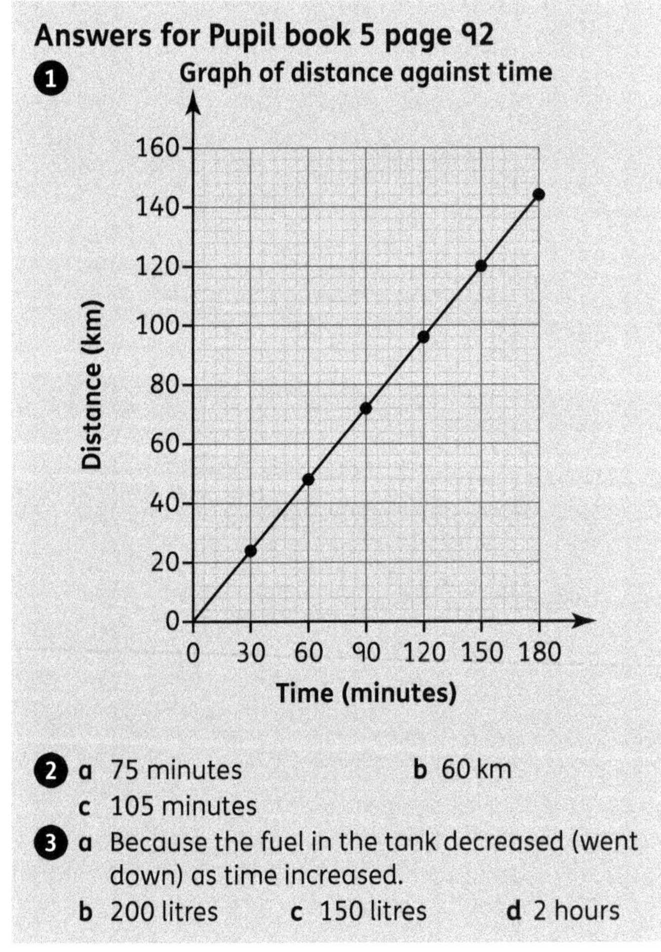

Answers for Workbook 5 page 58

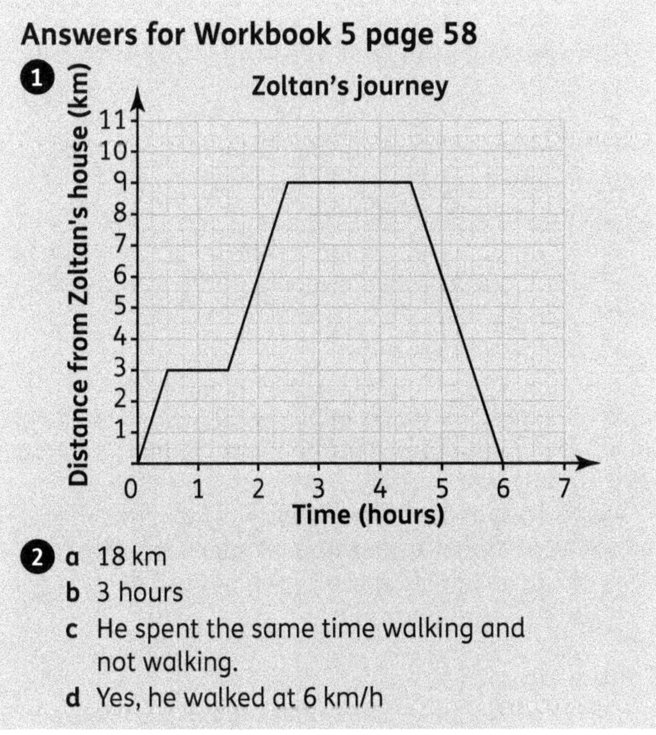

1

2
a 18 km
b 3 hours
c He spent the same time walking and not walking.
d Yes, he walked at 6 km/h

is, find the mean of the two values). The children could also draw number lines to determine the midpoint between two values.

- For question 1, check that the children remember to arrange the data in order so that they can easily find the different values. If they order the data before they try to find the mode, they will find it easier to see the groups of the same value. Note that the data in part a is already in order (descending order works just as well as ascending).
- Problem solving: For question 2, tell the children that they can act out, model or draw diagrams to solve problems like this.

Answers for Pupil book 5 page 93

1 a mode = 75; median = 77.5; range = 35
   b mode = 6; median = 6; range = 3
   c mode = 2 and 6; median = 3.5; range = 5

2 Possible answers: 134, 134, 135, 136, 140; 134, 134, 135, 137, 140; 134, 134, 135, 138, 140; 134, 134, 135, 139, 140

The mode, median and range

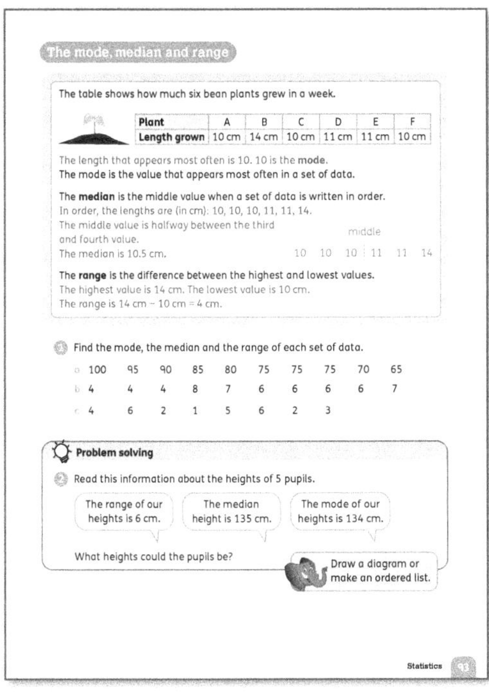

Frequency tables

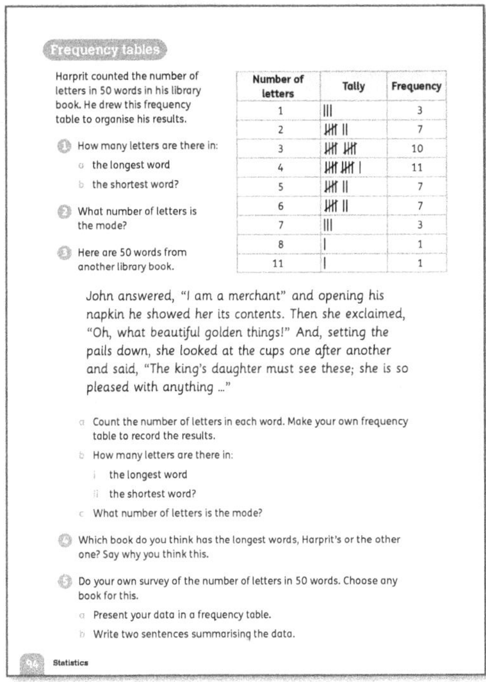

Warm-up

Practise finding the value halfway between two numbers as a starter for this lesson.

Focus

- Teach the children the term *mode* and work through the example on **Pupil book 5 page 93** with the class to show how the mode is found. Work through the rest of the example to explain how to find the *median* and the *range* of a set of data.
- Spend some time discussing how to find the median when there is an even number of values in the data set. Mathematically, you can add the two middle values and then divide by 2 to find the median (that

Warm-up

Do some activities involving counting on and back in fives from any number as a warm-up for this lesson.

Focus

- Revise using tallies with the class. Choose a section of text from the Pupil book and ask the children to use tallies to record the number of full stops or capital letters in the text. Make sure that the children mark every fifth tally with a line through a group of four tallies. Check that they can count tallies by skip counting in fives.
- Use questions 1–4 on **Pupil book 5 page 94** to revise the concept of frequency tables and to make sure the children can interpret and create them.

- For question 5, give the children time to do their own surveys and to present their results. They should include a copy or print-out of the text that they used with their frequency table and sentences.

Answers for Pupil book 5 page 94

1 a 11   b 1

2 4

3 a

Number of letters	Tally	Frequency
1	||	2
2	ⅢⅠ |	6
3	ⅢⅠ ⅢⅠ ||||	14
4	ⅢⅠ |||	8
5	||||	4
6	ⅢⅠ	5
7	||||	4
8	ⅢⅠ	5
9	||	2

  b i 9   ii 1   c 3

4 Harprit's; the modal size of word in Harprit's book is longer (4 letters rather than 3) and the longest word in Harprit's book is longer (11 letters rather than 9).

5 Individual answers.

Frequency tables with groups

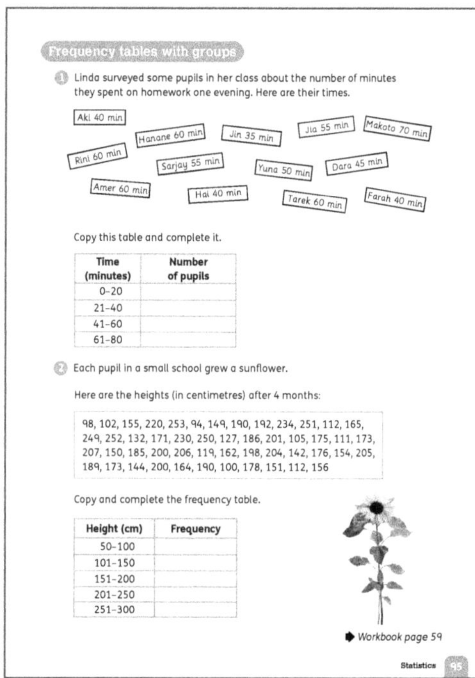

Warm-up

Ask each child to pick a number between 0 and 60 and write it on a piece of paper. Ask them to group themselves into five groups: 0–10, 11–20, 21–30, 31–40 and 41–50.

Focus

- Draw a frequency table to represent the number of children in each group from the warm-up activity.
- Ask the children why we need to group data in some cases. Test results are useful as an example to demonstrate this.
- Tell the class to imagine that each child got a different mark out of 100 for a test. Ask: *How can we record this on a frequency table?* If we record each mark, we will have a very long table with too many rows to fit on a page.
- Explain that we can make recording the data simpler by grouping the marks 0–20, 21–40, and so on. Point out that the groups must be the same size and that they must not overlap.
- Let the children work independently to complete question 1 and question 2 on **Pupil book 5 page 95**. Check their work to make sure they can create a grouped frequency table.

Follow-up

Use **Workbook 5 page 59** to check that the children can tally results and complete a frequency table. Use question 2 (which involves using grouped frequencies) to assess whether the children are able to interpret and work with grouped intervals.

Support

Encourage the children to use coloured dots to help them count the words with different numbers of letters on a photocopy of the text and revise how to use tallies to keep track of a count if necessary.

Interesting mistakes

The children may miss out numbers in grouped frequency tables or overlap the intervals. Spend some time talking about where a number on the margins should go. For example, ask: *Where does a height of 150 go?* This will help the children to make sense of the groups and see that the heights at the two ends of each interval should only be included in one group.

Answers for Pupil book 5 page 95

1

Time (minutes)	Number of pupils
0–20	0
21–40	4
41–60	8
61–80	1

2

Height (cm)	Frequency
50–100	3
101–150	12
151–200	22
201–250	10
251–300	3

①

Shape	Tally	Frequency
heart	卌 卌 IIII	14
star	卌 卌 卌 II	17
moon	卌 卌 卌 II	17
circle	卌 卌 I	11
rectangle	卌	5

②

Results	Tally	Frequency
1–2		0
3–4	IIII	4
5–6	卌 卌	10
7–8	卌 卌 I	11
9–10	卌 III	8

Do your own investigation

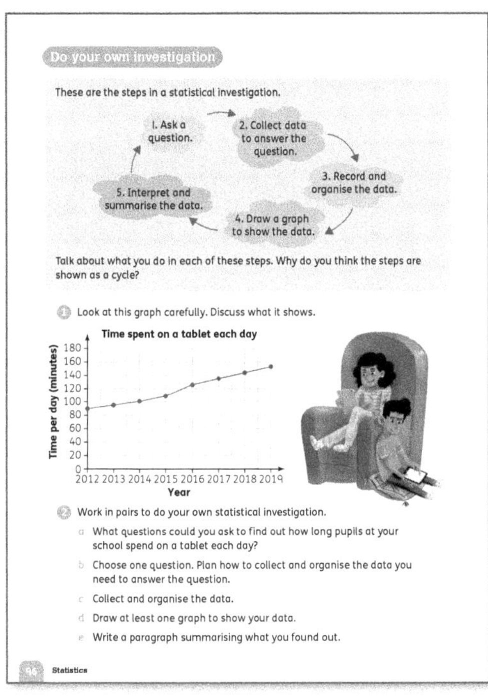

Materials
Graph paper or squared paper; Internet access (if needed)

Warm-up
Spend time discussing the steps in a statistical investigation at the top of **Pupil Book 5 page 96** with the class. Focus on step 1 'Ask a question' and let the children suggest good questions for investigations. A good question is any question that you can answer by collecting and analysing data. Also talk about different methods of collecting data, including surveys and research using secondary sources (data that has been collected by someone else).

Focus
- Discuss the steps in a statistical investigation and why they are needed. The question in the box asks the children to explain why this is called a cycle. (It's called a cycle because each step follows from

the previous one, and all the steps may need to be repeated several times as the results are improved.)
- Discuss the graph in question 1 on **Pupil book 5 page 96** and the trend it shows, with the class. (The graph shows an upward trend, showing that the time spent on a tablet each day is increasing each year.)
- Work with the class to list a set of questions that they could investigate. Then let the children work in groups to talk about the questions and to decide how they could answer them by collecting, organising and presenting data.
- As a whole class, work through the suggested questions one by one, asking groups to share their plans for answering each question. Once a group has answered, avoid other groups giving the same method by asking: *Who else has a similar method? Does any group have a different method?*
- Let the groups either choose one of the questions or make up their own question to investigate. Some children could carry out the investigation in question 2. Explain that they are going to plan and carry out their own investigation over the next few days. Let them write a plan for collecting, organising and presenting the data. Check the groups' plans before they start work to make sure that these are sensible and practical.
- Set aside some time over the next few lessons for the children to work on their investigations. Tell them when they need to be finished and what they need to present to you or the class as a final product.

Answers for Pupil book 5 page 96
① The average number of minutes that children spent on a tablet each day between 2012 and 2019.
② Individual answers. The summary should describe the trend shown by the data, that is, whether it is upwards, downwards or level.

Timetables

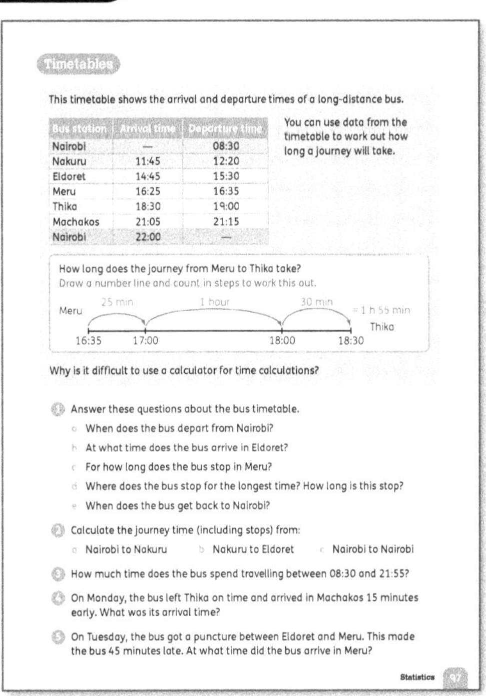

Materials

A range of different local timetables: bus, metro and train timetables, TV schedules, and so on (you can also use the timetables to reinforce 24-hour clock times during this lesson); Internet access (optional)

Warm-up

Use 'Finding times earlier and later than a given time' (page 28) as a starter for this lesson.

Focus

*   Ask the class to share what they remember about *timetables*. Make sure they can tell you that a timetable is a list of times with details of what happens at each time.
*   Revise working out *time intervals* (how much time passes from one point to another) and the duration of events (how long something takes) using the times on timetables. It is always useful to relate this work to everyday timetables that the children may use, such as bus or train timetables, tide tables, television schedules and so on.
*   Let the children develop a timetable of their own to show what they do on a typical day at the weekend. Before they start, explain that a timetable shows the time at which an event starts. It does not state the duration of an event, but you can work out the duration by subtracting the start time from the time the next event starts. This can cause confusion but dealing with it in this way often makes it easier for the children when they work with other timetables.
*   Use the timetable on **Pupil book 5 page 97** to make sure the children can read a simple timetable. Let the children read and work through the example to show how to work out how much time different journeys take. Remind them that units of time are not decimal units, so it is not practical to use a calculator to work out durations. (You cannot state a time of day in which the minutes are greater than 59 because a total of 60 minutes is equivalent to 1 hour.)
*   Answer the questions in question 1 orally with the class to make sure that they can interpret the timetable correctly.
*   Encourage the children to use timelines to work out the answers to questions 2–5.

Challenge

The bus stop names in this activity are all towns in Kenya. The travelling distance between towns can generally be found using the Internet. Let the children choose five towns in your own area and let them develop their own timetable for a round trip between the places.

Support

Provide a timeline like the one the top of the next column and get the children to model the movement of the bus to help them to see where the bus is at any time and to identify stops.

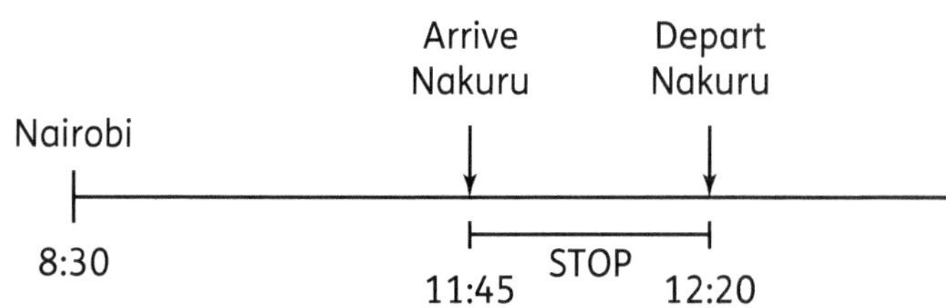

Interesting mistakes

*   Some children need additional support to answer questions such as: *Where is the bus at 16:30?* because only starting times are given in schedules. You will need to pose questions carefully and focus on the duration of intervals/events on timetables.
*   Use the school timetable to illustrate how this works. For example, if PE starts at 9:15 and continues until 10:00, you could ask: *What lesson do you have at 9:30?* (Make sure that you give a time between the lesson start and end times.) The children then need to work out that because the lesson starts at 9:15 and continues to 10:00, they have PE at 9:30, even though the actual time 9:30 does not appear on the timetable.

Answers for Pupil book 5 page 97

1 **a** 08:30   **b** 14:45
   c 10 minutes   **d** Eldoret; 45 minutes
   e 22:00
2 **a** 3 hours and 15 minutes
   b 2 hours and 25 minutes
   c 13 hours and 30 minutes
3 11 hours 15 minutes
4 20:50
5 17:10

End-of-unit check

Ask some or all of these questions to assess how well the children have understood the concepts in this unit.

*   *What is the best way to represent this data?* (Show the children a table of data.) *Should we use a pictogram or a bar graph? Why?* (Pictograms are useful when the number of each item is not very large. When the numbers are larger, bar charts are better.)
*   *What units should we use on the vertical axis?*
*   *Which is the longest bar on this bar graph? What does it show us?* (Show a bar chart.) (The longest bar shows the most popular item.)
*   *Which is the shortest bar?*
*   *What statements can you make about this graph just by looking at it quickly?*
*   *How can you tell from a bar chart which item is the least popular?* (It is the shortest bar.)
*   *In bar graphs, are the bars always vertical?* (No, the bars can be horizontal.)
*   *In a bar graph, a bar 1 cm high represents 8 people. How high is a bar that represents 24 people?* (3 cm)
*   *How can the interval we choose for the scale make a graph look different?* (It can make the differences between different bars look greater or less.)

- *How do you know what the symbols on a pictogram represent?* (There is a key.)
- *If a circle represents 8 people, how could you show 6 people?* ($\frac{3}{4}$ circle) *4 people?* ($\frac{1}{2}$ circle) *2 people?* ($\frac{1}{4}$ circle) (Show a pictogram.)
- *What is the mode/median/range?* (Give a data set.)
- *What do these tallies add up to?* (Show a tally chart.)
- *How could you group this data to make a smaller frequency table?* (Show a very large frequency table.)
- *Why should the groups in a grouped frequency table not overlap?* (So that each data value can only go in to one group.)
- *How many children got a mark of* (choose a mark within a group)*?* (Show the children a grouped frequency table.)
- *How many pieces of data were collected altogether?*
- *What is the heading of this graph?*
- *What scale is used on the vertical axis?*
- *How are these two graphs similar/different?* (Show two different types of graph or two graphs with different scales.)
- *What does this graph show?* (Show a line graph.)
- *Why does the line on this graph slope down to the right?* (As the values on the horizontal axis increase, the values on the vertical axis decrease.)
- *What is wrong with this graph?* (Show a graph with one or more errors, missing labels, etc.)
- *What steps do you follow to do an investigation? Why?* (Ask a question. Collect data to answer the question. Record and organise the data. Draw a graph to show the data. Interpret and summarise the data. The steps form a cycle so that you can change the question according to what the data shows.)
- *Give me an example of a good question for an investigation. What makes it a good question?* (For example: How much time do children spend playing sport each week? You can collect data on how much sport children do each week.)
- *Look at this TV guide. At what time does the news start? What is on TV at 7.30 p.m.?*
- Use bus or metro timetables. Show times on an analogue clock face and ask the children to say where the bus/train is at different times. Include some times that are between times shown on the timetable.

Unit 11 Fractions

Learning objectives

- Compare and order fractions whose denominators are all multiples of the same number

- Identify, name and write equivalent fractions of a given fraction, represented visually, including tenths and hundredths

- Find equivalent fractions and understand that they have the same value and the same position on a number line

- Recognise mixed numbers and improper fractions and convert from one form to the other and write mathematical statements > 1 as a mixed number

- Understand that proper fractions can act as operators and find non-unit fractions of quantities

- Add and subtract fractions with the same denominators and denominators that are multiples of each other

- Multiply fractions by a whole number

- Solve problems involving multiplication, including scaling by simple fractions

Key words

numerator   denominator   improper fraction
mixed number   equivalent fractions

Materials
Blank copies of an equivalent fraction wall (you could include 1 whole, halves, thirds, quarters, fifths, sixths, eighths, tenths and twelfths, as shown below)

Teaching guidance
Hand out a blank equivalent fraction wall to each child (or to each group).

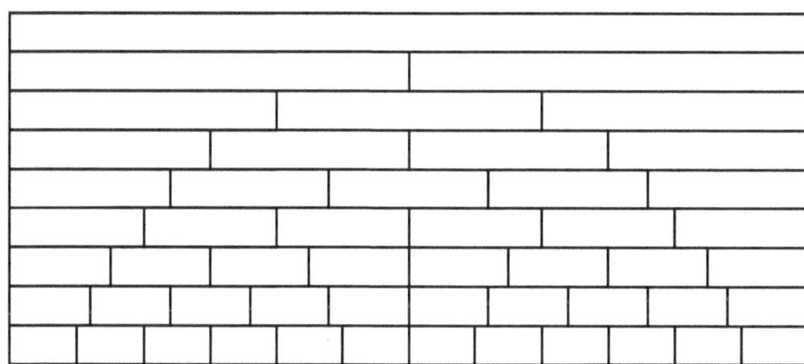

- Let the children label the fractions in each row. Check that they are able to do this.
- The children have worked with equivalent fractions previously, so they should understand the concept.
- Display a simple one-step addition pyramid with fractions, for example,

- Remind the children that the number in each block is the sum of the two numbers below it. Then ask: *How can we use the fraction wall to find the missing number?*
- Let the children work together to figure this out. The children will probably start by looking at the halves and the thirds. They will soon realise that this does not give them the answer, but they will be able to see that if they add a half and a third, the piece missing from the strip is $\frac{1}{6}$. This allows them to see that $\frac{1}{2} + \frac{1}{3}$ is equivalent to $\frac{5}{6}$.
- Some children will work with sixths and say that '$\frac{1}{2}$ is the same length as $\frac{3}{6}$ and $\frac{1}{3}$ is the same length as $\frac{2}{6}$ so together they make $\frac{5}{6}$'. Children who need additional support to do this could cut out each row of the table to make fraction strips. They can then move the strips around to compare and find equivalent fractions.
- Once the children have explained their thinking and found the answer, give them a slightly more challenging pyramid. For example:

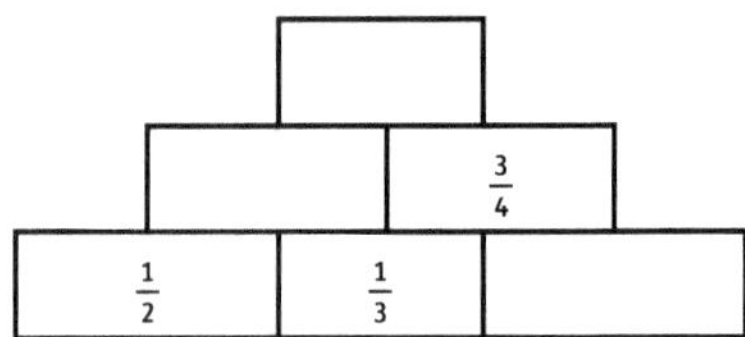

- The missing values are: top row $\frac{19}{12}$ or $1\frac{7}{12}$; middle row $\frac{5}{6}$; bottom row $\frac{5}{12}$.
- Challenge the children to make up their own numbers for a pyramid with 1 at the top. For example:

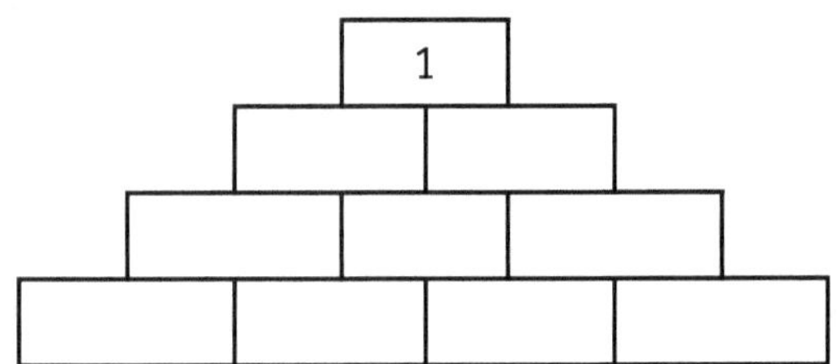

- At its simplest, this is a version of the fraction wall, but if you tell the children they cannot repeat fractions, then it becomes more challenging.
- Similarly, filling in $\frac{1}{10}$ or $\frac{1}{4}$ in the bottom row (any cell) means they cannot simply use halves, quarters and eighths to complete the pyramid.

Revisit fractions

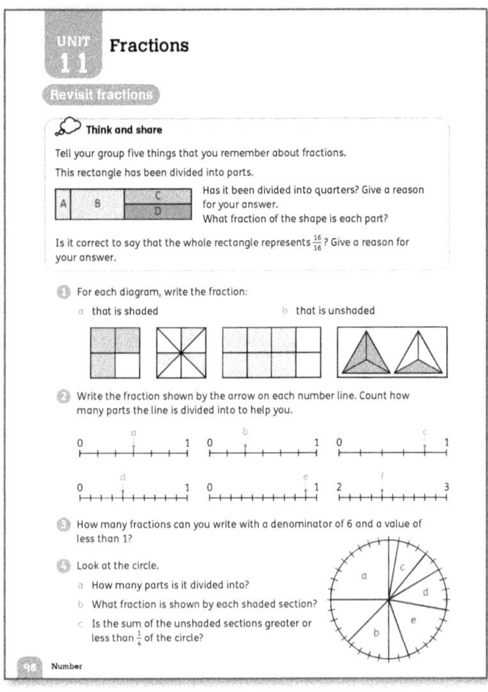

Materials
Coloured pencils or marker pens

Warm-up
Select any suitable 'Mental problem solving' activity from the Activity bank (pages 24–25) as a starter for this lesson.

Focus
Use **Pupil book 5 page 98** to formally revise fractions and to check that the children understand the basic concepts.
- <u>Think and share:</u> Let the children work on the questions in groups and take feedback from the groups. The first question asks whether all the rectangles have been divided into quarters. (No, because the parts are not all the same size. A is approximately one tenth, B is approximately two fifths, and C and D are approximately one quarter of the rectangle.) The second question asks whether the whole bar represents $\frac{16}{16}$. (Yes, because $\frac{16}{16}$. because this simplifies to 1.)
- Let the children work on their own or in pairs to complete questions 1–4.
- Make sure the children realise that the markings around the outside of the circle in question 4 divide the circle into 36 equal fractional parts and that the nine sections inside the circle are different fractional parts.

Challenge
Write on the board:

  3 out of 4 parts improper fraction
  mixed number 4 thirds

Then write these numbers:

$$\frac{7}{8} \quad \frac{3}{4} \quad \frac{4}{3} \quad \frac{6}{5} \quad 2\frac{1}{2}$$

Ask children to match each number to the correct description and write a description for the one left over.

Support

Work with children who need support in small groups to revise the basics and teach the vocabulary of fractions.

- Write a fraction on the board, for example, $\frac{2}{3}$ or $\frac{4}{7}$.
- Remind the children of the terms *numerator* and *denominator* if necessary. Ask the children to tell you what the numbers in the fraction mean. Use their responses to establish that the denominator is the number of parts into which the whole has been divided and the numerator tells us how many of these parts we have.
- Draw sets of shapes on the board divided into equal parts, for example, a shape divided into quarters and a copy of the same shape divided into eighths. Shade a fraction of one shape, for example, $\frac{1}{4}$, and ask the children how much of the other shape must be shaded to make the two diagrams look the same. Ask the children to work out where the fraction ($\frac{1}{4}$) belongs on a 0–1 number line with labels at 0, $\frac{1}{2}$ and 1. Let them explain why.

Interesting mistakes

Children often need additional support with fractions. Many misconceptions seem to arise from over-reliance on the part of a whole model and always presenting this using shaded diagrams.

- When children work with shaded diagrams, they sometimes express the shaded part using a part–part model rather than a part of the whole model. For example: they may say the fraction that is shaded in this diagram is $\frac{2}{3}$ rather than $\frac{2}{5}$.

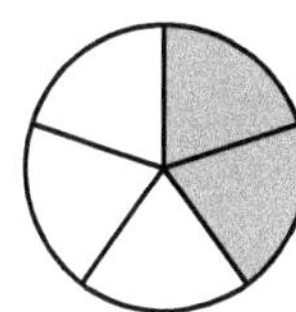

- Address this misconception by asking questions such as: *How many parts is the shape divided into?* (5) *How many of the 5 parts are shaded?* (2 of the 5) *How do we write 2 out of 5 as a fraction?* ($\frac{2}{5}$)
- It is important for the children to see fractions as numbers that they can order on a number line and compare. They also need to be aware that the 'whole' can be a quantity, so, for example, we can express 12 cm out of 24 cm as the fraction $\frac{12}{24}$ or $\frac{1}{2}$.

Answers for Pupil book 5 page 98

1 a $\frac{3}{4}, \frac{3}{8}, \frac{6}{8}, 1\frac{1}{3}$ b $\frac{1}{4}, \frac{5}{8}, \frac{2}{8}, \frac{2}{3}$

2 a $\frac{1}{2}$ b $\frac{2}{6} = \frac{1}{3}$ c $\frac{4}{5}$

     d $\frac{4}{10} = \frac{2}{5}$ e $\frac{9}{10}$ f $2\frac{2}{5}$

3 5

4 a 36

     b $a = \frac{1}{4}, b = \frac{1}{8}, c = \frac{1}{12}, d = \frac{1}{9}, e = \frac{1}{6}$

     c Unshaded portion is $\frac{9\frac{1}{2}}{36}$ or $\frac{19}{72}$ which is greater than $\frac{1}{4}$ ($\frac{18}{72}$).

Equivalent fractions

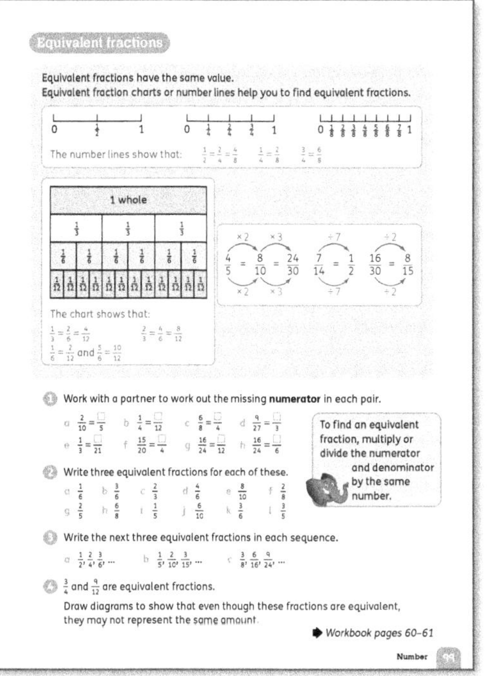

Materials

Large equivalent fraction wall for display in the classroom

Warm-up

As a mental warm-up for this lesson, display a fraction. Go round the class asking the children to say an *equivalent fraction* and record their answers. Challenge them to find as many as possible. Ask: *Are there any more? How do you know?*

Focus

- Use **Workbook 5 page 60** to revise equivalence and to check that the children can show and write equivalent fractions. Observe the children as they work and help anyone who needs support. Then check the answers.
- Let the children read through the examples at the top of **Pupil book 5 page 99**, which show how to use number lines to find equivalent fractions.
- Write a fraction on the board (for example, $\frac{1}{4}$). Ask the children to say some equivalent fractions. Continue with different fractions until you have written several pairs of equivalent fractions on the board.
- Ask the children to look at each pair and say what they notice about the numerator and denominator. Use the children's answers to work towards the idea that we can find equivalent fractions by multiplying or dividing the numerator and denominator by the same number.
- Work through the examples on **Pupil book 5 page 99** to show how this method works.
- Let the children work independently to complete questions 1–4.

Follow-up

Use **Workbook 5 page 61** to informally assess the children's ability to recognise and work with equivalent fractions and to compare and order fractions whose denominators are all multiples of the same number. Provide the answers and let the children mark their own (or each other's) work.

Interesting mistakes

- Some children may add the same number to the numerator and denominator rather than multiply to make equivalent fractions. For example, they might reason that '$\frac{3}{4} = \frac{7}{8}$ because 3 + 4 = 7 and 4 + 4 = 8'. (This is more likely to be a problem with missing number statements such as $\frac{3}{4} = \frac{\square}{8}$).
- Use the fraction wall to show the children that their answer cannot be correct. Also, show them how to find the equivalent numerators and denominators by multiplication and/or division.
- It is useful to start with $\frac{1}{2}$ because they can easily see the equivalences on the fraction wall and because they will find division and multiplication by 2 quite straightforward. Start by showing that $\frac{1}{2} = \frac{5}{10}$ so the children can see that you cannot just add 4 (or 8) to each number to find the equivalence.

Answers for Pupil book 5 page 99

1 a 1   b 3   c 3
d 1   e 7   f 3
g 8   h 4

2 Possible answers:

a $\frac{2}{12}, \frac{3}{18}, \frac{4}{24}$ b $\frac{1}{2}, \frac{2}{4}, \frac{4}{8}$ c $\frac{4}{6}, \frac{6}{9}, \frac{9}{12}$

d $\frac{2}{3}, \frac{6}{9}, \frac{9}{12}$ e $\frac{4}{5}, \frac{12}{15}, \frac{16}{20}$ f $\frac{1}{4}, \frac{3}{12}, \frac{4}{16}$

g $\frac{4}{10}, \frac{6}{15}, \frac{8}{20}$ h $\frac{3}{4}, \frac{18}{24}, \frac{30}{40}$ i $\frac{2}{10}, \frac{3}{15}, \frac{4}{20}$

j $\frac{3}{5}, \frac{9}{15}, \frac{12}{20}$ k $\frac{1}{2}, \frac{2}{4}, \frac{4}{8}$ l $\frac{6}{10}, \frac{9}{15}, \frac{12}{20}$

3 a $\frac{4}{8}, \frac{5}{10}, \frac{6}{12}$ b $\frac{4}{20}, \frac{5}{25}, \frac{6}{30}$ c $\frac{12}{32}, \frac{15}{40}, \frac{18}{48}$

4 Individual answers, for example: $\frac{3}{4}$ of a bag of 24 sweets is 18 sweets, but $\frac{3}{4}$ of a bag of 48 sweets is 36 sweets.

Answers for Workbook 5 page 60

1 a $\frac{1}{2} = \frac{2}{4} = \frac{3}{6} = \frac{4}{8}$

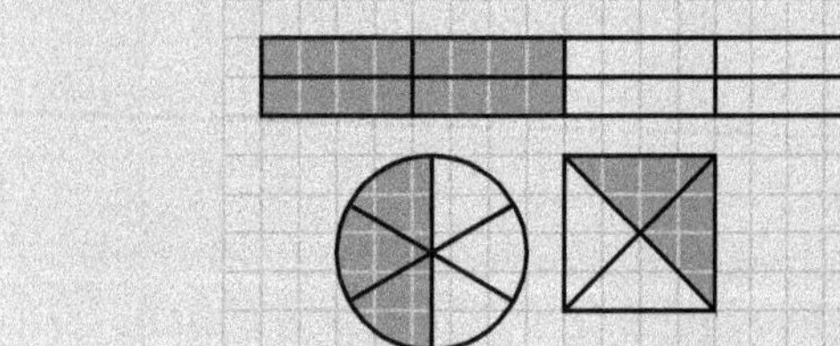

b $\frac{2}{10} = \frac{1}{5} = \frac{3}{15} = \frac{6}{30}$

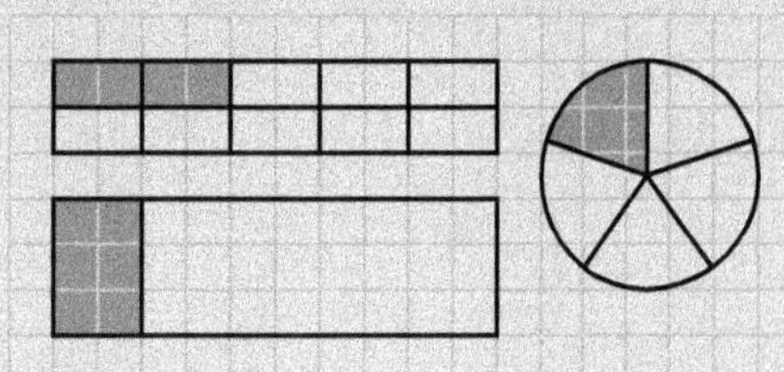

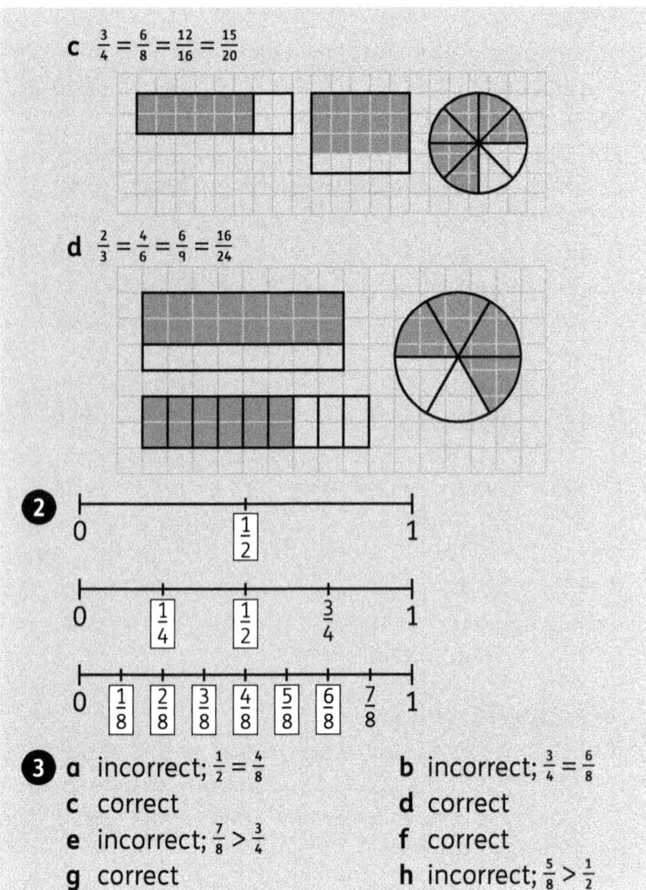

c $\frac{3}{4} = \frac{6}{8} = \frac{12}{16} = \frac{15}{20}$

d $\frac{2}{3} = \frac{4}{6} = \frac{6}{9} = \frac{16}{24}$

2

3 a incorrect; $\frac{1}{2} = \frac{4}{8}$ b incorrect; $\frac{3}{4} = \frac{6}{8}$
c correct   d correct
e incorrect; $\frac{7}{8} > \frac{3}{4}$ f correct
g correct   h incorrect; $\frac{5}{8} > \frac{1}{2}$

Answers for Workbook 5 page 61

1 a $\frac{4}{8}$ b $\frac{5}{10}$ c $\frac{6}{8}$ d $\frac{5}{10}$ e $\frac{3}{5}$ f $\frac{2}{6}$

2 a $\frac{2}{5}$ b $\frac{3}{10}$ c $\frac{8}{10}$

3 $\frac{3}{10}, \frac{2}{8}, \frac{1}{6}, \frac{3}{8}, \frac{4}{10}, \frac{1}{4}, \frac{2}{5}$

4 a $\frac{1}{4}, \frac{1}{2}, \frac{3}{4}$ b $\frac{3}{10}, \frac{1}{2}, \frac{4}{5}, \frac{9}{10}$ c $\frac{3}{8}, \frac{1}{2}, \frac{5}{8}, \frac{3}{4}$

d $\frac{1}{10}, \frac{3}{10}, \frac{2}{5}, \frac{1}{2}, \frac{3}{5}$ e $\frac{1}{10}, \frac{1}{8}, \frac{1}{5}, \frac{1}{3}, \frac{1}{2}$

5 a correct   b incorrect $\frac{1}{2} < \frac{3}{4}$ c incorrect $\frac{2}{8} = \frac{1}{4}$
d correct   e incorrect $\frac{5}{8} > \frac{1}{2}$ f correct
g correct   h correct   i incorrect $\frac{5}{8} < \frac{3}{4}$

Improper fractions and mixed numbers

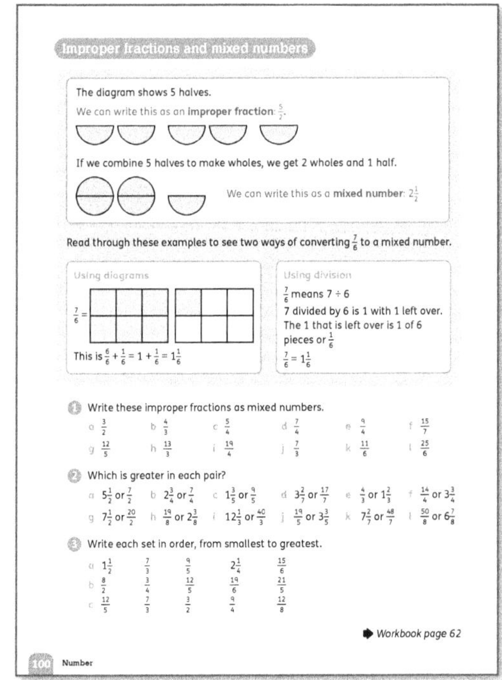

The diagram shows 5 halves.
We can write this as an **improper fraction**: $\frac{5}{2}$.

If we combine 5 halves to make wholes, we get 2 wholes and 1 half.

We can write this as a **mixed number**: $2\frac{1}{2}$

Read through these examples to see two ways of converting $\frac{7}{6}$ to a mixed number.

Using diagrams

$\frac{7}{6} =$

This is $\frac{6}{6} + \frac{1}{6} = 1 + \frac{1}{6} = 1\frac{1}{6}$

Using division

$\frac{7}{6}$ means 7 ÷ 6
7 divided by 6 is 1 with 1 left over. The 1 that is left over is 1 of 6 pieces or $\frac{1}{6}$
$\frac{7}{6} = 1\frac{1}{6}$

1 Write these improper fractions as mixed numbers.

a $\frac{3}{2}$ b $\frac{4}{3}$ c $\frac{5}{4}$ d $\frac{7}{4}$ e $\frac{9}{4}$ f $\frac{15}{7}$

g $\frac{12}{5}$ h $\frac{13}{3}$ i $\frac{19}{4}$ j $\frac{7}{3}$ k $\frac{11}{6}$ l $\frac{25}{6}$

2 Which is greater in each pair?

a $5\frac{1}{2}$ or $\frac{7}{2}$ b $2\frac{3}{4}$ or $\frac{7}{4}$ c $1\frac{3}{5}$ or $\frac{9}{5}$ d $3\frac{2}{7}$ or $\frac{17}{7}$ e $\frac{4}{3}$ or $1\frac{2}{3}$ f $\frac{14}{4}$ or $3\frac{3}{4}$

g $7\frac{1}{2}$ or $\frac{20}{2}$ h $\frac{19}{8}$ or $2\frac{3}{8}$ i $12\frac{1}{3}$ or $\frac{40}{3}$ j $\frac{19}{5}$ or $3\frac{3}{5}$ k $7\frac{2}{7}$ or $\frac{48}{7}$ l $\frac{50}{8}$ or $6\frac{7}{8}$

3 Write each set in order, from smallest to greatest.

a $1\frac{1}{2}$ $\frac{7}{3}$ $\frac{9}{5}$ $2\frac{1}{4}$ $\frac{15}{6}$

b $\frac{8}{2}$ $\frac{3}{4}$ $\frac{12}{5}$ $\frac{19}{6}$ $\frac{21}{5}$

c $\frac{12}{5}$ $\frac{7}{3}$ $\frac{3}{2}$ $\frac{9}{4}$ $\frac{12}{8}$

➡ Workbook page 62

Materials

Sets of shapes, with each shape cut into equal parts (For example, you could include a set of circles cut into halves, a set of squares cut into quarters (diagonally), a set of rectangles cut into thirds and so on. Each set should also include one whole shape.)

Warm-up

Revise dividing numbers to give a remainder, for example $8 \div 3 = 2$ remainder 2, as a starter for this lesson.

Focus

- Divide the class into groups. Give each group a set of shapes cut into fractional parts and one whole shape for comparison.
- Ask the children to take a handful of parts and say how many they have, for example, 'I have three halves', 'I have ten thirds' and so on.
- Explain how to write each amount as an improper fraction. Let the children make different groups using the parts they have and discuss how to write these as (improper) fractions. Remind the class that any fraction with a numerator greater than its denominator is called an *improper fraction*.
- Using the same fractional parts, let the children work out how many parts make a whole (for example, three thirds make one whole rectangle and so on).
- Then ask them to take another handful of pieces, express the result as an improper fraction and say how many wholes and parts this is.
- Remind the children that a number with a whole number part and a fraction part is called a *mixed number*. Discuss the relationship between improper fractions and mixed numbers and let the children share the methods they used to convert from an improper fraction to a mixed number.
- Read through the examples on **Pupil book 5 page 100** with the class to formalise and consolidate what they learnt during the practical activities. Then let them complete questions 1–3.

Follow-up

Use **Workbook 5 page 62** for additional practice and to make sure that the children can position mixed numbers on a number line.

Challenge

Let the children work in pairs to design a game involving equivalent mixed numbers and improper fractions. (Dominoes and card games work well, but also allow the children to explore their own ideas.) Let each pair show their game to the class and explain how it works. If you have time, let the children play the different games and rate them.

Support

- At this level, the children may find it easier to understand improper fractions and mixed numbers as concrete amounts. Use examples such as these:
  - *I cut some pizzas in half, I ate 1 half and now I have 7 halves left. How many whole pizzas did I cut into halves?* (4)
  - *I cut pieces of card into quarters. I gave each group 9 quarters. How many whole cards can they make?* (2)
- Concrete examples like these help the children to understand that a numerator can be greater than a denominator and that they can rearrange improper fractions to make equivalent mixed numbers.

Interesting mistakes

Some children will grasp that we can divide to convert from improper fractions to mixed numbers. Other children may find using division too abstract and need to use diagrams and jottings. Continue to ask the children to share their thinking and explain their methods to expose them to different ways of working and thinking about converting fractions.

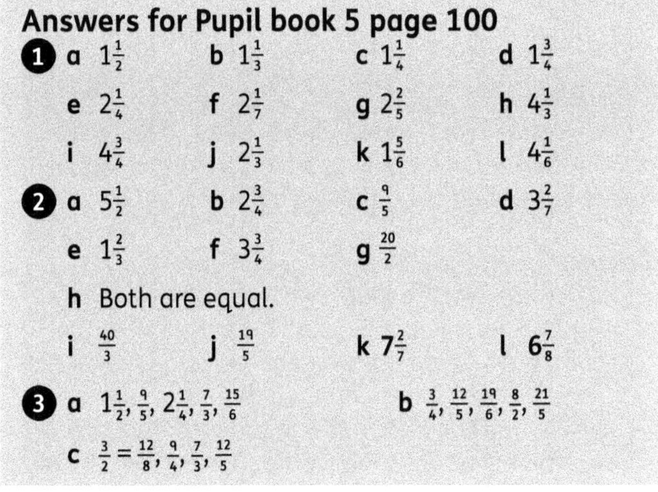

Answers for Pupil book 5 page 100

1.
   a $1\frac{1}{2}$ b $1\frac{1}{3}$ c $1\frac{1}{4}$ d $1\frac{3}{4}$
   e $2\frac{1}{4}$ f $2\frac{1}{7}$ g $2\frac{2}{5}$ h $4\frac{1}{3}$
   i $4\frac{3}{4}$ j $2\frac{1}{3}$ k $1\frac{5}{6}$ l $4\frac{1}{6}$

2.
   a $5\frac{1}{2}$ b $2\frac{3}{4}$ c $\frac{9}{5}$ d $3\frac{2}{7}$
   e $1\frac{2}{3}$ f $3\frac{3}{4}$ g $\frac{20}{2}$
   h Both are equal.
   i $\frac{40}{3}$ j $\frac{19}{5}$ k $7\frac{2}{7}$ l $6\frac{7}{8}$

3.
   a $1\frac{1}{2}, \frac{9}{5}, 2\frac{1}{4}, \frac{7}{3}, \frac{15}{6}$ b $\frac{3}{4}, \frac{12}{5}, \frac{19}{6}, \frac{8}{2}, \frac{21}{5}$
   c $\frac{3}{2} = \frac{12}{8}, \frac{9}{4}, \frac{7}{3}, \frac{12}{5}$

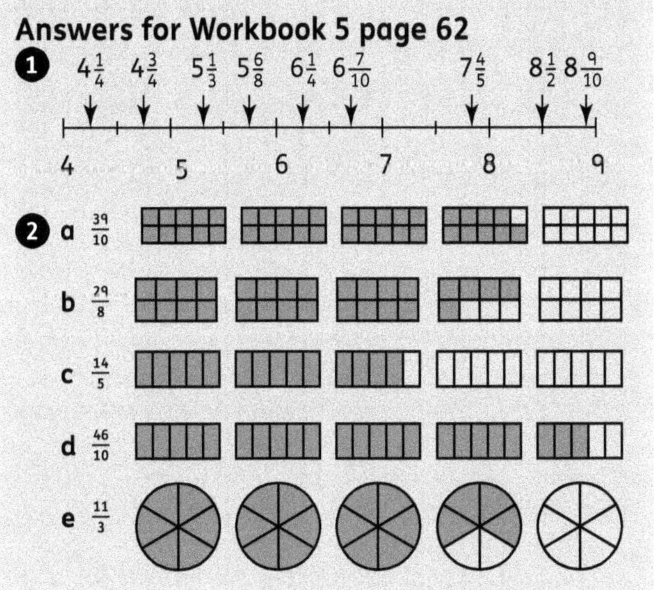

Answers for Workbook 5 page 62

1. $4\frac{1}{4}$ $4\frac{3}{4}$ $5\frac{1}{3}$ $5\frac{6}{8}$ $6\frac{1}{4}$ $6\frac{7}{10}$ $7\frac{4}{5}$ $8\frac{1}{2}$ $8\frac{9}{10}$

2.
   a $\frac{39}{10}$
   b $\frac{29}{8}$
   c $\frac{14}{5}$
   d $\frac{46}{10}$
   e $\frac{11}{3}$

Compare and order fractions

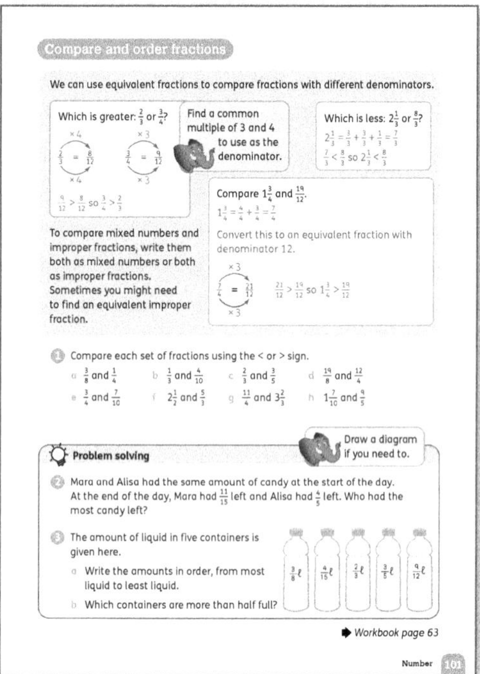

Materials
Calculators

Warm-up
Do any comparing and ordering activity using whole numbers and/or decimals as a starter for this lesson. Revise the use of the < and > symbols as you do so.

Focus
- Revise ordering fractions with the same denominators (such as $\frac{2}{5}$ and $\frac{4}{5}$) by looking at the numerators. Then ask the children to compare $\frac{2}{3}$ and $\frac{7}{10}$ and say which is greater.
- Remind the children that we can write fractions as equivalent fractions with different denominators and that this does not change the value of the fraction. Explain that we can use this principle to compare and order fractions.
- Let the children suggest how to rewrite $\frac{2}{3}$ and $\frac{7}{10}$ to make them easier to compare. It is likely that, in this case, the children will use the lowest common multiple of 3 and 10 and write the fractions as thirtieths. However, they can use any common multiple to find a common denominator; it does not have to be the lowest common denominator.
- Next, display a mixed number such as $3\frac{1}{5}$ and an improper fraction such as $\frac{18}{7}$. Ask the children which number is greater. Give them time to think about this and let them share their answers as well as the method they used.
- Draw a number line like the one in the next column. Explain that we can locate $3\frac{1}{5}$ quite easily on this number line and mark it on the line.
- Then say: *What about eighteen sevenths? I know that $\frac{7}{7}$ is 1, so I can count up in ones.* (Draw the jumps as shown on the number line.) *Seven sevenths is 1, another seven sevenths makes 2 and I still have*

4 sevenths left. That takes me to $2\frac{4}{7}$. Now I can compare the numbers and see that $3\frac{1}{5}$ is greater than $2\frac{4}{7}$.

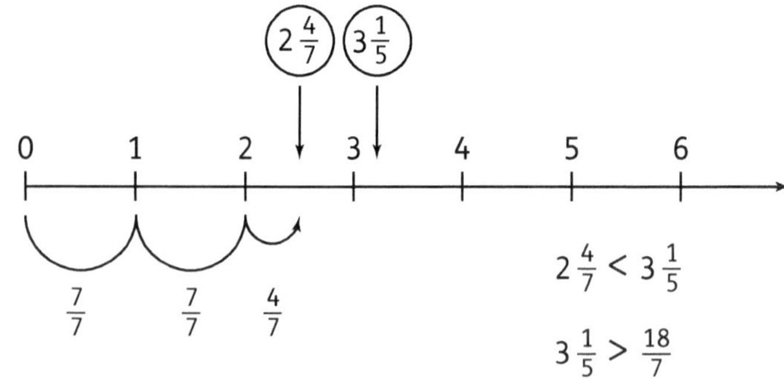

$$2\frac{4}{7} < 3\frac{1}{5}$$

$$3\frac{1}{5} > \frac{18}{7}$$

- Turn to **Pupil book 5 page 101**. Work through the first example with the class to reinforce how to compare fractions with different denominators.
- Then look at the second example (comparing a mixed number and an improper fraction). Explain that drawing a number line takes time and that we can use more efficient methods to compare the numbers.
- Work through the example with the class, asking questions to make sure they understand.
- The children can answer question 1.
- Problem solving: Then the children can complete question 2 and question 3.

Follow-up
Use **Workbook 5 page 63** as an informal assessment task to make sure that the children can compare and order fractions and position them correctly on a number line. Have the children complete the work in 'test' conditions, but you could mark the work together as a class, addressing any mistakes.

Challenge
- Let the children investigate how to use cross multiplication to see whether or not fractions are equivalent. To do this, they multiply the numerator of each fraction by the denominator of the other. If the products are equal, the fractions are equivalent. For example,

$$\frac{3}{5} \diagdown \frac{24}{40} \qquad 5 \times 24 = 120 \qquad \checkmark \quad \text{Fractions are} \\ 3 \times 40 = 120 \qquad \text{equivalent.}$$

$$\frac{4}{7} \diagdown \frac{7}{10} \qquad 7 \times 7 = 49 \qquad \times \quad \text{Fractions are} \\ 4 \times 10 = 40 \qquad \text{not equivalent.}$$

- Once the children have explored this method, they can use the same method to compare fractions. To do this, they write the products next to the numerator they are multiplying by. (But you can let the children work that out for themselves.) For example:

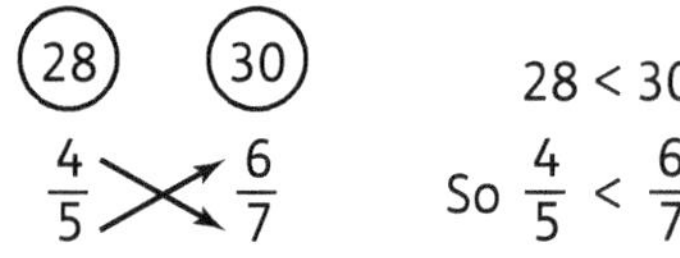

$$28 < 30$$
$$\text{So } \frac{4}{5} < \frac{6}{7}$$

- Let the children discuss why they think this method works. (It is a shortcut method of using common denominators.) The children can find more information about this method online by searching for 'order-fractions-from-least-to-greatest'.

Support

Some children may have a good understanding of place value and decimal comparison but need additional support with fractions. It may be useful to show them how to use a calculator to express fractions as decimals (by dividing the numerator by the denominator) and to use the decimals to compare and order the fractions. For example, in question 3, the decimal amounts are 0.375, 0.266, 0.667, 0.6, 0.75. This method allows the children to order the amounts without having to find lots of equivalent fractions.

Interesting mistakes

Children often decide that a fraction is greater than another by comparing just the denominators or just the numerators. This mistake is common when children do not understand the relative sizes of fractions. It is useful to give the children a benchmark to compare the fractions with. For example, ask them to compare each fraction with an appropriate fixed point such as 0, $\frac{1}{2}$ or 1. Then ask them to explain and justify their thinking and get them to show you what they are doing so that you can address mistakes using real objects or number lines.

Answers for Pupil book 5 page 101

1 a $\frac{3}{8} > \frac{1}{4}$　　b $\frac{1}{3} < \frac{4}{10}$　　c $\frac{2}{3} > \frac{3}{5}$

d $\frac{19}{8} < \frac{12}{4}$　　e $\frac{3}{4} > \frac{7}{10}$　　f $2\frac{1}{2} > \frac{5}{3}$

g $\frac{11}{4} < 3\frac{2}{3}$　　h $1\frac{7}{10} < \frac{9}{5}$

2 Alisa

3 a $\frac{9}{12}\,\ell,\ \frac{2}{3}\,\ell,\ \frac{3}{5}\,\ell,\ \frac{3}{8}\,\ell,\ \frac{4}{15}\,\ell$

b The containers with $\frac{9}{12}\,\ell$, $\frac{2}{3}\,\ell$, and $\frac{3}{5}\,\ell$ are more than half full.

Answers for Workbook 5 page 63

1 a $\frac{1}{2} > \frac{1}{6}$　　b $\frac{1}{3} < \frac{1}{2}$　　c $\frac{1}{2} = \frac{4}{8}$

d $\frac{1}{4} > \frac{1}{5}$　　e $\frac{2}{3} > \frac{2}{5}$　　f $\frac{3}{7} > \frac{3}{8}$

g $\frac{3}{4} = \frac{9}{12}$　　h $\frac{6}{12} = \frac{4}{8}$　　i $\frac{14}{15} > \frac{4}{5}$

j $\frac{9}{10} > \frac{4}{5}$　　k $\frac{3}{16} > \frac{1}{8}$　　l $\frac{3}{4} > \frac{7}{24}$

2

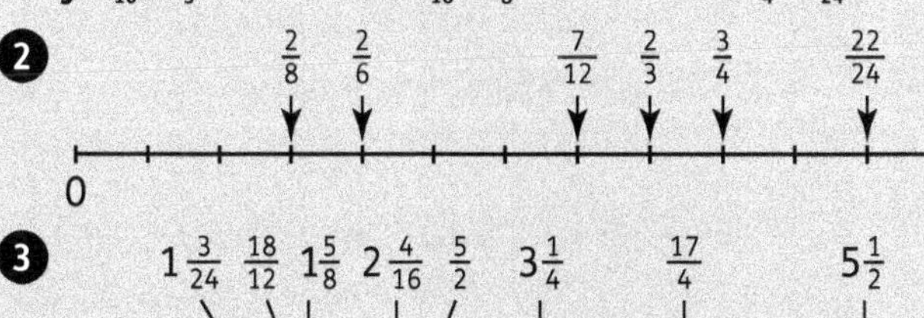

3

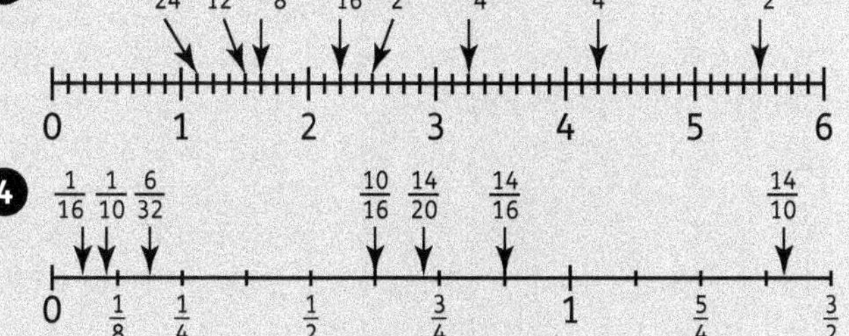

4

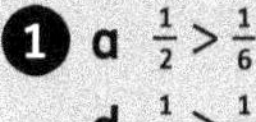

Add and subtract fractions

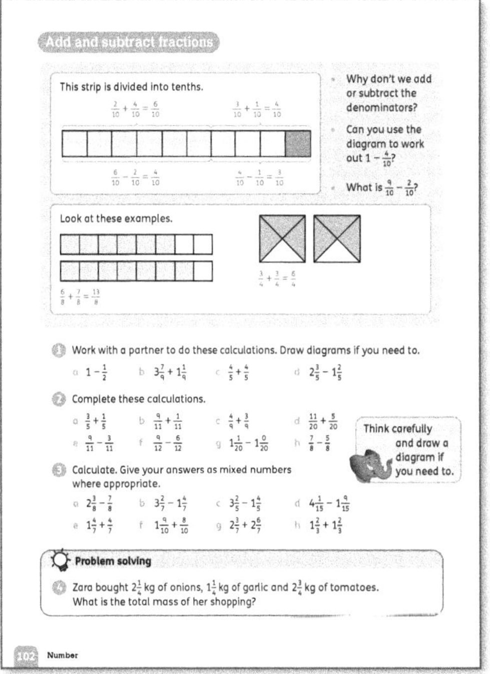

Materials

Shapes cut into fractional pieces; number lines or fraction strips

Warm-up

Use any suitable mental adding and subtracting activities as a starter for this lesson.

Focus

- Use fraction pieces and fraction number lines to start a discussion about adding and subtracting fractions with the same denominators. (The next lesson covers adding and subtracting fractions with different denominators.) For example, give the children a set of circles cut into sixths. Instruct the children to take 2 sixths and 3 sixths from the set. Ask: *What is 2 sixths add 3 sixths?* (5 sixths)
- Then say: *Make a set of 5 sixths. How many sixths are left if you subtract 4 sixths?* ($\frac{1}{6}$)
- Include some additions where the sum is greater than 1 and leave the answers as improper fractions. For example, 4 *fifths add 3 fifths is 7 fifths.*
- Working with practical apparatus at this stage is important because it allows the children to see that you are adding sixths (or any other fractional amount) and that they remain sixths when you add or subtract them. Using the names of the fractions also helps the children understand what you are doing.
- Display a number line (or fraction strip) like this one:

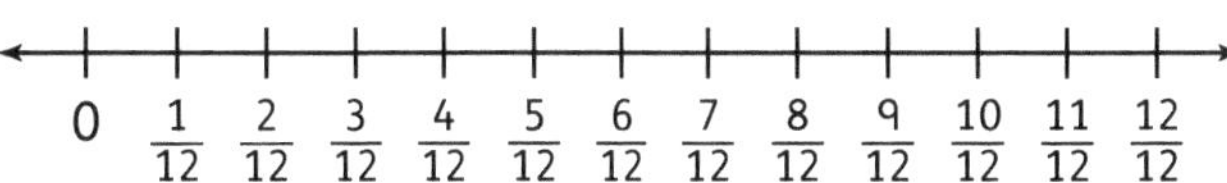

- Do some activities that involve counting on and counting back in fractions. For example:
- *Start at 8 twelfths. Count back 3 twelfths. Where do you land?* ($\frac{5}{12}$)
- *Start at 1 twelfth, count on 10 twelfths. Where do you land?* ($\frac{11}{12}$)

- Then ask: *How can we write counting on and counting back as addition or subtraction?* Let the children write down the calculations.
- Once you can see that the children are able to add and subtract the fractions, turn to **Pupil book 5 page 102**.
- Work through the questions and examples with the class. This will help the children formalise what they have learnt and show them how to move from real items to diagrams and then to written work.
- For the questions that involve mixed numbers in question 1, the children can leave their answers as mixed numbers. If the answer is an improper fraction, they do not need to convert it.
- Children with a good grasp of fractions and equivalence can simplify their answers to question 2 and give them in simplest form.
- Ask the children how they could solve the calculations in question 3. They should realise that changing the mixed numbers to improper fractions will make it easier to subtract or add and allow them to avoid 'carrying a whole'.
- <u>Problem solving:</u> For question 4, remind the children that a diagram can help them see what is involved.

Interesting mistakes

- The most common mistake that children make at this level is to assume that if they are adding the numerators, they must also add the denominators. For example, they may say that '$\frac{2}{5} + \frac{2}{5} = \frac{4}{10}$'. This indicates that the children are simply applying rules they already know and do not fully understand the concept of a *denominator*.
- Use fraction pieces and number lines to help them see that the denominator stays the same. In other words, when we add one type of fraction the answer is the same type of fraction. Using the names as you model the operations helps make this clear.
- The children may need additional support with calculations involving mixed numbers. Using a number line and counting on or back can help them to make sense of the calculation. Remind them that rewriting mixed numbers as improper fractions can also help. For example, the number line below shows that the calculation $1\frac{3}{5} - \frac{4}{5}$ is equivalent to $\frac{5}{5} + \frac{3}{5} - \frac{4}{5} = \frac{8}{5} - \frac{4}{5} = \frac{4}{5}$.

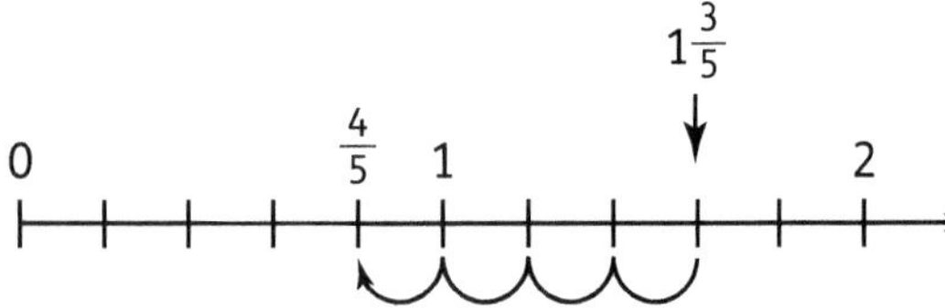

Answers for Pupil book 5 page 102

1 a $\frac{1}{2}$ b $\frac{44}{9}$ or $4\frac{8}{9}$ c $\frac{8}{5}$ or $1\frac{3}{5}$ d $\frac{6}{5}$ or $1\frac{1}{5}$

2 a $\frac{4}{5}$ b $\frac{10}{11}$ c $\frac{7}{9}$ d $\frac{16}{20}$ or $\frac{4}{5}$

    e $\frac{6}{11}$ f $\frac{3}{12}$ or $\frac{1}{4}$ g $\frac{1}{20}$ h $\frac{2}{8}$ or $\frac{1}{4}$

3 a $1\frac{1}{2}$ b $1\frac{5}{7}$ c $1\frac{3}{5}$ d $2\frac{7}{15}$

    e $2\frac{1}{7}$ f $2\frac{7}{10}$ g $5\frac{2}{7}$ h $3\frac{1}{3}$

4 $6\frac{1}{4}$ kg

More adding and subtracting

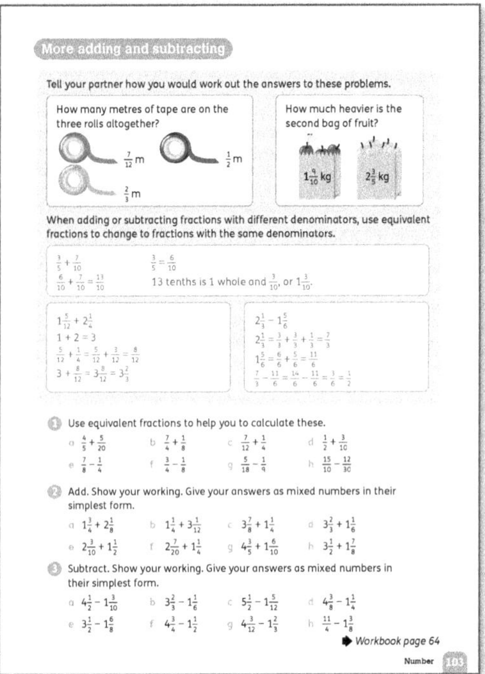

Materials

Large equivalent fraction walls for reference; fraction tiles or rods (if available); identical shapes cut into fractions that are multiples of each other

Warm-up

As a mental warm-up for this lesson, ask the children to find the lowest common multiple (LCM) of different pairs or sets of numbers (for example: 2 and 3; 3 and 4; 3 and 5; 3 and 6; 2 and 8; 3 and 9; 5 and 10; 4 and 12). Include some sets with three numbers (for example: 2, 3 and 8; 2, 4 and 12).

Focus

- Use fraction pieces to show that $\frac{1}{3} + \frac{1}{3} = \frac{2}{3}$. Then show $\frac{1}{3} + \frac{1}{6}$. Ask the children how they could add these two fractions. Let them share their ideas.
- The children should be able to relate thirds to sixths and say that $\frac{1}{3}$ is equivalent to $\frac{2}{6}$. $\frac{2}{6} + \frac{1}{6} = \frac{3}{6}$, which can be simplified to $\frac{1}{2}$.
- Emphasise that we can only add fractions that have the same denominator, so we use equivalent fractions to make the denominators the same. (Generally, we use the lowest common denominator, but it is not incorrect to use any common denominator. For example, we could write $\frac{1}{3} + \frac{1}{6}$ as $\frac{4}{12} + \frac{2}{12} = \frac{6}{12}$, which can be simplified to $\frac{1}{2}$.)
- Repeat the procedure for a different pair of fractions, such as $\frac{1}{3} + \frac{5}{12}$.
- Turn to **Pupil book 5 page 103**. Let the children discuss the problems at the top of the page in pairs and give different children a chance to show how they would work out the solutions.
- Work step-by-step through the examples with the class. As you show each step, verbalise what you are doing. Make sure the children understand that mixed numbers can be converted to improper fractions if it will make their calculation easier.
- The children can then complete questions 1–3.

Follow-up

Use **Workbook 5 page 64** to give the children additional practice in adding and subtracting fractions. They can complete question 1 on their own and then work in pairs on the magic squares in question 2.

Support

Let the children continue to use the fraction wall or other apparatus to model the fractions and find equivalents if necessary. You could also give them a set of equivalent fraction number lines to use.

Interesting mistakes

Some children may not realise that they can simplify the fractional part of a mixed number without affecting its value. Remind them that a number such as $3\frac{4}{10}$ means 3 wholes $+ \frac{4}{10}$, so the fraction part can be treated separately and simplified to $\frac{2}{5}$.

Answers for Pupil book 5 page 103

1
a $\frac{21}{20} = 1\frac{1}{20}$ b $\frac{15}{8} = 1\frac{7}{8}$ c $\frac{5}{6}$

d $\frac{4}{5}$ e $\frac{5}{8}$ f $\frac{5}{8}$

g $\frac{1}{6}$ h $\frac{11}{10} = 1\frac{1}{10}$

2
a $1\frac{3}{4} + 2\frac{1}{8} = \frac{7}{4} + \frac{17}{8} = \frac{14}{8} + \frac{17}{8} = \frac{31}{8} = 3\frac{7}{8}$

b $1\frac{1}{4} + 3\frac{1}{12} = \frac{5}{4} + \frac{37}{12} = \frac{15}{12} + \frac{37}{12} = \frac{52}{12} = 4\frac{1}{3}$

c $3\frac{7}{8} + 1\frac{1}{4} = \frac{31}{8} + \frac{5}{4} = \frac{31}{8} + \frac{10}{8} = \frac{41}{8} = 5\frac{1}{8}$

d $3\frac{2}{3} + 1\frac{1}{6} = \frac{11}{3} + \frac{7}{6} = \frac{22}{6} + \frac{7}{6} = \frac{29}{6} = 4\frac{5}{6}$

e $2\frac{3}{10} + 1\frac{1}{2} = \frac{23}{10} + \frac{3}{2} = \frac{23}{10} + \frac{15}{10} = \frac{38}{10} = 3\frac{4}{5}$

f $2\frac{7}{20} + 1\frac{1}{4} = \frac{47}{20} + \frac{5}{4} = \frac{47}{20} + \frac{25}{20} = \frac{72}{20} = 3\frac{3}{5}$

g $4\frac{3}{5} + 1\frac{6}{10} = \frac{23}{5} + \frac{16}{10} = \frac{46}{10} + \frac{16}{10} = \frac{62}{10} = 6\frac{1}{5}$

h $3\frac{1}{2} + 1\frac{7}{8} = \frac{7}{2} + \frac{15}{8} = \frac{28}{8} + \frac{15}{8} = \frac{43}{8} = 5\frac{3}{8}$

3
a $4\frac{1}{2} - 1\frac{3}{10} = \frac{9}{2} - \frac{13}{10} = \frac{45}{10} - \frac{13}{10} = \frac{32}{10} = 3\frac{1}{5}$

b $3\frac{2}{3} - 1\frac{1}{6} = \frac{8}{3} - \frac{7}{6} = \frac{16}{6} - \frac{7}{6} = \frac{9}{6} = 1\frac{1}{2}$

c $5\frac{1}{2} - 1\frac{5}{12} = \frac{11}{2} - \frac{17}{12} = \frac{66}{12} - \frac{17}{12} = \frac{32}{10} = \frac{49}{12} = 4\frac{1}{12}$

d $4\frac{3}{8} - 1\frac{1}{4} = \frac{35}{8} - \frac{5}{4} = \frac{35}{8} - \frac{10}{8} = \frac{25}{8} = 1\frac{1}{8}$

e $3\frac{1}{2} - 1\frac{6}{8} = \frac{7}{2} - \frac{14}{8} = \frac{28}{8} - \frac{14}{8} = \frac{14}{8} = 1\frac{3}{4}$

f $4\frac{3}{4} - 1\frac{1}{2} = \frac{19}{4} - \frac{3}{2} = \frac{19}{4} - \frac{6}{4} = \frac{13}{4} = 3\frac{1}{4}$

g $4\frac{3}{12} - 1\frac{2}{3} = \frac{51}{12} - \frac{5}{3} = \frac{51}{12} - \frac{20}{12} = \frac{31}{12} = 2\frac{7}{12}$

h $\frac{11}{4} - 1\frac{3}{8} = \frac{11}{4} - \frac{11}{8} = \frac{22}{8} - \frac{11}{8} = \frac{11}{8} = 1\frac{3}{8}$

Answers for Workbook 5 page 64

1
a $\frac{4}{6} = \frac{2}{3}$ b $\frac{5}{8}$ c $\frac{13}{12} = 1\frac{1}{12}$

d $\frac{5}{12}$ e $\frac{5}{12}$ f $\frac{4}{10} = \frac{2}{5}$

2

$1\frac{2}{10}$	$\frac{2}{10}$	$1\frac{6}{10}$
$\frac{4}{10}$	1	$\frac{6}{10}$
$\frac{4}{10}$	$\frac{18}{10}$	$\frac{8}{10}$

$\frac{1}{12}$	$\frac{11}{48}$	$\frac{3}{48}$
$\frac{5}{48}$	$\frac{1}{8}$	$\frac{7}{48}$
$\frac{3}{16}$	$\frac{1}{48}$	$\frac{1}{6}$

$\frac{14}{100}$	$\frac{9}{100}$	$\frac{1}{10}$
$\frac{7}{100}$	$\frac{11}{100}$	$\frac{15}{100}$
$\frac{12}{100}$	$\frac{13}{100}$	$\frac{8}{100}$

$5\frac{3}{5}$	$6\frac{1}{2}$	$4\frac{2}{5}$
$4\frac{3}{10}$	$5\frac{1}{2}$	$6\frac{7}{10}$
$6\frac{3}{5}$	$4\frac{1}{2}$	$5\frac{2}{5}$

Multiply fractions by whole numbers

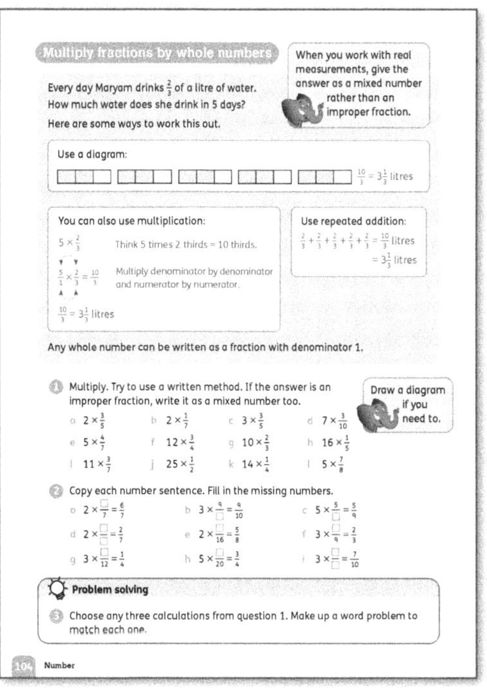

Warm-up

Revise multiplication table facts as a mental starter for this lesson.

Focus

- Display a diagram like this one:

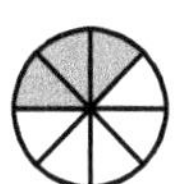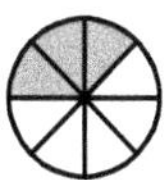

- Ask the class: *What does this diagram show?*
- Take responses, asking the children to explain their answers. For example, they might give one of these responses:
  - 'It shows three lots of 3 eighths.'
  - 'It shows three circles with $\frac{3}{8}$ shaded on each and $\frac{5}{8}$ not shaded.'
  - 'It shows 1 and $\frac{1}{8}$.'
  - 'It shows $\frac{9}{8}$.'
- Ask the children to write an addition sum and a multiplication to match the diagram. They might wrile: $\frac{3}{8} + \frac{3}{8} + \frac{3}{8}$ and $3 \times \frac{3}{8}$ or $\frac{3}{8} \times 3$.
- Ask the class how they could use the diagram below to work out $3 \times \frac{7}{8}$ and give the answer as a mixed number.
- Take suggestions and then demonstrate shading 3 lots of 7 eighths in different ways. Continue to say the fraction parts as you demonstrate and explain.

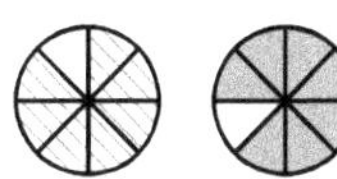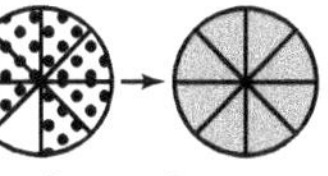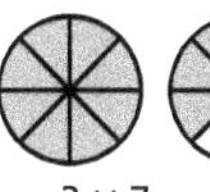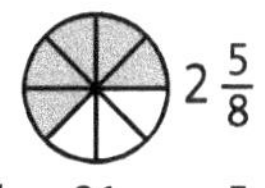 $2\frac{5}{8}$

7 striped + 7 grey + 7 dotted   Check: $\frac{3 \times 7}{8} = \frac{21}{8} = 2\frac{5}{8}$

- Say: *We have shaded 2 wholes and $\frac{5}{8}$. $2\frac{5}{8}$ is the same as $\frac{8}{8} + \frac{8}{8} + \frac{5}{8}$, which gives us $\frac{21}{8}$. We can check by multiplying. 3 times 7 eighths is 21 eighths. We write $\frac{21}{8}$.*

- Turn to **Pupil book 5 page 104**. Let the children work in pairs or small groups to read the problem and work through the solutions. Once they have done this, ask them to show $5 \times \frac{2}{3}$ using circle diagrams.
- The children can work on their own to complete question 1. If they are not yet confident with a written method they can use the visual method they prefer.
- The children can work out the missing values in question 2 using inverse operations. However, if they have not learnt how to divide by fractions they will not be able to do this. Encourage them to use a 'guess, check and improve' strategy. The answers are in simplest form, so they need to work back from those using the numbers they are given. The answers given below also show the answer before simplifying for reference.
- <u>Problem solving</u>: When the children have completed question 3, they can swap their word problems with a partner to check that they make sense and can be worked out.

Interesting mistakes

The children may need support with finding the missing numbers in question 2. If so, have a number talk with the class or with smaller groups to think about the different types of questions. For example:

- $2 \times \dfrac{\square}{7} = \dfrac{6}{7}$. The denominator in the answer is the same as the one in the fraction, so the missing number is worked out by thinking: *2 × what gives me 6?*

- $3 \times \dfrac{9}{\square} = \dfrac{9}{10}$ The numerators are 3 and 9. $3 \times 9 = 27$. You can find the denominator by thinking: *I know that* $\dfrac{9}{10} = \dfrac{27}{\square}$. *9 × 3 = 27, so the denominator must be 10 × 3.*

Answers for Pupil book 5 page 104

1

a $\frac{6}{5}$ or $1\frac{1}{5}$ b $\frac{2}{7}$ c $\frac{9}{5}$ or $1\frac{4}{5}$

d $\frac{21}{10}$ or $2\frac{1}{10}$ e $\frac{20}{7}$ or $2\frac{6}{7}$ f 9

g $\frac{20}{3}$ or $6\frac{2}{3}$ h $\frac{16}{5}$ or $3\frac{1}{5}$ i $\frac{33}{7}$ or $4\frac{5}{7}$

j $\frac{25}{2}$ or $12\frac{1}{2}$ k $\frac{14}{4}$ or $3\frac{1}{2}$ l $\frac{35}{8}$ or $4\frac{3}{8}$

2

a $2 \times \frac{3}{7} = \frac{6}{7}$ b $3 \times \frac{9}{30} = \frac{27}{30} = \frac{9}{10}$

c $5 \times \frac{5}{45} = \frac{25}{45} = \frac{5}{9}$ d $2 \times \frac{2}{14} = \frac{4}{14} = \frac{2}{7}$

e $2 \times \frac{5}{16} = \frac{10}{16} = \frac{5}{8}$ f $3 \times \frac{2}{9} = \frac{6}{9} = \frac{2}{3}$

g $3 \times \frac{1}{12} = \frac{3}{12} = \frac{1}{4}$ h $5 \times \frac{3}{20} = \frac{15}{20} = \frac{3}{4}$

i $3 \times \frac{7}{30} = \frac{21}{30} = \frac{7}{10}$

3 Individual answers.

Multiply with mixed numbers

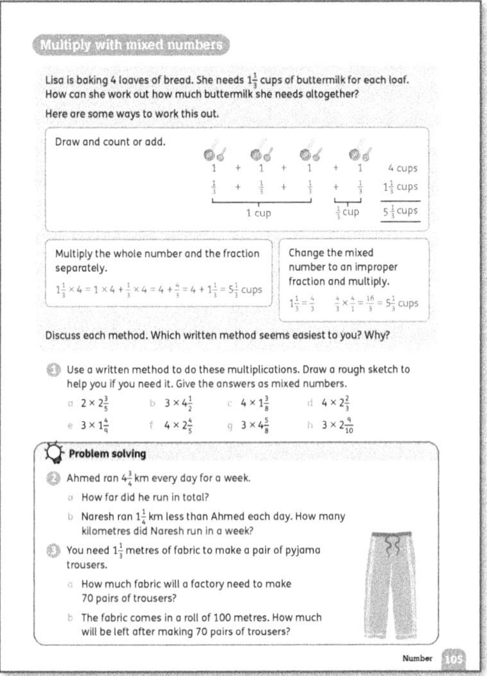

Materials

Large copy of the ingredients for a simple recipe with mixed-number measurements (You could copy and enlarge this example.)

> **Choco-chip cookies**
>
> For 12 cookies, you will need:
>
> - $2\frac{1}{3}$ cups sugar
> - $1\frac{5}{8}$ cups flour
> - 1 cup butter
> - $1\frac{1}{2}$ teaspoons of baking powder
> - $\frac{3}{4}$ teaspoon of salt
> - $2\frac{1}{4}$ cups of chocolate chips
> - 1 egg

Warm-up

Use any activity that involves finding/recalling multiplication and division facts as a mental starter for this lesson.

Focus

- Display the list of ingredients for choco-chip cookies (or any other recipe) and ask the children how much of each ingredient they need to make three times this number of cookies.
- Give them time to discuss this and to list the new quantities. Then, ask different children to explain how they worked out the new amounts. Ask whether anyone worked differently and let them show or explain their method.
- Remind the class that they already know how to multiply fractions and that they can apply what they know to work with mixed numbers.
- Turn to **Pupil book 5 page 105** and work through the examples with the class. Make sure the children realise that multiplying the whole number part and fraction

part separately will give the same result as converting the mixed number to an improper fraction and then multiplying it by the whole number. Encourage the children to try all three methods before they explain which they prefer. The method they choose may depend on the numbers in the calculation.

- The children can then work independently to complete question 1, either using the method they prefer throughout or mixing methods to suit the numbers in the questions.
- Problem solving: Give the children similar instructions for answering questions 2 and 3.

Answers for Pupil book 5 page 105

1 a $5\frac{1}{5}$ b $13\frac{1}{2}$ c $5\frac{1}{2}$

d $10\frac{2}{3}$ e $4\frac{1}{3}$ f $11\frac{1}{5}$

g $13\frac{7}{8}$ h $8\frac{7}{10}$

2 a $33\frac{1}{4}$ km   b $24\frac{1}{2}$ km

3 a $93\frac{1}{3}$ m   b $6\frac{2}{3}$ m

Fractions of an amount

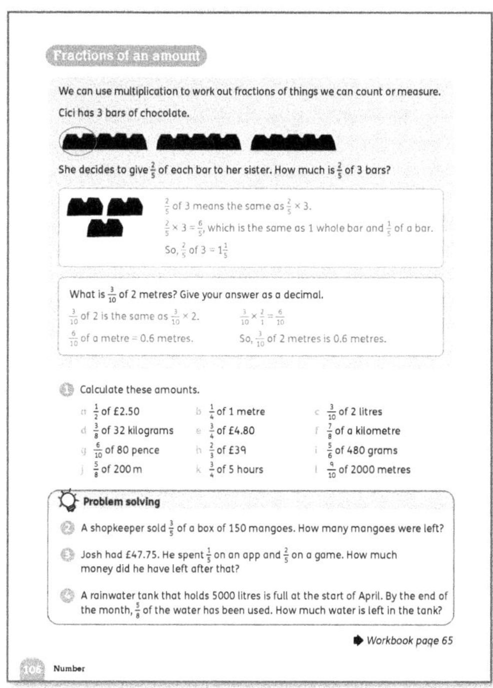

Warm-up

Practise writing one number as a fraction of another. Choose 5 children to stand up. Ask 3 of them to stand on one leg. What fraction are standing on one leg? ($\frac{3}{5}$) Repeat with different sized groups and activities.

Focus

- Use **Workbook 5 page 65** to check that the children can express the number of each set of shapes as a fraction of the whole set.
- Check the Workbook answers, then turn to **Pupil book 5 page 106** and work through the example to show the children what it means to find a fraction of an amount using real objects (chocolate bars) and measurements.

- Let the children work on their own to calculate the amounts in question 1. Check their answers and make sure that they include units. The children could use bar models to visualise these questions.
- Problem solving: The children can then work in pairs to discuss the problems in questions 2–4 before solving them on their own.

Challenge

Let the children explore how to use a calculator to find a fraction of an amount. If they realise that the fraction (for example, $\frac{2}{5}$) can be expressed as 2 divided by 5, they will be able to find methods of working this out. Let them share what they find out with the class.

Answers for Pupil book 5 page 106

1 a £1.25   b 25 cm   c 600 ml

d 12 kg   e £3.60   f 875 m

g 48p   h £26   i 400 g

j 125 m   k 3 hours and 45 minutes

l 1800 m

2 60   **3** £19.10   **4** 1875 litres

Answers for Workbook 5 page 65

1 square, triangle, circle, hexagon, triangle, square, circle, circle, hexagon, square, triangle, circle

2 a $\frac{3}{12} = \frac{1}{4}$ b $\frac{1}{12}$ c $\frac{2}{12} = \frac{1}{6}$

d $\frac{4}{12} = \frac{1}{3}$ e $\frac{2}{12} = \frac{1}{6}$ f $\frac{2}{12} = \frac{1}{6}$

g $\frac{8}{12} = \frac{2}{3}$ h $\frac{3}{12} = \frac{1}{4}$ i $\frac{1}{12}$

j $\frac{9}{12} = \frac{3}{4}$ k $\frac{2}{12} = \frac{1}{6}$ l $\frac{5}{12}$

m $\frac{7}{12}$ n $\frac{3}{12} = \frac{1}{4}$ o $\frac{5}{12}$

p $\frac{1}{12}$

More work with fractions

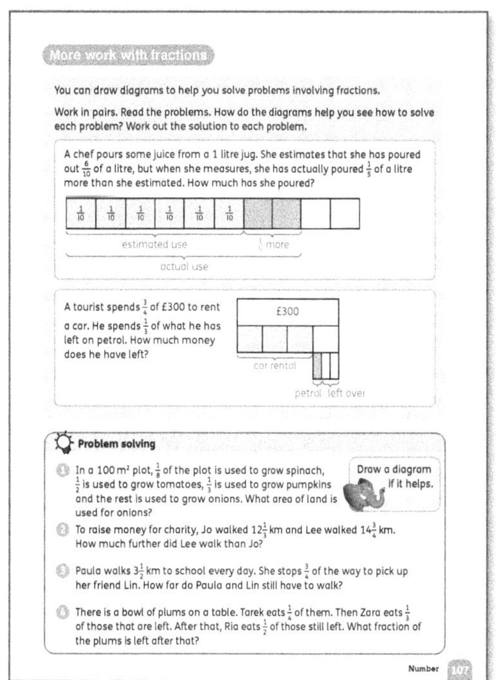

Materials

Rulers; squared paper; marker pens

Warm-up

Choose any 'Mental problem solving' activity from the Activity bank on pages 24–25 as a starter for this lesson.

Focus

The focus of this lesson is to extend the children's use of bar models and other diagrams to include fractions.

- Place the children in pairs to work through and discuss the examples on **Pupil book 5 page 107**.
(First example: $\frac{6}{10} + \frac{1}{5} = \frac{6}{10} + \frac{2}{10} = \frac{8}{10} = \frac{4}{5}$, so the chef has poured $\frac{4}{5}$ litre.
Second example: $1 - \frac{3}{4} = \frac{1}{4}$;
$\frac{1}{4}$ or £300 $= \frac{1}{4} \times \frac{300}{1} = \frac{300}{4} = £75$;
$\frac{1}{3}$ of £75 $= \frac{1}{3} \times \frac{75}{1} = \frac{75}{3} = £25$, so the tourist has £25 left.)
- Remind them that a two-step or multi-step problem may need two or more bars to show the different steps. The answers to the problems are $\frac{8}{10}$ or $\frac{4}{5}$, and £150.
- Problem solving: If necessary, talk through the problems in question 1 and question 2, asking the children to show the diagram(s) they would use to help them visualise/show each problem situation.
- Question 3 and question 4 involve fractions of fractions. However, the children can use diagrams to solve these without knowing formal methods for multiplying fractions by fractions.

Challenge

Let the children make sets of dominoes in which fractions of amounts are matched with the resulting values. Provide a template for the children to use. The children can work on their own to create their set of dominoes, and then cut them up. They can exchange their dominoes with a partner and play each other's games.

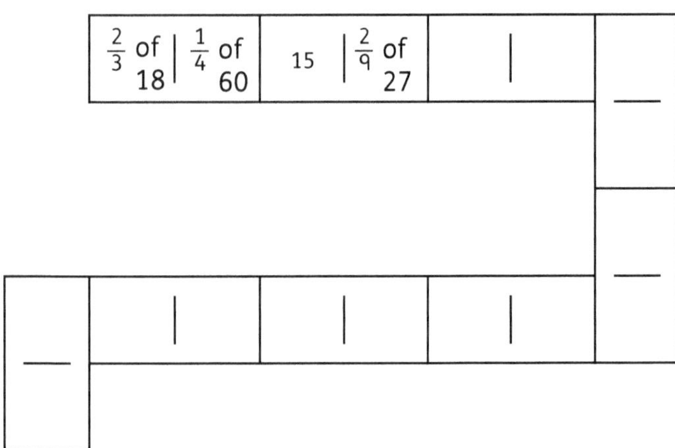

Answers for Pupil book 5 page 107

1 $4\frac{1}{6}$ m

2 $2\frac{5}{12}$ km

3 $\frac{7}{8}$ km

4 $\frac{1}{4}$

Ask some or all of these questions to assess how well the children have understood the concepts in this unit:

- *What did you find easiest in this unit? What made it easy?*
- *What did you find most challenging? Why was it challenging?*
- *What are three things that are important to remember when you work with fractions? Why are these things important?* (For example: To find equivalent fractions, multiply or divide the numerator and denominator by the same number. To add or subtract fractions, the denominators must be the same. In an improper fraction, the numerator is larger than the denominator, and the fraction can be converted to a mixed number.)
- *Give me a fraction that is equivalent to $\frac{3}{5}$ but has a denominator of 10. How did you work it out?* ($\frac{6}{10}$ Multiply both the numerator and denominator by 2.)
- *Give me a fraction that is equivalent to $\frac{2}{3}$.* (For example, $\frac{4}{6}, \frac{6}{9}, \frac{20}{30}$)
- *How would you order these fractions from smallest to greatest? $\frac{3}{5}, \frac{3}{4}, \frac{5}{8}, \frac{7}{10}$* (Write as equivalent fractions with the same denominator. $\frac{3}{5}, \frac{5}{8}, \frac{7}{10}, \frac{3}{4}$)
- *Into how many sixths can you cut 8 cakes?* (48)
- *What fraction of the children in our class wear glasses/do not wear glasses/have long hair/do not have long hair?* (Ask other questions like this.)
- *What fraction of this group of shapes are triangles?* (Show a group of shapes.)
- *How would you change $1\frac{1}{5}$ to an improper fraction?* ($1 = \frac{5}{5}, \frac{5}{5} + \frac{1}{5} = \frac{6}{5}$)
- *Look at the number $2\frac{7}{5}$ What is wrong with this number? Why?* (The numerator in the fraction part is greater than the denominator and there is also a whole number part. The number should be either $3\frac{2}{5}$ or $\frac{17}{5}$)
- *What is $\frac{2}{7} + \frac{3}{7}$?* ($\frac{5}{7}$) (and similar)
- *If I subtract $\frac{3}{8}$ from $\frac{7}{8}$, what am I left with?* ($\frac{4}{8} = \frac{1}{2}$) (and similar)
- *What is $\frac{4}{5} + \frac{1}{3}$?* ($1\frac{2}{15}$) (and similar)
- *How would you subtract $\frac{3}{8}$ from $\frac{3}{4}$?* (Write both fractions with denominator 8 and subtract the numerators: $\frac{6}{8} - \frac{3}{8} = \frac{3}{8}$ (and similar)
- *What is $2\frac{3}{4} - 1\frac{7}{8}$?* ($\frac{7}{8}$) (and similar)
- *What is $3 \times \frac{2}{9}$?* ($\frac{6}{9} = \frac{2}{3}$) (and other multiplications of fractions by a whole number)
- *How can you draw a diagram to show four lots of $\frac{2}{7}$?* (For example, four 7-part bars, each with 2 parts shaded.)
- *What is $3 \times 1\frac{1}{2}$?* ($4\frac{1}{2}$) (and other multiplications of mixed numbers by a whole number)
- *What is $\frac{1}{3}$ of 12?* (4) (and other fractions of amounts)
- *Make up a two- or three-step problem using fractions. Show me how you would solve it.*

Position, direction and movement

Learning objectives

- Compare the relative position of coordinates, with or without a grid
- Use knowledge of 2D shapes and coordinates to plot points to form lines and shapes in the first quadrant
- Identify, describe and represent the position of a shape following a reflection or translation, using the appropriate language, and know that the shape has not changed
- Reflect 2D shapes in both horizontal and vertical mirror lines to create patterns on square grids
- Use knowledge of reflective symmetry to identify and complete symmetrical patterns

Key words

coordinates   reflection   mirror line   translation
symmetrical   coordinate grid

Unit introduction

Materials

'Gridlock' game board (shown below); spinners labelled 1 to 6 or sets of 1–6 digit cards; coloured marker pens

Teaching guidance

- Play a game of 'gridlock' to revise working with coordinates in the first quadrant.
- Divide the class into pairs (these can change after the first game). Each pair will need two spinners or two sets of 0–6 digit cards, two different-coloured marker pens and a copy of the game board.
- The game board shown here allows the children to play two games, but you may want to add more grids.

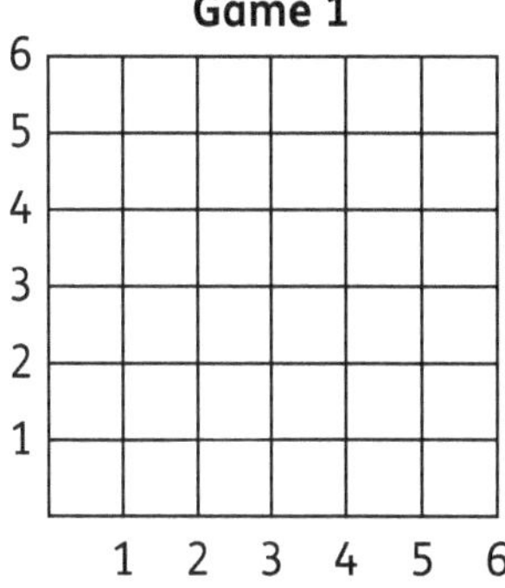

Game 1

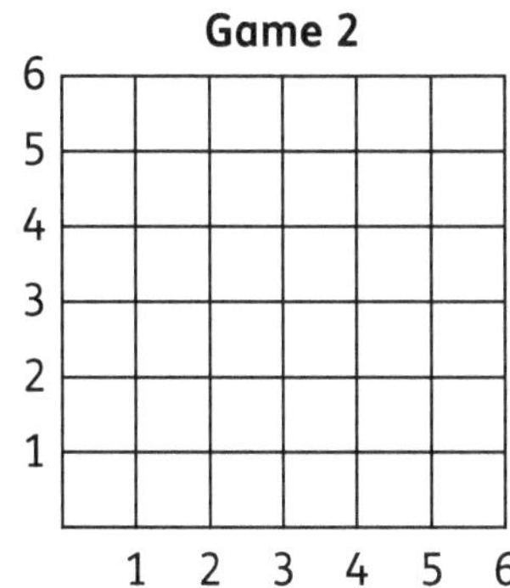

Game 2

Explain the rules to the class.

- Players take turns to spin both spinners, or pick a digit card from each set without looking. They use the numbers (in any order) to form a coordinate pair. So, if Player 1 gets 2 and 3, they make (2, 3) or (3, 2).
- Player 1 then marks a dot at that position on the grid.
- Player 2 spins the spinners or picks digit cards and makes a coordinate pair with the two numbers they get.
- If there is already a dot at both possible positions, the player cannot use their turn.
- The winner is the first player to get three dots in a row (horizontal, vertical or diagonal).
- If the game ends and neither player has three in a row, it is a tie.

Give the children time to play one or two games. Have a class discussion about the strategies they used to try to get three in a row while trying to prevent their partner from getting three in a row.

Coordinates on a grid

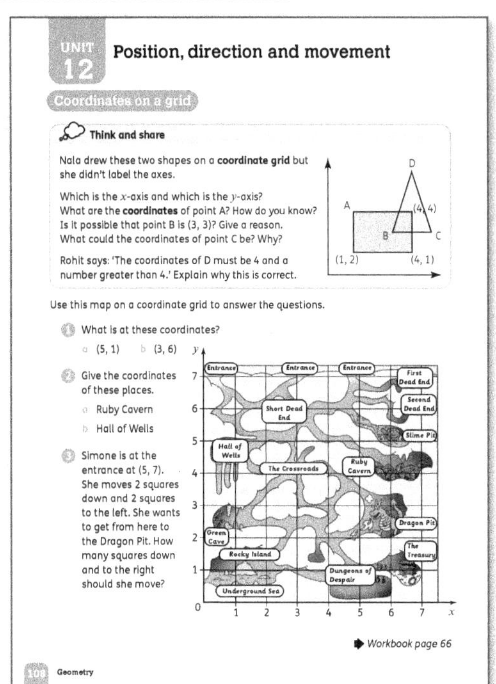

Materials

Small counters; rulers

Warm-up

Select any suitable activity from the 'Place value and number sense' section of the Activity bank (pages 22–23) as a starter for this lesson.

Focus

- <u>Think and share:</u> Use the activity at the top of **Pupil book 5 page 108** to revise coordinates and to assess how well the children understand and can use them. Discuss the answers to the questions. (The x-axis is the horizontal line and the y-axis is the vertical line. Point A has coordinates ((1, 4). Yes, B could be (3, 3): the x-coordinate is between 1 and 4 and is closer to 4 than to 1, and the y-coordinate is half way between 2 and 4. C could have coordinates

(5, 3) because the x-coordinate is one bigger than 4 and the y-coordinate is the same as for B. D has the same x-coordinate as the bottom right-hand corner of the rectangle, which is 4, and a greater y-coordinate than the top right-hand corner.)

- For question 1, encourage the children to use a small counter or other small object to show a position on the grid if they need to.
- The places given in question 2 cover more than one point on the *coordinate grid*. Make sure that the children realise at they need to find the intersection of two grid lines at each place. Allow the children to check each other's work.
- For question 3, encourage the children to use a small counter to physically move around the grid if they need to.

Follow-up

Use **Workbook 5 page 66** to check that the children can plot points given the coordinates and accurately write the coordinates of points on a grid.

Challenge

Let the children investigate how people in different countries who speak different languages are able to play chess against each other by stating their moves using different systems. They can report back on what they find out.

Support

Demonstrate to the children how to draw a set of axes.

- If possible, use a projection of grid paper or stick a piece of squared paper onto the board.
- Emphasise that the vertical axis (the line going up and down) is called the *y-axis*, and that the horizontal axis is called the *x-axis*.
- Show that the numbers along the axes increase from left to right and bottom to top. 0 is always at the intersection of the x- and y-axes.
- Invite different children to plot *coordinates* on your drawn graph so that you can see what they are doing and correct any mistakes.

Interesting mistakes

- The children may mix up the two axes and/or give coordinate pairs in the wrong order. Keep reminding them that the x-axis is the horizontal line and that the y-axis is the vertical line, and that the coordinates are given in the order (x, y). The first number represents the point along the x-axis, and the second represents the point along the y-axis.
- The children may also identify the coordinate on the x- or y-axis and forget about the second coordinate. If necessary, they can use two rulers (or any two straight edges, such as books or sheets of paper) to form a corner so that they can easily find the coordinates.

Answers for Pupil book 5 page 108

1 a Dungeon   b Short Dead End
2 a (5, 4)   b (1, 5)
3 4 squares to the right and 3 squares down

Answers for Workbook 5 page 66

1 a 170-180 km^2.

b

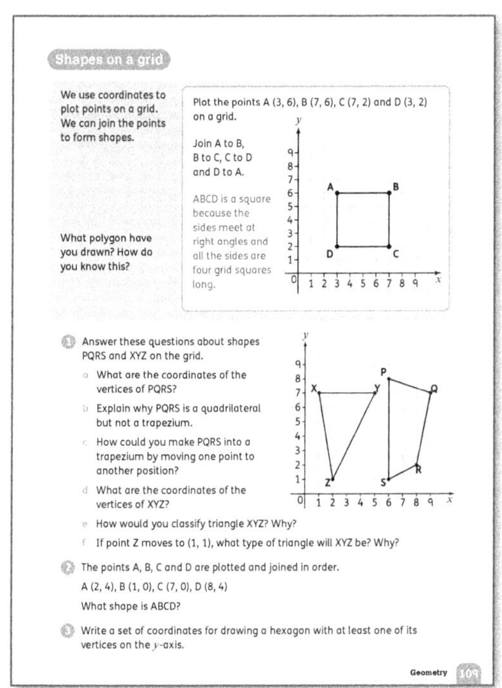

c Individual answers.

Shapes on a grid

Materials

Squared paper or prepared coordinate grids

Warm-up

Revise properties of shapes as the warm-up for this lesson. For example, you could name a polygon and ask the children to state the properties of its sides and/or angles. You could also give a set of properties and ask the children to name the polygon you are describing.

Focus

- Turn to **Pupil book 5 page 109**. Work through the example with the class. (The polygon is a square. It has four sides which are equal in length and four angles of 90°.)
- Ask them why it is important to plot the points in the given order. Take their responses and, if necessary, show that joining A to C, then C to B and B to D would give a different result.

- You can do question 1 orally with the class or you can let the children discuss it and answer the questions in pairs.
- Give the children squared paper or a coordinate grid to work on for question 2. It is likely that they will need to plot the coordinates to find the shape.
- Again, the children can do question 3 on a coordinate grid. The children will give different answers and they should check each other's work by plotting the points to see what shape results. Note that the hexagon does not have to be regular.

Answers for Pupil book 5 page 109
1 a P(6, 8), Q(9,7), R(8,2), S(6,1)
  b It has four sides but none of them are parallel.
  c Move Q to (8, 7) (other answers are possible)
  d X(1, 7)), Y(5, 7), Z(2,1)
  e scalene, as no sides are the same length
  f right-angled as it would have a 90° angle
2 a parallelogram **3** Individual answers.

Reflect shapes on a grid

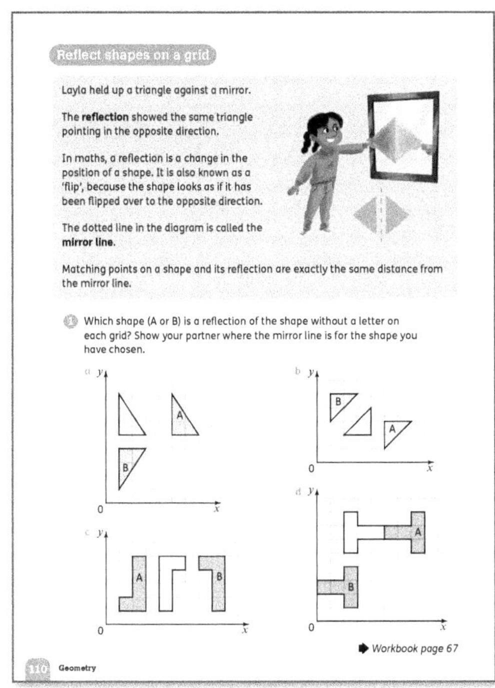

Materials
Small mirrors (such as mirror tiles); squared paper or 10 × 10 coordinate grids

Warm-up
Select suitable activities from the 'Place value and number sense' section of the Activity bank (pages 22–23).

Focus
- Write the word *transform* on the board. Explain that when we move a shape to a new position on a grid, we call this change a *transformation*. (Changing the size of a shape (enlargement) is also a type of transformation; the children will not be dealing with enlargements of shapes at primary school.)

- Demonstrate a *reflection* on a grid by drawing a shape (for example, a scalene triangle) and a *mirror line*, and then drawing the shape's reflection on the other side of the mirror line.
- Ask the children to draw a shape anywhere on their squared paper and then to draw a mirror line. Let them use a mirror to work out where the reflection of the shape will be and then draw it.
- Let the children experiment with different mirror lines – horizontal, vertical and diagonal – including some that are touching the shape and some that are at a distance from the shape. Have a class discussion about the shapes and their reflections. Try to elicit these facts: the reflection is the same size as the original shape but it is flipped to face in the opposite direction; the corresponding vertices of the shape and its reflection are the same distance from the mirror line. The children can use a ruler to measure and check these facts.
- Use **Workbook 5 page 67** to check that the children can draw the reflection of a shape on squared paper.
- Turn to **Pupil book 5 page 110**. Read through the information with the class.
- Let the children work on their own to identify the reflections in question 1. They can use a mirror if they need to. Let them place a ruler on the grid to show their partner where the mirror lines are.

Interesting mistakes
When reflecting shapes, some children do not understand that the *image* (the reflected shape) must be the same distance away from the mirror line as the original shape. They might simply copy the shape on the other side of the line without thinking about the orientation of the image. Using a mirror to check their work can help overcome this error.

Answers for Pupil book 5 page 110
1 a B **b** B **c** B **d** A

Answers for Workbook 5 page 67
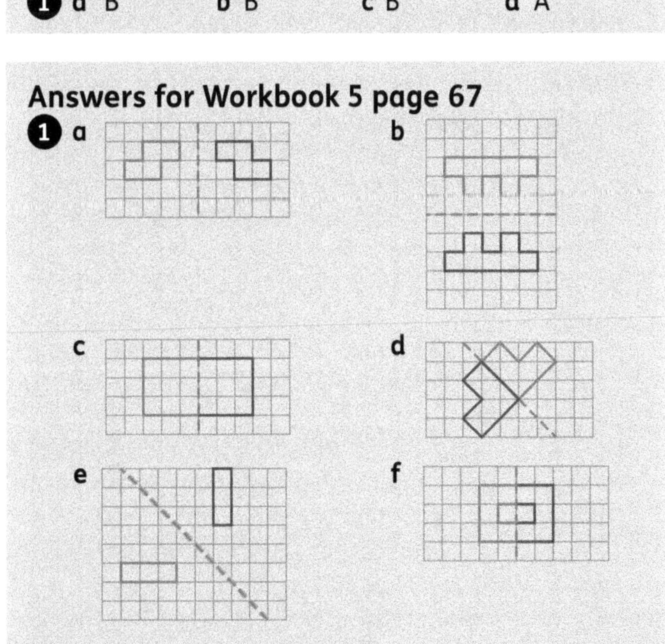
1 a ... b ... c ... d ... e ... f

2 Individual answers.

Translate shapes on a grid

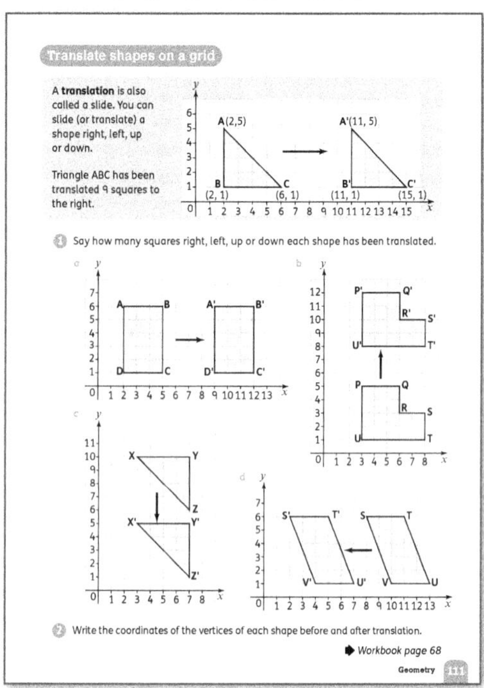

Materials

Squared paper; 10 × 10 coordinate grids

Warm-up

- Display a large grid. Draw a shape on the grid and then perform a translation on it and draw the new shape. Ask the children to describe the transformation as moves left/right and up/down.
- Draw another shape on the grid.
- Ask the children to take turns to describe different ways of moving the original shape until it lies on top of the second shape. Ask: *What is the fewest number of moves you can make?*

Focus

- Introduce the term *translation* and define it as a movement up or down and/or left or right. Translations are also called *slides* because the shape moves without changing its orientation in any way.
- Use **Workbook 5 page 68** to give the children an opportunity to translate shapes. Spend some time discussing how the shapes move. Make sure the children understand that the shapes remain the same size and in the same orientation, but move to a different position. Also point out that a shape may seem to move diagonally, but that this is the result of a movement up/down and then left/right, not a diagonal slide.
- Use the example at the top of **Pupil book 5 page 111** to demonstrate that the vertices of the shape move to a new set of coordinates on the grid.
- The symbol ' is used to indicate the vertices of the shape after a translation so that we can tell the difference between the original position and the position after translation.
- The children can do question 1 orally in groups.

- For question 2, remind the children to use the notation ' for the translated vertices (for example: A', B', and so on).

Challenge

- Ask pairs of children to sit opposite each other with a screen, such as an open file, between them. Give each child a 10 × 10 coordinate grid.
- Each child draws a shape on their grid and translates it to a new position.
- The children then take turns to give their partner the coordinates of the original shape and to describe the translation to move it to its new position.
- Once they have both followed each other's instructions, they can reveal their work to check that they have both drawn the same original shape and translated shape.

Answers for Pupil book 5 page 111

1. **a** 7 squares right **b** 7 squares up
   c 5 squares down **d** 6 squares left
2. **a** Before: A(2, 6), B(5, 6), C(5, 1), D(2, 1)
   After: A'(9, 6), Bv(12, 6), C'(12, 1), D'(9, 1)
   b Before: P(3, 5), Q(6, 5), R(6, 3), S(8, 3), T(8, 1), U(3, 1)
   After: P'(3, 12), Q'(6, 12), R'(6, 10), S'(8, 10), T'(8, 8), U'(3, 8)
   c Before: X(3, 10), Y(7, 10), Z(7, 6)
   After: X'(3, 5), Y'(7, 5), Z'(7, 1)
   d Before: S(8, 6), T(11, 6), U(13, 1), V(10, 1)
   After: S'(2, 6), T'(5, 6), U'(7, 1), V'(4, 1)

Answers for Workbook 5 page 68

1. **a** Provided as an example.

b 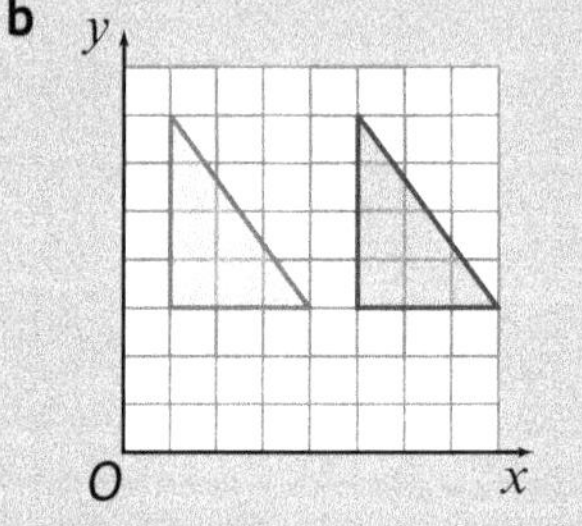 **c** 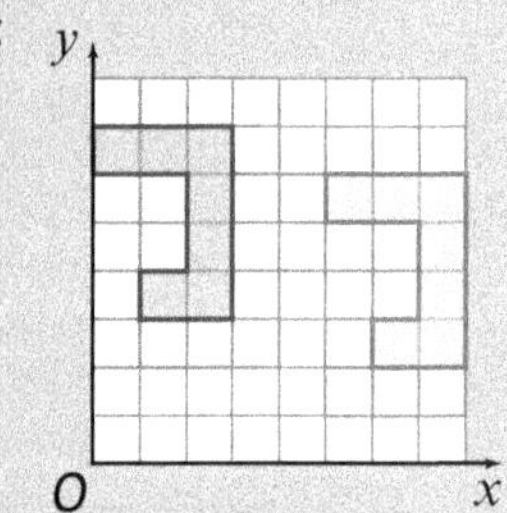

d 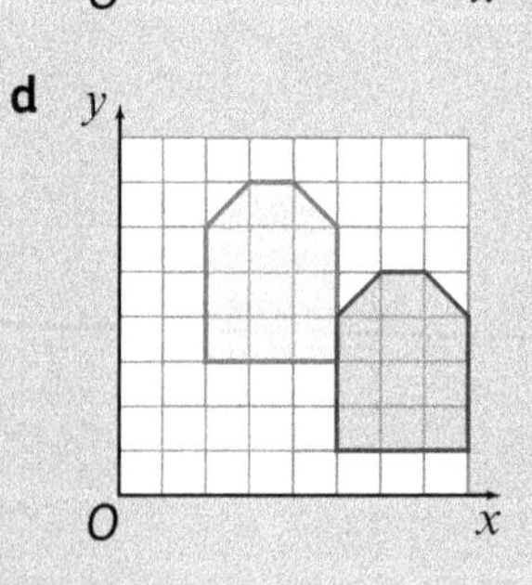 **e** 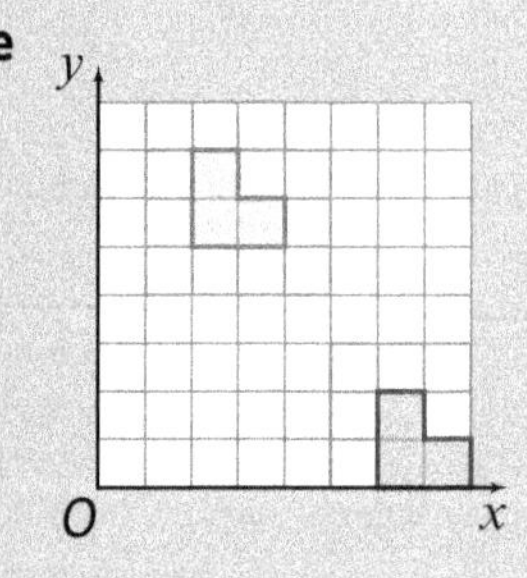

f 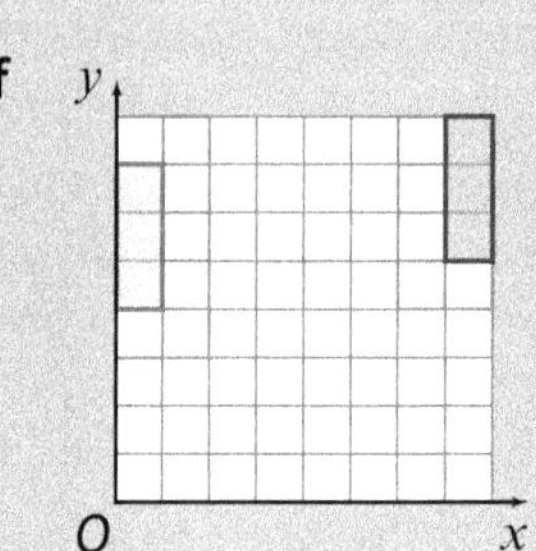

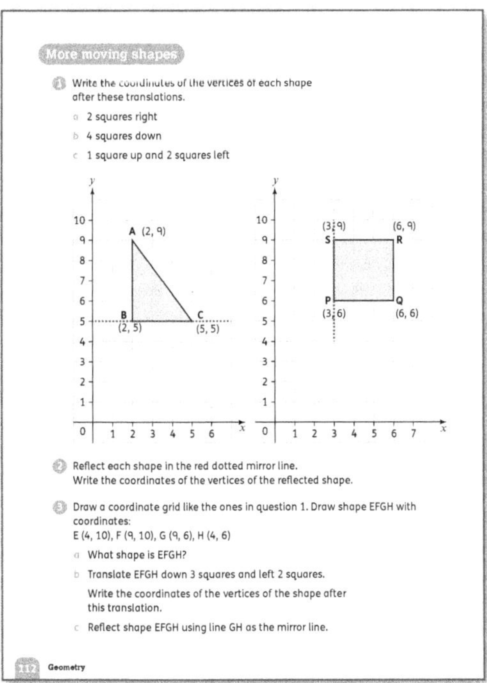

Materials

Squared paper or 10 × 10 coordinate grids; tracing paper

Warm-up

Choose any 'Mental problem solving' from the Activity bank (pages 24–25) that involves adding and subtracting small numbers as a warm-up for this lesson.

Focus

This lesson involves translations and reflections.

- Give each child squared paper or a coordinate grid, then ask the children to work on their own to complete questions 1–3 on **Pupil book 5 page 112**.
- The children can then check each other's diagrams.
- Spend some time talking about what the children found easy and what they found challenging in the lesson.

Challenge

- Let the children investigate how coordinates change with different translations and try to find patterns that will allow them to work out the coordinates without moving or drawing the shapes.
- For example, when a shape moves 1 square to the right, all the x-coordinates increase by 1 but the y-coordinates stay the same.

Support

Some children might need support to visualise the movements. Encourage them to trace the shape and move it physically on the grid rather than redrawing the shape and its image on a grid; this will save time and effort.

Answers for Pupil book 5 page 112

1   a   A(4, 9), B(4, 5), C(7, 5); P(5, 6), Q(8, 6), R(8, 9), S(5, 9)

    b   A(2, 5), B(2, 1), C(5, 1); P(3, 2), Q(6, 2), R(6, 5), S(3, 5)

    c   A(0, 10), B(0, 6), C(3, 6); P(1, 7), Q(4, 7), R(4, 10), S(1, 10)

2   A(2, 1), B(2, 5), C(5, 5); P(3, 6), Q(0, 6), R(0, 9), S(3, 9)

3   a   rectangle

    b   E'(2, 7), F'(7, 7), G'(2, 3), H'(7, 3)

    c

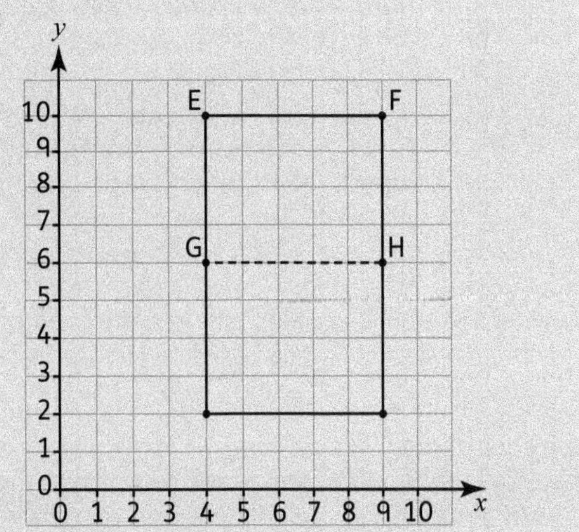

Reflections and symmetrical patterns

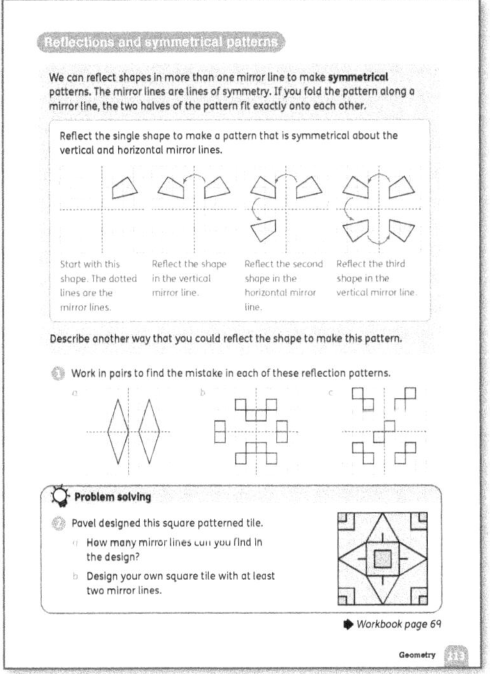

Materials

Large paper shape with more than one line of symmetry, such as a circle or a rectangle; examples and pictures of symmetrical patterns from real life (for example, fabrics, tiles, paving, buildings and so on); small mirrors; squared paper; small cut-out shapes or shape stencils; rulers; coloured marker pens; colouring pencils; squared paper

Warm-up

Choose any 'Mental problem solving' activity in the Activity bank (pages 24–25) as a starter for this lesson.

Focus

Briefly revise the concept of *symmetry* with the class.

- Show the children a large cut-out 2D shape that has more than one line of symmetry. Remind the children that if you fold the shape in half, the fold line is called the *line of symmetry*.
- Position a mirror on the fold line to show how the other half of the shape is reflected in the mirror.
- Fold the shape in a different way to show another line of symmetry. Explain that some shapes have more than one line of symmetry.
- Ask the children to look at shapes around the room, for example, doors, windows, floor tiles, brick patterns, ventilation grids and so on and identify which are symmetrical and where the lines of symmetry are. Use a mirror to demonstrate the symmetry if the children need support to see it.
- Show the class examples of symmetrical patterns from real life. Let them suggest where the mirror lines are in each pattern.

Give each child a sheet of squared paper and a cut out shape or stencil. Work through the example at the top of **Pupil book 5 page 113** with the class. Ask them to draw the lines of symmetry on their own shapes as you explain the steps. (Another way of making this pattern would be to reflect the original shape in the x-axis and then both shapes could be reflected in the y-axis.)

- The children can do question 1 orally in pairs before sharing their answers with the class.
- <u>Problem solving</u>: For the second part of question 2, the children will need colouring pencils and squared paper to design their own tile patterns. Make sure they realise that they should not show the mirror lines in their pattern. Let them check and comment on each other's work.

Follow-up

Use **Workbook 5 page 69** to informally assess how well the children can apply what they have learnt about reflections and symmetry. The children should complete the work on their own. Either check their work or let them check each other's work.

Support

Let the children do a small project to find examples of symmetrical patterns in the local environment. They could record these photographically and prepare a slide show. Alternatively, they could draw the symmetrical patterns they find and make a poster to show the designs.

Interesting mistakes

The children may have difficulty visualising the reflections in symmetrical patterns with diagonal lines of symmetry. Allow them to physically move shapes to reflect them. Also encourage them to use a mirror and to fold the patterns along the mirror lines to check that the shapes fit onto each other.

Answers for Pupil book 5 page 113

1 These are the correct reflection patterns.

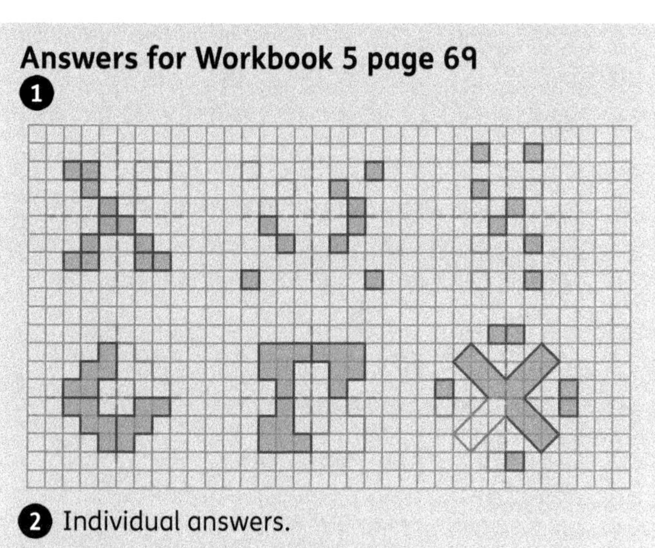

2 a The tile has 4 lines of symmetry.
b Individual answers.

Answers for Workbook 5 page 69

1

2 Individual answers.

End-of-unit check

Ask some or all of these questions to assess how well the children have understood the concepts in this unit.

- *What do we call the horizontal line in a pair of axes?* (x-axis)
- *What do we call the vertical line in a pair of axes?* (y-axis)
- *At which number do the axes intersect?* (Clarify that 'intersect' means 'cross' if necessary.) (0)
- *Show me where point (4, 5) is on your grid.* (and similar)
- *What is at (2, 3)?* (and similar)
- *Look at the two positions of this shape* (Show a grid with a shape in two positions.). *Can you describe the movement of the shape from position 1 to position 2?*
- *Which two translations are used to move this shape from A to B?* (Show a grid with a translated shape.)
- *Where does this shape end up when you move it following these directions?* (Show a shape on a grid and give some directions.)

- *What is the fewest possible number of movements to move this shape from A to B?* (Show a grid with a translated shape.)
- *Is this pattern symmetrical?* (Show a pattern.) *Why or why not?*
- *What does 'symmetry' mean?* (If you fold a pattern along a mirror line, the two halves of the pattern fit exactly onto each other.)
- *What is a 'line of symmetry'?* (the mirror line, such that the pattern is reflected in the line)
- *Where is the line of symmetry on this pattern?* (Show a pattern with at least one line of symmetry.) *Is there only one line of symmetry?*

Mixed practice 2

Answers to Mixed practice 2 Pupil book 5 pages 113–114

1 a 64 b 64 c 121
2 c
3 $3^3 = 27$ or $3^2 = 9$
4 Possible answers (other fractions are possible):
  a $\frac{3}{4}$ 75% = 0.75 = $\frac{6}{8}$ b 20% = 0.2 = $\frac{2}{10} = \frac{1}{5}$
  c 0.25 = 25% = $\frac{1}{4} = \frac{25}{100}$ d $\frac{4}{10} = \frac{2}{5} = 0.4 = 40\%$
5 a 45 minutes b £3 c 25 cm

6 a Zagreb 20 miles; Split 40 miles; Trieste 45 miles; Ljubljana 30 miles; Dubrovnik 50 miles
  b 16 km
  c Possible answer: The graph shows that 50 km is 31.25 miles. If you multiply this by two, then 100 km is 62.5 miles.
7 A: 8 pupils; B: 7 pupils; total = 15 pupils
8 a $2540 b $1500
9 Individual answers.
10 Individual answers.
11 a (2, 1)
  b A
  c Square; all the sides are 4 units long and all the angles are right angles.
  d The line AD goes vertically upwards so there is no change in x-coordinate.
  e isosceles
  f i perimeter = 80 mm
    ii area = 400 mm²
  g a parallelogram.
12 Marc wants to double the area so he could *either* double the length to 80 mm *or* double the width to 60 mm (but not both!)

UNIT 13 Multiplication and division 2

Learning objectives

- Use known facts and laws of arithmetic to simplify multiplication and division calculations
- Estimate and multiply numbers with up to 4 digits by 1-digit or 2-digit whole numbers, including using long multiplication for the 2-digit multipliers
- Estimate and divide numbers with up to 4 digits by a 1-digit number using the formal method of short division
- Solve problems involving addition, subtraction, multiplication and division and a combination of these, including interpreting remainders appropriately for the context
- Solve problems involving multiplication and division, including problems involving simple rates

Key words

place value   inverse   short division   remainder   rate

Unit introduction

Teaching guidance

- Display a grid like this one for the class to consider.

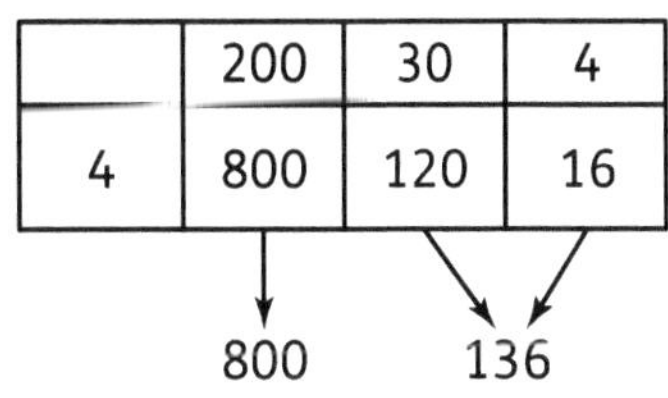

	200	30	4
4	800	120	16

800   136

- Ask: *What do you notice? What do you wonder?*
- Give the children a few minutes to write down things they notice and things they wonder. Then ask different children to share things they noticed.
- As the children are sharing, record their observations for the class to see. Allow as many children as possible to contribute. Record all the children's suggestions. Avoid praising, restating, clarifying or asking questions.
- Next, ask the children to share what they wondered. Encourage them to use a question format, for

example, *Would this work for bigger numbers?* List all their questions on the board.

- Choose some questions for the children to investigate, either as a class or in small groups. You could add some more starter questions to get the children thinking and talking. For example:
  ○ *What calculation is this? What is the answer?*
  ○ *What is the biggest number that could go in each block? Why?*
  ○ *What words can we use to talk about this calculation?*
  ○ *How else could I model this calculation?*
  ○ *How would we do this calculation using a written method?*
- Discuss the questions and the answers that the children suggest.

Written multiplication

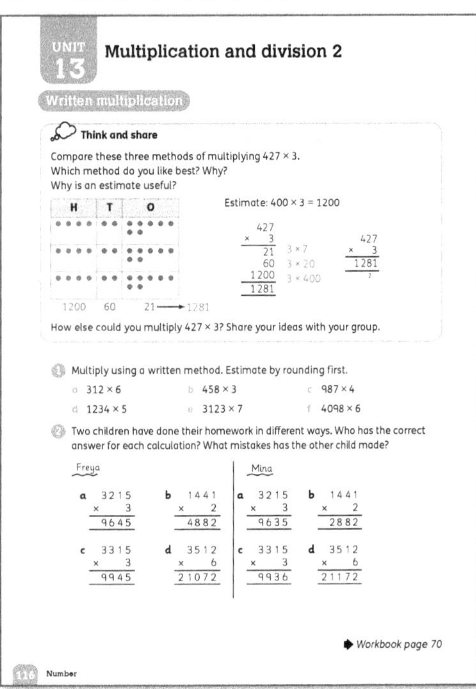

Materials

Place-value tables (page 21); place-value counters

Warm-up

Choose any of the multiplication activities from the 'Calculation skills' section of the Activity bank (pages 25–28) as a starter for this lesson.

Focus

- Think and share: Turn to **Pupil book 5 page 116**. Let the children work in groups to discuss the questions and the different representations of 427 × 3.
- Treat this as a continuation of the number talk from the 'Unit introduction' lesson. Once the children have discussed and answered the questions, take feedback from them.
- Observe the children as they work through question 1 to make sure they write down their estimates. Take every opportunity to help them get into the habit of estimating.
- The children can do question 2 orally in pairs.

Follow-up

Use **Workbook 5 page 70** to provide additional practice and to consolidate multiplication by a 1-digit number. Check the answers as a class. Discuss any interesting mistakes and their answers to question 3.

Challenge

- Write some criteria for creating word problems on the board. For example:
  Write a word problem that ...
  ○ *is solved by multiplying a 3-digit number by a 1-digit number*
  ○ *has a product less than 1200* (and so on).
- Ask the children to create a word problem that meets each criterion. They should share the problems they develop and make sure they meet the criteria before asking other children to find the solutions.

Support

Revise written multiplication methods with any children who need this.

- Write some 2-digit by 1-digit multiplications on the board. Next to each multiplication, write a choice of three answers, only one of which is correct.
- Ask the children to look at the answers and, without actually doing the calculation, say which answer is correct. Discuss how they decided (for example, by estimating, using a known fact or finding multiples of the multiplier).
- Continue to allow mental strategies and different approaches to multiplication, including informal methods and jottings. However, at the same time, introduce and model more compact and efficient written methods where possible. The children should not be forced to use these methods until they are comfortable and confident with them.

Answers for Pupil book 5 page 116

1 a 1872   b 1374   c 3948
   d 6170   e 21 861   f 24 588

2 a Freya is correct; Mina has forgotten to add '1' to the tens column from the '15' in the units column.

   b Mina is correct; In the thousands column Freya has incorrectly written 1 × 2 = 4.

   c Freya is correct; Mina has multiplied 3 × 5 and written 6 instead of writing 5 in the ones column and adding 1 to the tens column.

   d Freya is correct; Mina has incorrectly entered '1' in the hundreds column. This should be '0', since 5 × 6 = 30 and the '3' is carried over to the thousands column.

Answers for Workbook 5 page 70

1 a 956   b 2324   c 1395
   d 2814   e 1988   f 1791

2 a $1026   b $5593   c $83.25   d $71.92

3 Individual answers.

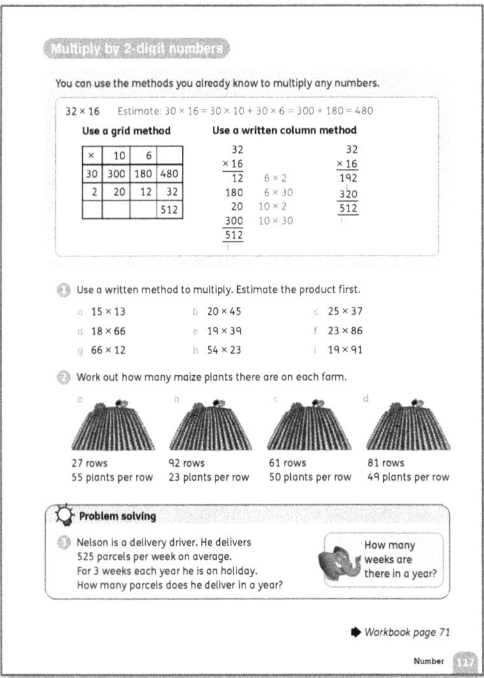

- When children estimate first, they can quickly see when the product is not reasonable. For example, in the second example, estimating would give 40 × 4 = 160 and, in the third example, estimating 30 × 10 would give 300.
- Modelling the calculations using place-value counters will help the children to see that this method is not mathematically correct. Writing the multiplication on a grid will also help to overcome this.

Answers for Pupil book 5 page 117

1. **a** 195   **b** 900   **c** 925
      d 1188   **e** 741   **f** 1978
      g 792   **h** 1242   **i** 1729
2. **a** 1485   **b** 2116   **c** 3050   **d** 3969
3. 25 725

Answers for Workbook 5 page 71

1. **a** 2576   **b** 5681   **c** 14 455
      d 69 608   **e** 32 071   **f** 92 313
2. 1601 × 15: the other calculations give the answer 24 024 but this has answer 24 015
   1088 × 10: the other calculations give the answer 11 880 but this has answer 10 880

Materials
0–9 digit cards; squared paper

Warm-up
Choose multiplication activities from the 'Calculation skills' section of the Activity bank (pages 25–28) as a starter for this lesson.

Focus
- Pick digit cards to generate two 2-digit numbers. Ask the children to find the product using whatever methods they like. Give different children a chance to explain how they found the answers.
- Explain that the children can use these methods and any other methods that they have learnt to multiply any numbers.
- Turn to **Pupil book 5 page 117** and work through the examples with the class. Spend some time talking about the different methods and why it is useful to be able to use the most efficient column method.
- Use **Workbook 5 page 71** to practise multiplying 3- and 4-digit numbers by 2-digit numbers in a grid. Then check the answers.
- Turn back to **Pupil book 5 page 117** and ask the children to work independently to complete question 1 and question 2.
- <u>Problem solving</u>: The children should work independently to complete question 3.

Interesting mistakes
Some children may apply what they have learnt about 1-digit multiplication to column multiplication so that they multiply each column as a separate 1-digit multiplication.
- For example, they might work out 32 × 4 as 4 × 2 and 4 × 3. This incorrect thinking works in some cases but fails when there is a number to carry and when multiplying by a 2-digit number (as shown below).

32		44		32	
× 4		× 4		× 13	
128	✓	1616	✗	36	✗

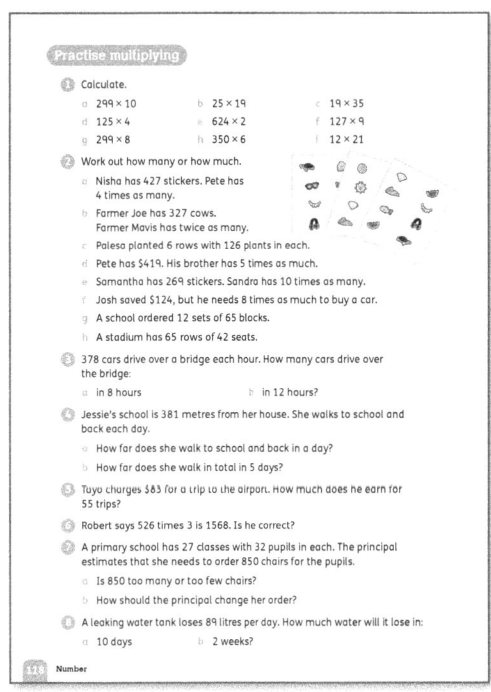

Materials
Calculators

Warm-up
Use multiplication activities from the 'Calculation skills' section of the Activity bank (pages 25–28) as a starter for this lesson.

Focus
Pupil book 5 page 118 provides additional multiplication practice and provides word problems involving multiplication.

- It is a good idea to break up the work so the children do not feel overwhelmed. You can do this by: doing some of the questions and then stopping to check them; setting some activities as homework tasks; or splitting the work over two or more lessons.
- Remind the children that they do not always need to do column multiplication and that if they can use a mental strategy, then they should do so. For example, for question 1, part a, the children could apply what they know about place value to work out that 299 × 10 is 2990 without doing any calculation.

Challenge

Give the children problem-solving activities that involve finding the missing digits.

Ask: *What are the missing digits in these multiplications? How did you work this out?*

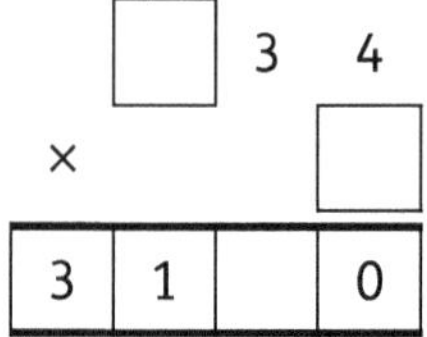 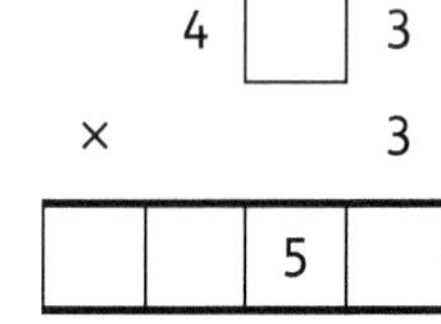

Solution: 634 × 5 = 3170   Solution: 453 × 3 = 1359

- The children can then try this activity:
  The digits 0–9 are used once each in this calculation. Can you work out where each digit goes? (Point out that 3 has already been used, so the children shouldn't use 3 again. A possible answer is 6819 × 3 = 20 457.)

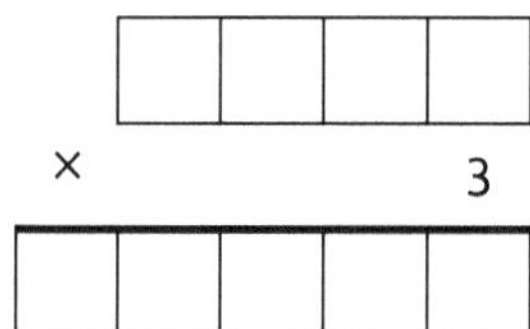

- Search online for more challenging missing-digit problems.

Support

Ask the children to make a poster showing different methods of multiplying, with one or two examples of each method. This will help them to reflect on what they have learnt. You could also ask them to include a set of hints for solving word problems involving multiplication.

Answers to Pupil book 5 page 118

1. a 2990   b 475   c 665   d 500
   e 1248   f 1143   g 2392   h 2100
   i 252
2. a 1708   b 654   c 756   d $2095
   e 2690   f $992   g 780   h 2730
3. a 3024   b 4536
4. a 762 m   b 3810 m
5. $4565
6. No; the correct answer is 1578
7. a too few
   b She must order 864 chairs in total.
8. a 890 ℓ   b 1246 ℓ

Revisit division

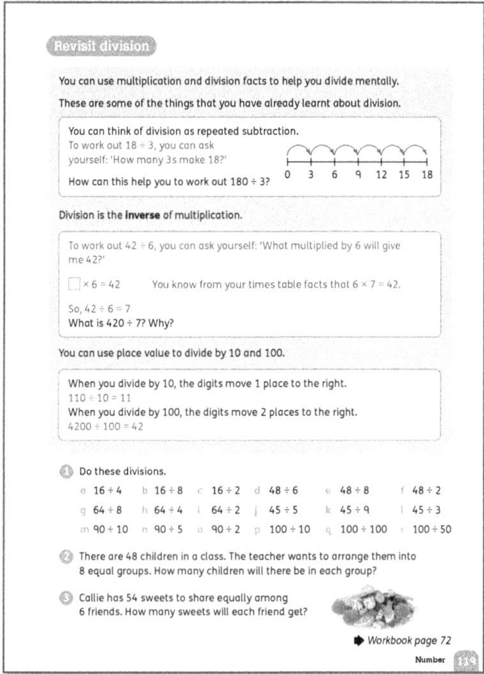

Materials

Calculators; large wall chart with multiplication facts for reference

Warm-up

Give the children times table multiplication facts and ask for the related division facts, as a starter for this lesson.

Focus

- Write the word *division* on the board and ask the class to think about what this word means in maths. Take feedback from the class and write down their ideas.
- If they get stuck, ask questions such as: *How do you know you have to divide? How do you know if you can divide a number by 2? Does splitting mean the same thing as division? How do you know what share everyone gets if you share something equally? What do you do when the number of items does not give you an equal share?*
- Explain to the class that you are going to start by revising mental strategies for division because these strategies will be useful when they move on to written division and work with larger numbers in the next few lessons.
- Demonstrate these mental division strategies: to divide by 4, halve and halve again; to divide by 5, divide by 10, then multiply by 2. Work through some examples to confirm that these give the correct answers. Ask the children: *How could I use factors of 6 to divide mentally by 6?* (Divide by 2, then by 3.)
- Place the children in pairs and let them read through the text and examples on **Pupil book 5 page 119**. There are no new concepts involved here, so this is just a refresher for them.
- The children can then complete questions 1–3 independently. The children can use calculators to check their own work.

- Spend time talking about any difficulties they had or mistakes they made and let them say what they think went wrong.

Follow-up
- **Workbook 5 page 72** provides a challenging problem-solving activity to reinforce division facts. The children can compare answers and use calculators to check each other's work.

Challenge
Put the children into groups. Assign each group a different 1-digit number and give them some time to investigate and develop a rule for testing whether a large number can be divided by that number. For example, different groups can investigate rules for numbers that are divisible by 4, 5, 7 and so on. As a class, ask each group for their rule and use some different numbers to test that the rule works.

Interesting mistakes
The children may get incorrect answers to calculations due to mental arithmetic errors. Types of errors might include incorrect times tables facts and errors in addition or subtraction. Encourage the children to always check their calculations and to make an estimate to check whether their answer is reasonable.

Answers for Pupil book 5 page 119

1 a 4   b 2   c 8   d 8
e 6   f 24   g 8   h 16
i 32   j 9   k 5   l 15
m 9   n 18   o 45   p 10
q 1   r 2

2 6   **3** 9

Answers for Workbook 5 page 72

1
$$100 \div 10 = 10$$
$$\div \qquad \div$$
$$50 \div 5 = 10$$
$$= \qquad =$$
$$2 \div 2 = 1$$

$$144 \div 6 = 24$$
$$\div \qquad \div$$
$$24 \div 3 = 8$$
$$= \qquad =$$
$$6 \div 2 = 3$$

2 a 1.25 m   b 2.5 m
c 0.5 m   d 0.625 m

Written division

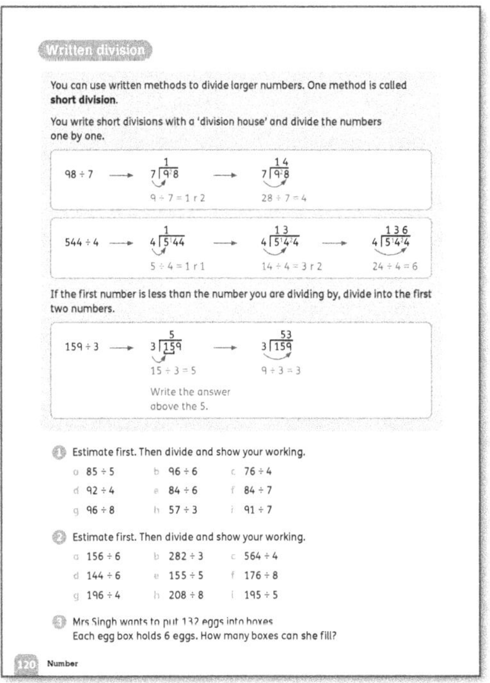

Materials
Squared paper; concrete resources, for example counters, interlocking cubes (page 20), base-ten blocks (pages 20–21); marker pens

Warm-up
Use the 'Division grid' activity from the Activity bank (page 28) as a starter for this lesson.

Focus
- Display these two problems for the class:
  A group of 96 children is divided into 3 equal groups to travel on a school outing to a museum. How many children are in each group? (32)
  At the museum, the children are put into groups of 4. How many groups are there? (24)
- Ask the children to work in pairs to solve each problem. They should use any strategy that makes sense to them. Encourage the children to use concrete resources and diagrams if these will help them.
- Give each pair a sheet of squared paper and marker pens. Let them divide the paper into two halves and record their solutions to the problems, one in each part.
- Tell the children that they should be prepared to share their thinking and the strategies they used with the class.
- Observe the children as they work to see what strategies they use. Help anyone who is stuck. If any children finish early, ask them to show a different method or to suggest how they could divide the 96 children into 8 equal groups to travel to the museum and into groups of 6 at the museum.
- Choose a few pairs who used different strategies to present their methods to the class. Then ask questions to help them think critically about what they did. For example: *Did your strategy make it easy to find a solution? What worked well? What didn't work so well? What would you change if you had to solve this problem again? Why?*

- Explain that one strategy is not necessarily better than another, but that in maths we aim to find a compact and efficient way of recording the work that we do.
- Turn to **Pupil book 5 page 120**. Work through the examples with the class, talking about what you are doing at each step. For example: *I'm going to start by dividing 7 into 9. How many times does 7 go into 9? I'm writing the 1 above the 9 and carrying the 2 that remains to the next digit.*
- Spend some time demonstrating *short division* because this method can be used for almost any division at primary school level.
- The children can then work independently to complete the written divisions in questions 1–3 on **Pupil book 5 page 120**. Let them use squared paper if they need it to keep the digits aligned in the correct places.

Challenge

Challenge the children to work in pairs to find 2-, 3-, 4-, 5- and 6-digit numbers that are divisible by 2. Repeat this for other single-digit divisors up to 9. Ask the children to share how they worked and how they checked that the numbers they chose were divisible by the given digit.

Interesting mistakes

Some children may make mistakes because they do not understand how the written methods work or may remember the stages incorrectly.

- Encourage the children to 'have a conversation in their head' about what they are doing at each stage of the calculation and model how to do this with example divisions.
- Give the children examples of correct and incorrect worked examples and ask them to mark them and identify where mistakes have been made. Allow and encourage informal methods and jottings to support the children's understanding as they work towards formal, written methods.

Answers for Pupil book 5 page 120

1	a 17	b 16	c 19
	d 23	e 14	f 12
	g 12	h 19	i 13
2	a 26	b 94	c 141
	d 24	e 31	f 22
	g 49	h 26	i 39
3	22 boxes		

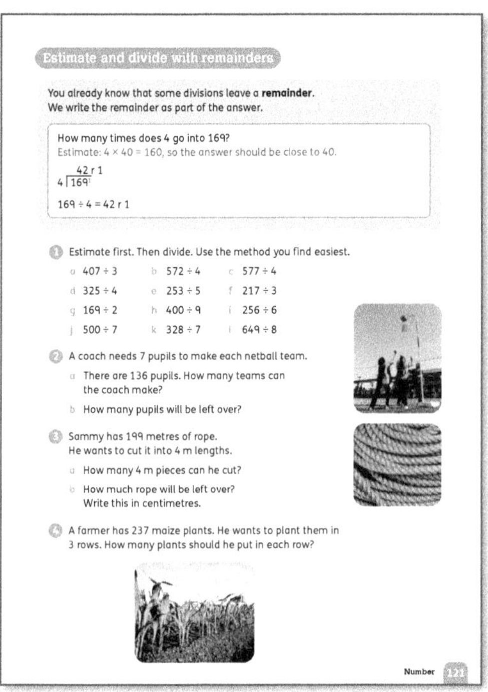

Materials

Several sets of at least 20 small objects (The sets do not need to have the same number of objects.)

Warm-up

Use 'Multiplication, factors and divisibility rules' from the Activity bank (page 27) as a starter for this lesson.

Focus

- Divide the class into groups and give each group a set of objects. Ask the groups to divide their sets into two equal groups.
- Ask: *Does anyone have any objects left over? How many?* (The maximum answer will be 1.) *Could you have any other amount left over?* (No), *Why not?* (If you had more than 1 object, you could put them into the two groups.) *What do we call this leftover amount when we divide?* (a remainder)
- Repeat the warm-up activity with a few different divisors (stick to low numbers such as 3, 4 and 5) to revise the concept of a *remainder* and to help the children understand that the remainder has to be less than the number they are dividing by (the *divisor*).
- Turn to **Pupil book 5 page 121**. Discuss remainders with the class. Remind them that when a number is divided by a number that is not a factor of that number, there will be a remainder.
- Work through the example with the class, reminding them that they can use written short division even when there is going to be a remainder. Establish how you want the class to record remainders. The Pupil book uses 'r' for remainder, as this is the shortest and quickest way for the children to write the remainder.
- For question 1, ask the children to explain how they can use *inverse* operation facts to estimate the answer to a division.
- The children can work through questions 2–4 on their own or in pairs.

Can you divide the remainder?

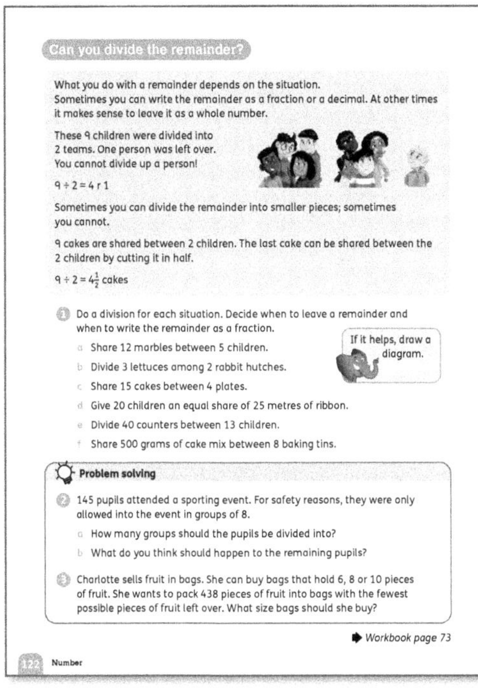

Previously, the children have only been expected to write a remainder. Now they have to decide whether or not to round up or down depending on the context of the problem.

Materials

Base-ten blocks (page 20); envelopes or boxes containing 6 counters each

Warm-up

Tell the children that each envelope contains 6 counters, and ask how many envelopes you need to be able to give each child one counter. Use that number of envelopes to see if their answers were correct.

Focus

- Set this problem in which the answer needs to be rounded up: *A teacher wants to put 18 calculators into boxes. Each box can hold 5 calculators. How many boxes does the teacher need?* The children need to realise that in this case the answer is 4. The teacher cannot use 3 boxes, as there will be 3 calculators left over. These 3 calculators need to go in a box, even if the box is not full.
- Set this problem in which the remainder needs to be divided further: *I have 3 cakes to share between 2 families. How much cake will each family get?*

The answer of 1 remainder 1 does not make sense here, because it makes sense to cut the third cake in half and give each family $1\frac{1}{2}$ cakes. Use concrete examples to show this where possible.

- Let the children work on their own to read and discuss the examples on **Pupil book 5 page 122**. When they have done this, ask them to think of some examples where we need to divide the remainder.
- Let the children work in pairs to do question 1. They should work out each division and talk about what to do with the remainder. Check the answers to make sure they realise that physical objects like food, which can be cut up, can be shared and that measurements often give a remainder that can be divided.
- <u>Problem solving:</u> The children can then complete the problems in question 2 and question 3.

Follow-up

Use **Workbook 5 page 73** to informally assess the children's ability to do written short division and to deal with the remainder sensibly in context. Check the answers as a class, discussing any differences between the children's answers.

Challenge

- The children can work with calculators to explore remainders and how the calculator deals with these. They will begin to understand that when a question requires a whole number remainder but their calculator gives a decimal answer, they have to work backwards to find the remainder.
- For example, on a calculator, the answer to $53 \div 4$ is 13.25. The children can use the fact that $13 \times 4 = 52$ to deduce that the remainder is 1. Alternatively, they can think about the fact that the remainder is a fraction of the divisor. Therefore, in this calculation 0.25 means $\frac{1}{4}$ of 4, which is 1. Let the children share what they discover in groups.

Support

You may wish to use base-ten blocks to help the children understand and model division work. Use 10-rods and ones cubes to represent the number being divided.

- This example models $43 \div 3$:

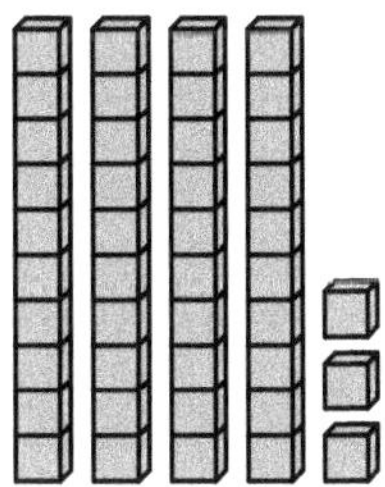

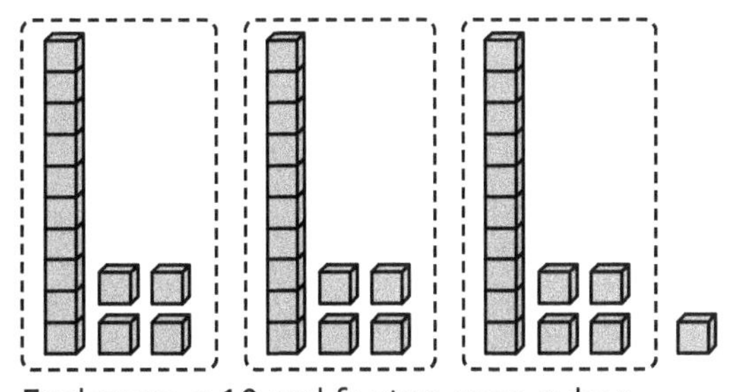

Exchange a 10-rod for ten ones cubes.

Answers for Pupil book 5 page 122

1 **a** 2 r 2   **b** $1\frac{1}{2}$ **c** $3\frac{3}{4}$
   d $1\frac{1}{4}$ m   **e** 3 r 1   **f** $62\frac{1}{2}$ g

2 **a** 18 groups of 8 pupils and 1 pupil left over
   b Either create 17 groups of 8 pupils and 1
     group of 9 pupils (if this is allowed)
     or (for example) make 17 groups of 8 pupils,
     1 group of 5 pupils and 1 group of 4 pupils.

3 The bag that holds 6 pieces.

Answers for Workbook 5 page 73

1 27 pieces, with 7 m left over

2 $57\frac{1}{8}$

3 **a** 25 cm
   b 12, with 10 cm left over

4 $18.75

5 Individual answers.

More work with remainders

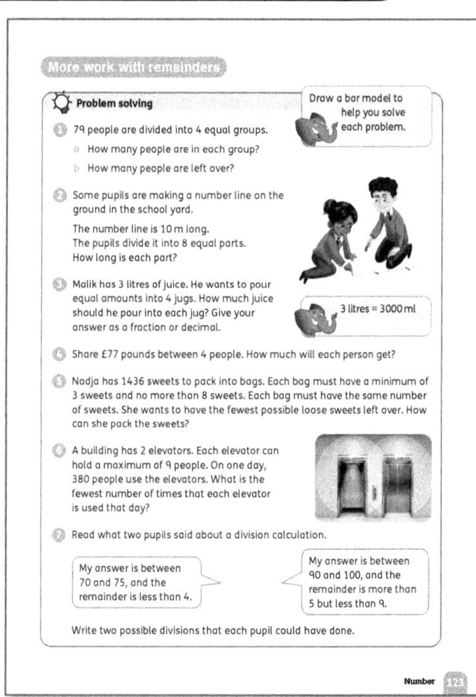

Materials
Calculators

Warm-up
Ask the children to give you division calculations that have the answer 6. Repeat for different numbers, then include remainders.

Focus
- Problem solving: There are no new concepts in this lesson. The children can work through questions 1–7 on **Pupil book 5 page 123** on their own to consolidate what they have learnt.

- Observe the children as they work and assist anyone who asks for help. Encourage the children to use a calculator to check their work if they are not sure about their answers.

Challenge
Ask the children to find all the possible whole-number solutions for the second pupil in question 7. Encourage them to use tables to organise their solutions.

Answers for Pupil book 5 page 123

1 **a** 19   **b** 3   **2** 1.25 m

3 0.75 litres   **4** £19.25

5 Packaging in bags of 4 sweets means 359 bags with no sweets left over.

6 22

7 Individual answers for example, 363 ÷ 5 = 72 remainder 3; 512 ÷ 7 = 73 remainder 1; 395 ÷ 4 = 98 remainder 3; 845 ÷ 9 = 93 remainder 8.

More division

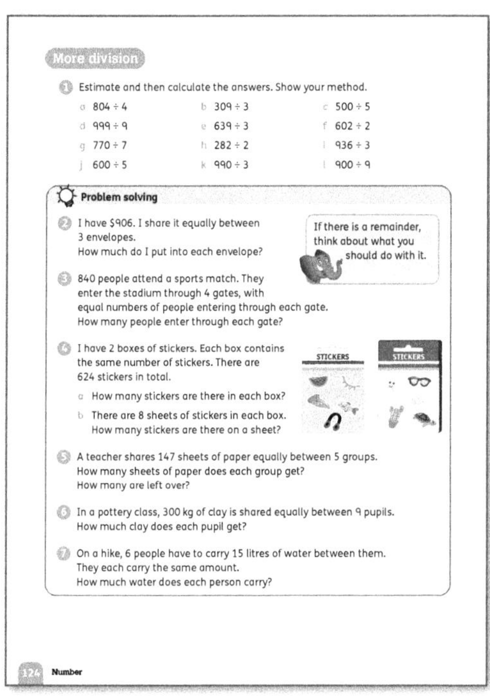

Warm-up
Use question 1 on **Pupil book 5 page 124** as a starter for this lesson. The children can use jottings like the ones below if they need to, but they should be able to use known facts to do these divisions mentally. Tell them in advance that there are no remainders.

a 804 ÷ 4   **b** 309 ÷ 3   **c** 500 ÷ 5
   201   103   100

Focus
- Problem solving: Spend some time talking about the word problems in questions 2–7 on **Pupil book 5 page 124**. Let the children share their ideas for solving them.
- After a class discussion, let the children work on their own to provide worked written solutions.

Solve division problems

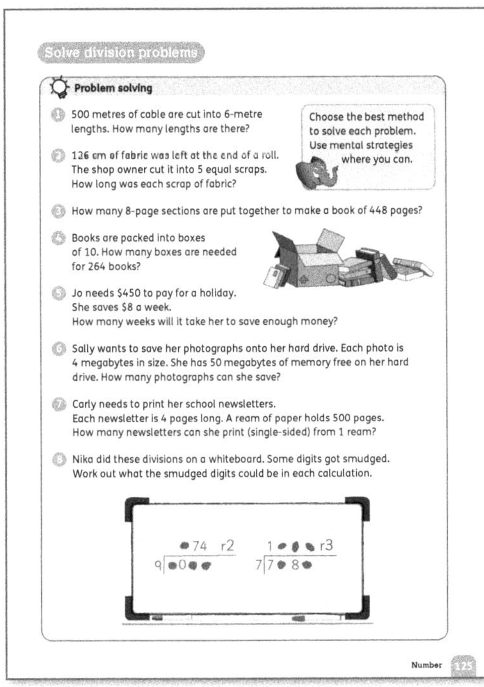

Materials

Poster paper and marker pens; grid and number cards (see below)

Warm-up

As a mental warm-up for this lesson, ask the children to work out where the number cards go on this grid.

48	÷		=	
÷		÷		÷
	÷		=	
=		=		=
	÷		=	

Number cards:

8	8	6	6
4	3	2	2

- Show the class the number cards and explain that the children must use each number once. Explain that the divisions must work both across and down.
- Let the children think about the problem and indicate when they have some answers.
- Select different children to give their solutions and explain their strategies and thinking. (A possible answer is: top row: 48 ÷ 6 = 8; middle row: 8 ÷ 2 = 4; bottom row: 6 ÷ 3 = 2. Vertically, this gives: 48 ÷ 8 = 6; 6 ÷ 2 = 3; 8 ÷ 4 = 2.)

Focus

Problem solving:
- Put the children into groups of three or four. Give each group a piece of poster paper and some coloured marker pens.
- Turn to **Pupil book 5 page 125**. Tell the groups that each group member can choose one of the problems (from questions 1–8). They must each choose a different problem.
- Each child should then work out the solution, asking the other group members for help if they need it.
- Then the group should use the poster paper to show how they solved each problem. Tell them that they should be prepared to explain their work to the class.
- Give the groups time to work and then ask different groups to show their solution posters. Give each child in the group the chance to say why they chose their problem and to explain how they solved it.

Follow-up

If any problems are left unsolved, you could allocate them for homework or use them at the end of the unit to check understanding.

Challenge

If any children finish quickly, give them some blank division grids like the one shown in the 'Warm-up' section above and ask them to choose a number to write in the top left corner and to produce a set of cards for the division grid. Challenge them to make the most difficult puzzle they can. You can use the completed puzzles as additional problem-solving activities as the children work through the rest of this unit.

Mixed problems

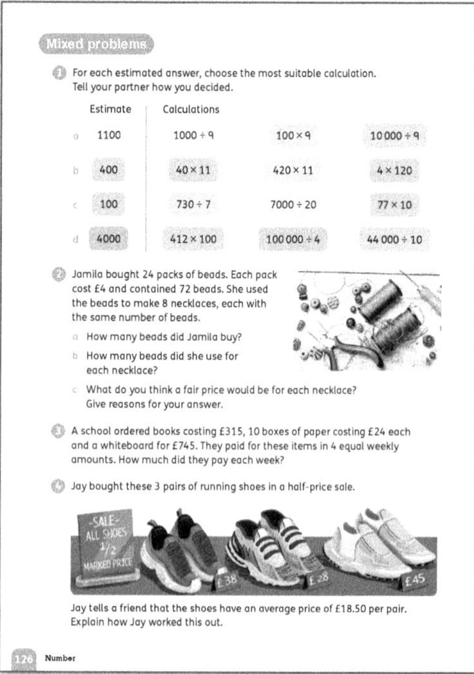

Mixed problems

1. For each estimated answer, choose the most suitable calculation. Tell your partner how you decided.

Estimate	Calculations		
a 1100	$1000 \div 9$	100×9	$10\,000 \div 9$
b 400	40×11	420×11	4×120
c 100	$730 \div 7$	$7000 \div 20$	77×10
d 4000	412×100	$100\,000 \div 4$	$44\,000 \div 10$

2. Jamila bought 24 packs of beads. Each pack cost £4 and contained 72 beads. She used the beads to make 8 necklaces, each with the same number of beads.
   a How many beads did Jamila buy?
   b How many beads did she use for each necklace?
   c What do you think a fair price would be for each necklace? Give reasons for your answer.

3. A school ordered books costing £315, 10 boxes of paper costing £24 each and a whiteboard for £745. They paid for these items in 4 equal weekly amounts. How much did they pay each week?

4. Jay bought these 3 pairs of running shoes in a half-price sale.

Jay tells a friend that the shoes have an average price of £18.50 per pair. Explain how Jay worked this out.

126 Number

What is a rate?

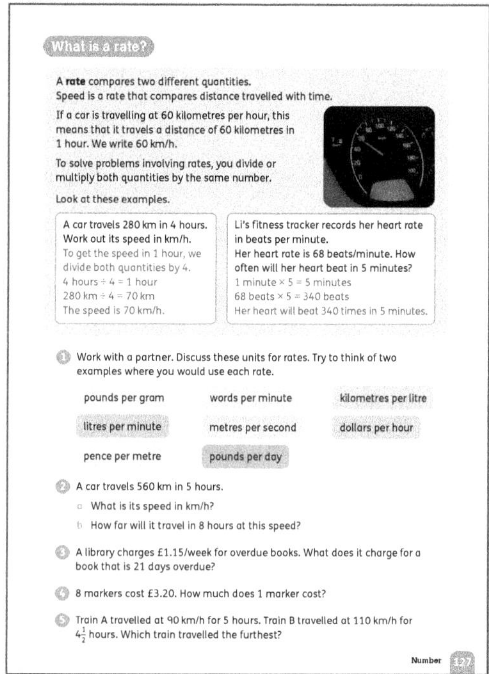

What is a rate?

A **rate** compares two different quantities.
Speed is a rate that compares distance travelled with time.

If a car is travelling at 60 kilometres per hour, this means that it travels a distance of 60 kilometres in 1 hour. We write 60 km/h.

To solve problems involving rates, you divide or multiply both quantities by the same number.

Look at these examples.

A car travels 280 km in 4 hours. Work out its speed in km/h.
To get the speed in 1 hour, we divide both quantities by 4.
4 hours ÷ 4 = 1 hour
280 km ÷ 4 = 70 km
The speed is 70 km/h.

Li's fitness tracker records her heart rate in beats per minute.
Her heart rate is 68 beats/minute. How often will her heart beat in 5 minutes?
1 minute × 5 = 5 minutes
68 beats × 5 = 340 beats
Her heart will beat 340 times in 5 minutes.

1. Work with a partner. Discuss these units for rates. Try to think of two examples where you would use each rate.

pounds per gram	words per minute	kilometres per litre
litres per minute	metres per second	dollars per hour
pence per metre	pounds per day	

2. A car travels 560 km in 5 hours.
   a What is its speed in km/h?
   b How far will it travel in 8 hours at this speed?

3. A library charges £1.15/week for overdue books. What does it charge for a book that is 21 days overdue?

4. 8 markers cost £3.20. How much does 1 marker cost?

5. Train A travelled at 90 km/h for 5 hours. Train B travelled at 110 km/h for $4\frac{1}{2}$ hours. Which train travelled the furthest?

Number 127

Warm-up

Choose a suitable activity from the 'Calculation skills' section of the Actvity Bank on pages 25–28 as a starter for this lesson.

Focus

This lesson provides the children with the opportunity to work with two-step problems involving different operations.

- The children can work through question 1 orally in pairs to reinforce estimating skills.
- For question 2 and question 3, encourage the children to show the problems visually by drawing bar models or other diagrams and to choose the most efficient strategies.
- For question 4, even though the children have not worked with the mean as an average at this stage, they should be able to see that dividing the total by 3 will give an average price of £18.50.

Answers for Pupil book 5 page 126

1. a $10\,000 \div 9$ b 40×11
   c $730 \div 7$ d $44\,000 \div 10$
2. a 1728   b 216
   c Individual answers. (Should take into account that cost of materials for each necklace is £12, and that some cost could be added for the time Jamila spent making the necklace.)
3. £325
4. Added the prices of the three pairs of shoes and then divided the total by 3.

In this lesson, the children will learn to work with simple unit rates. A unit rate is one quantity expressed for every 1 unit of another. For example: speed is measured in kilometres per hour (km/h) or heartrate is measured in beats per minute (bpm).

Materials

Pictures of speed limit signs; fitness tracker or health app on a smartphone (if possible); calculators

Warm-up

- Introduce the concept of a *rate* using real-life examples and activities that the children can do. For example, tell the children that you are going to time them for 1 minute to see how many times they can write (or type if you are using computers) their first name.
- At the end of the activity, ask the children to count how many times they wrote their name.
- Ask a child for their total, for example, Zayn, 23 times. Say: *Zayn wrote his name 23 times in 1 minute. His name-writing rate is 23 names per minute. What is your name-writing rate?* Then ask different children to share their rates in names per minute.
- Ask the class: *Why do you think different children gave different rates?* (Some children write faster than others, some names are longer and so on.)

Focus

- Discuss how the children can calculate their writing rate in letters per minute using the data they already have. (They can multiply the number of letters in their name by the name-writing rate.)
- Compare the children's letter-writing rates. Then ask the children: *Is there more similarity between your letter-writing rates than your name-writing rates?*

- Explain that a rate compares two things that are measured in different ways. The rate of names per minute compares the number of names written with the time in minutes.
- Display a speed sign and ask the children what it means. Make sure they realise that it gives a rate in kilometres (or miles) per hour.
- Use a fitness app on a phone or a fitness tracker and ask the children to find the heartrate function. Let them share their ideas of how this is tracked and measured. They should be able to tell you that heartrate is measured in beats per minute (bpm). Let them describe other rates that are tracked, for example, steps per day, hours of sleep per night, cycling or running speed and so on.
- Let the children work in pairs to read the information and to go through the worked examples on **Pupil book 5 page 127**. Then ask the children to complete questions 1–5.
- For question 1, give the children time to discuss the rates and think of examples. Then they can share their ideas as a class.

Follow-up

Show the children an example of when knowing how to work out rates could help them save money. Display the following information:

Fruit Juice Special

2-bottle pack for $3

3-bottle pack for $4

4-bottle pack for $5

$1.60 for a single bottle

- Ask the children how they can decide which pack is the best value for money. Let them work on this in groups and then take feedback, for example: working out the cost per bottle for each one. (The 4-bottle pack is the best value.)
- Ask further questions such as: *Mpho bought two 3-bottle packs. Nomsa bought a 4-bottle pack and a single bottle. Who paid the higher price per bottle? What was it?* (Mpho paid $1.33 per bottle and Nomsa paid $1.10 per bottle. So Mpho paid the higher price per bottle.).

Interesting mistakes

Children who need additional support with either multiplication or division may find it difficult to solve problems involving rate. Encourage them to use a calculator so that they can focus on the concept and solve the problems without having to deal with operations they find challenging.

End-of-unit check

Ask some or all of these questions to assess how well the children have understood the concepts in this unit.

- *Is multiplying by 3 and then multiplying by 10 the same as multiplying by 30?* (It is the same. $3 \times 10 = 30$)
- *What are the factors of* (give a number)?
- *Is dividing by 3 and then by 2 the same as dividing by 6?* (Yes)
- Give a problem, then ask: *What method did you use to find your answer?*
- *In this money problem, my calculator gives me an answer of 3.2. What does this mean?* (£3.20)
- Give a division problem with a remainder, then say: *In this division problem, there is a remainder of* (give a number). *What should I do with it?*
- Give the children some correct and some incorrect calculations. Ask: *Which of these calculations are correct and which are incorrect? What has this person done wrong in this calculation? How can you help them to correct it?*
- *What clues can you look for in word problems to help you know whether to multiply or divide?* (For example: Words such as product, altogether, lots of, per may mean multiply. Words such as share or groups may mean divide.)
- *Make up a two-step word problem that must be solved using multiplication and division.*
- *It took Josh 2 hours to cycle 40 km. What was his speed in kilometres per hour?* (20 km/h)
- *Carin walks at a speed of 5.5 km/h. How far does she walk in $1\frac{1}{2}$ hours?* (8.25 km)
- Give the children a situation such as: *Maria cleans her phone screen every 4 days.* Then give them a set of statements and ask them to say which are true. For example: *She cleans her phone 4 times per day.* (False) *She cleans her phone at a rate of 0.25 per day.* (True) *Every day that passes, $\frac{1}{4}$ of the phone will be cleaned.* (False) *After 12 days, she will have cleaned her phone 3 times.* (True)

Work with negative numbers

Learning objectives

- Interpret negative numbers in context
- Count backwards and forwards with positive numbers, including through zero
- Add and subtract integers, including where one integer is negative

Key words

positive numbers   negative numbers
minus sign   thermometer   degrees Celsius

Unit introduction

Teaching guidance

- Display a set of numbers represented in different ways. For example:

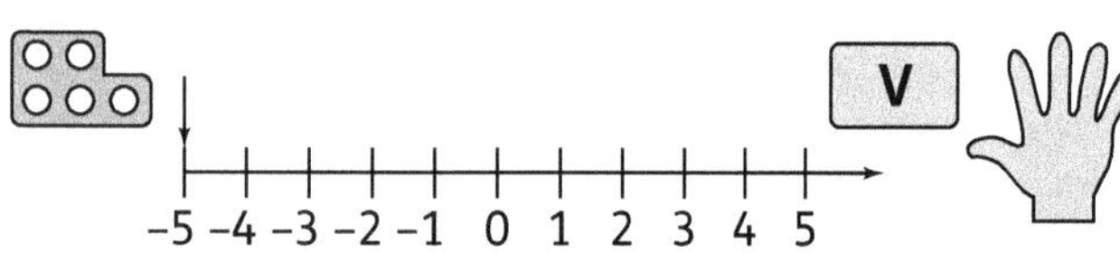

- Ask the class: *What do you notice? What do you wonder?* Give the children time to think and then ask them to share some of their ideas.
- Ask the class to look at the set again to decide which one does not belong and why.
- Then let different children share their ideas. Note that they could choose any of the representations, as long as they have a reason for their choice. For example, they could say that 'V does not belong because it is a Roman numeral and does not belong in the decimal number system we use'.
- Point to the number line and ask: *What do we call the numbers that are less than 0?* (negative numbers)
- Explain that we use a *minus sign* to show that a number is negative.
- Discuss with the children where negative numbers are used in real life and what they are used for.

Positive and negative numbers

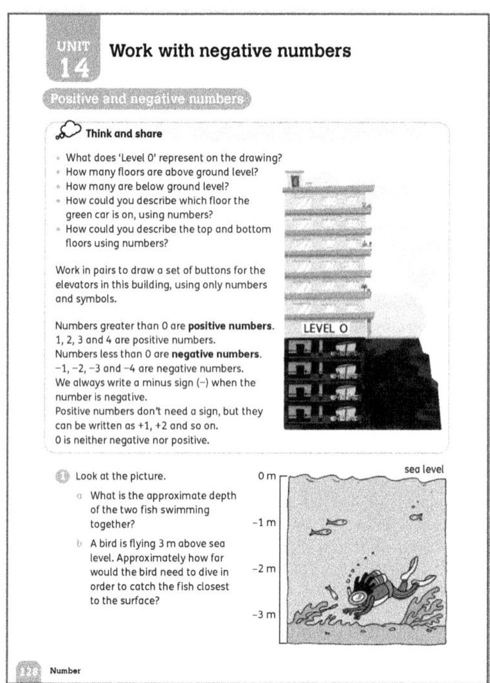

The children have already worked with negative numbers in context and they should also be confident counting back through zero to negative values. In this lesson, the idea of directed (negative and positive) numbers is revised and extended.

Materials

100 chart, showing the numbers 1 to 100; play money notes

Warm-up

As a mental starter, display a 100 chart. Play games in which the children have to find numbers by counting back and/or on in given amounts.

Focus

- <u>Think and share:</u> Turn to **Pupil book 5 page 128.** Let the children work in pairs to answer the questions about the building. (Level 0 represents ground level. There are six floors above ground level. There are four floors below ground level. The green car is on level −3. The top floor could be called level +6 and the bottom floor could be called level −4.)
- Before the children design a set of elevator buttons, take feedback and ask some additional questions to make sure they understand and can work with *negative numbers* as well as *positive numbers*. Ask: *How many floors are there altogether?* (12) *What do we call the first floor under ground level?* (−1) *What do we call the first floor above ground level?* (+1) *How many floors do you need to go up from the deepest underground floor to the top floor?* (11) and so on.

- Then let the children design the set of buttons. Display these in the classroom and discuss any interesting or creative ideas.
- The children can complete question 1 on their own.

Support
Reinforce the ideas in this lesson by playing a banking game using play money notes. Let the children take turns to be the banker who records all transactions. Give different children different amounts of play money. Instruct the children to deposit, withdraw and borrow money from the bank. Discuss what happens to the bank balance when someone has no money in their account and they borrow from the bank. Make sure that the children realise that the bank balance will be a negative amount of money.

Answers for Pupil book 5 page 128
1 a −1 m   b 4 m

Count on and back through 0

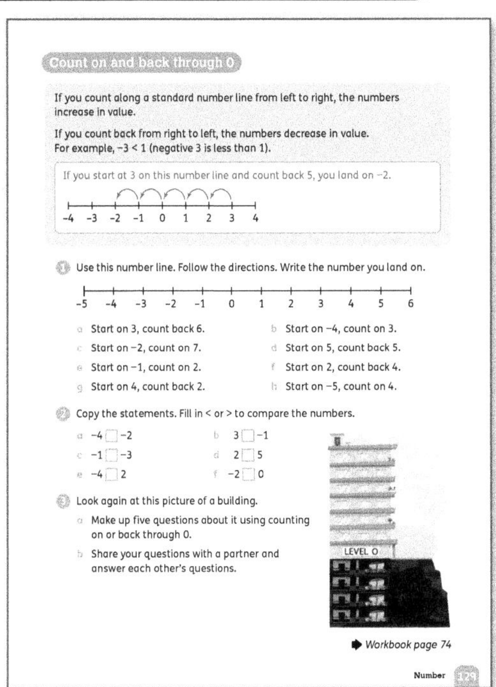

Number 129

Materials
Number lines that show both positive and negative values

Warm-up
As a mental starter, display a positive number line and play games in which the children find numbers by counting back and/or on in given amounts.

Focus
- Display a large number line extending to negative numbers and demonstrate how to count on and back bridging zero.
- Read through the explanation box and example at the top of **Pupil book 5 page 129** with the children. Practise comparing positive and negative numbers, and two negative numbers.
- Then ask the children to complete questions 1–3 on their own.

Follow-up
Use **Workbook 5 page 74** to assess whether the children can work with, order and compare positive and negative numbers.

Interesting mistakes
Children sometimes get confused by the fact that −12 is smaller than −2, but 12 is greater than 2. Remind them that numbers get smaller as you move to the left on a number line. Demonstrate this with positive numbers first, then extend it to negative numbers as well. It may be helpful to remind the children that someone who owes $12 (that is, has −$12) has less money than a person who owes only $2 (i.e. has −$2).

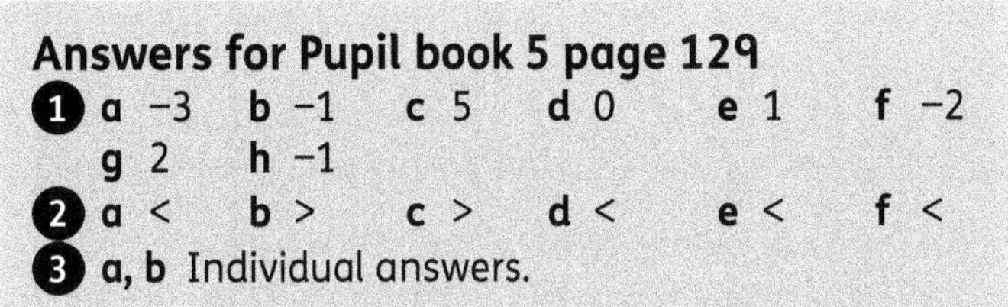

Answers for Pupil book 5 page 129
1 a −3   b −1   c 5   d 0   e 1   f −2
    g 2   h −1
2 a <   b >   c >   d <   e <   f <
3 a, b Individual answers.

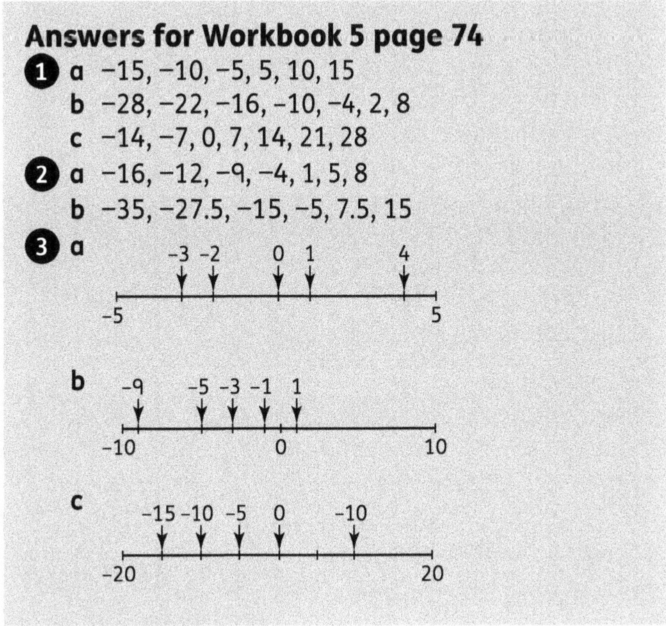

Answers for Workbook 5 page 74
1 a −15, −10, −5, 5, 10, 15
    b −28, −22, −16, −10, −4, 2, 8
    c −14, −7, 0, 7, 14, 21, 28
2 a −16, −12, −9, −4, 1, 5, 8
    b −35, −27.5, −15, −5, 7.5, 15
3 a

b

c

Count in steps

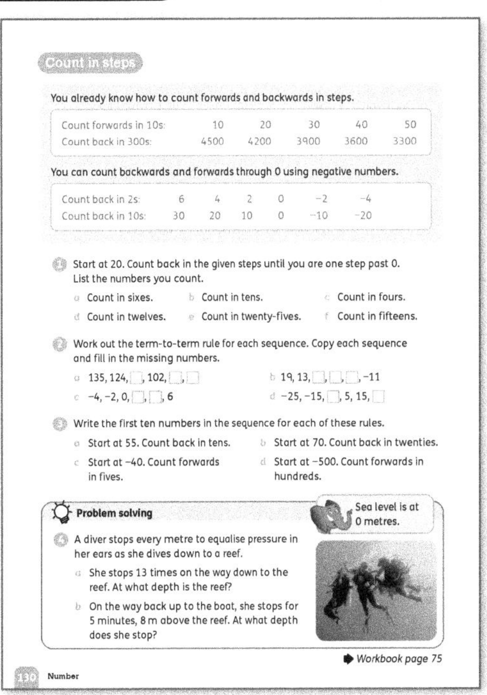

Workbook page 75

230 Number

The concept of step counting is not new to the children, but the aim here is for the children to develop more efficient methods of counting on by 'chunking' to reduce the number of 'jumps' and applying the mental strategies they have already mastered (such as counting in 100s, using multiples of 10, and so on) to include numbers less than 0 (negative numbers).

Materials
Blank number lines; calculators

Warm-up
Use any step-counting activity as a mental starter for this lesson.

Focus
- Work through the examples at the top of **Pupil book 5 page 130** as a class, allowing time for discussion of the strategies and methods that the children use to find the answers by counting on and back.
- Do question 1 orally as a class. Use a number line to show how to bridge 0 if anyone needs additional support to do this.
- Ask the children to identify the rules for the sequences in question 2 orally as a class. Then the children can copy and complete each sequence.
- The children can use number lines to help them with question 3 if they wish.
- Problem solving: For question 4 they can use a vertical number line.

Follow-up
When the children can count on and back in steps through 0 from any number, let them work independently to complete **Workbook 5 page 75**. They can check the answers in pairs using a calculator, and discuss how they answered questions 2a–c.

Interesting mistakes
The children may need additional support when they have to bridge 0, as they may forget to include 0 in the count. Encourage them to use number lines and to check their answers using a calculator. Make sure that they know how to enter negative values into the calculator using the +/− key if they are starting from a negative number.

Answers for Pupil book 5 page 130
1 a 20, 14, 8, 2, −4   b 20, 10, 0, −10
   c 20, 16, 12, 8, 4, 0, −4   d 20, 8, −6
   e 20, −5   f 20, 5, −10
2 a count back in 11s; missing numbers are 113, 91, 80
   b count back in sixes; missing numbers are 7, 1, −5
   c count on in twos; missing numbers are 2, 4
   d count on in 10s; missing numbers are −5, 25
3 a 55, 45, 35, 25, 15, 5, −5, −15, −25, −35
   b 70, 50, 30, 10, −10, −30, −50, −70, −90, −110
   c −40, −35, −30, −25, −20, −15, −10, −5, 0, 5
   d −500, −400, −300, −200, −100, 0, 100, 200, 300, 400
4 a −14 m   b −6 m

Answers for Workbook 5 page 75
1 a −56, −44, −32, −20, −8, 4, 16
   b 36, 24, 12, 0, −12, −24, −36
   c −20 −10, 0, 10, 20, 30, 40
   d 88, 77, 66, 55, 44, 33, 22
   e 86, 75, 64, 53, 42, 31, 20
   f −205, −190, −175, −160, −145, −130, −115
   g 60, 45, 30, 15, 0, −15, −30
2 a yes   b yes   c no
   d The numbers he counts must end in a 2 or a 7.

Number sequences

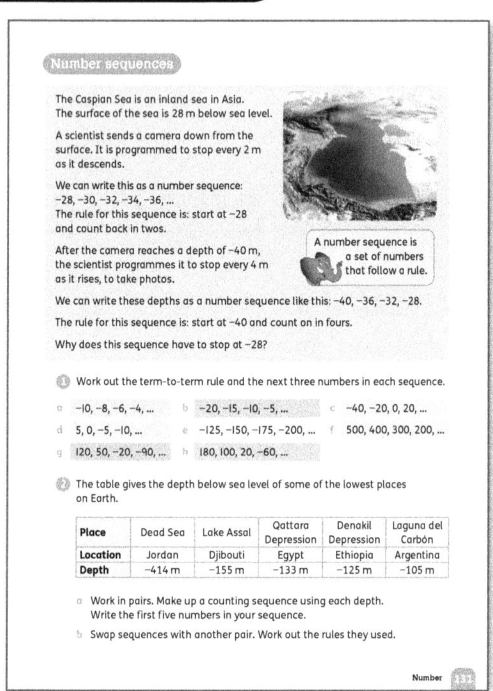

Materials
World map or atlas; blank positive and negative number lines; calculators

Warm-up
Start by finding the Caspian Sea on a world map or in an atlas. Discuss what it means when we say that the surface is 28 m below sea level. Make sure the children understand that sea level here means the general sea level in Asia, not the level of the Caspian Sea itself.

Focus
- Work through the examples of sequences of negative numbers on **Pupil book 5 page 131** to show how real-life activities can generate number sequences. If necessary, support the work by showing the sequences on a number line.
- Make sure the children understand that the increasing sequence stops at −28 m because that is the level of the surface.
- Let the children work in pairs to find the rules for the sequences in question 1 before completing the sequences on their own.

- Before the children do question 2 in pairs, explain that some places on Earth are below the level of the surrounding land and/or sea. Find the countries and places listed in the table on a world map. Then let the children work in pairs to generate sequences. Encourage them to include both ascending and descending sequences. Let the children use calculators to check each other's sequences.

Challenge
Let the children find out more about places on Earth that are below sea level. The Netherlands is a good starting point, as almost $\frac{1}{3}$ of the land area is below sea level. The lowest part of the country is almost 7 m below sea level. The highest part of the country is only about 300 m above sea level (so it is a very flat country). The children can create a presentation with numbers to show what they find out.

Interesting mistakes
Some children find it easy to work with number sequences when they create their own but need additional support to describe a sequence or pattern that is given to them. Let them model the sequence as jumps on a number line to help them work out the term-to-term rule.

Answers for Pupil book 5 page 131
1 a Add 2; −2, 0, 2   b Add 5; 0, 5, 10
   c Add 20; 40, 60, 80   d Subtract 5; −15, −20, −25
   e Subtract 25; −225; −250, −275
   f Subtract 100; 100, 0, −100
   g Subtract 70; −160, −230, −300
   h Subtract 80; −140, −220, −300
2 a, b Individual answers.

Temperature changes

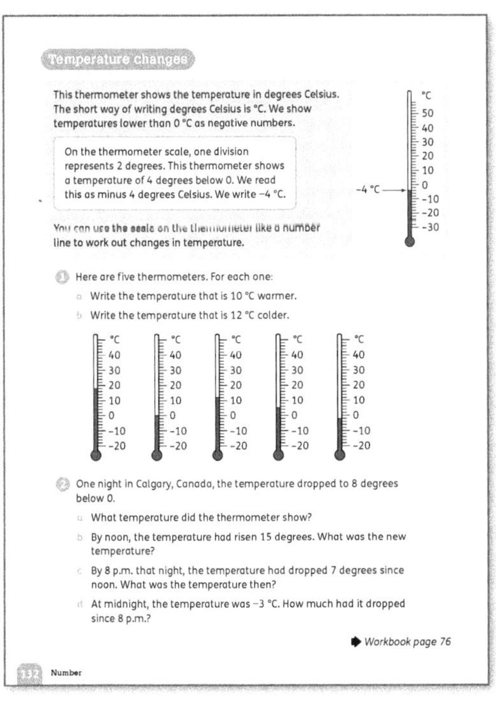

Materials
Large thermometer; pointer or counter; blank thermometer scales or number lines; calculators

Warm-up
Use 'Directed numbers' from the Activity bank (page 23) as a starter for this lesson.

Focus
Temperature changes are a good way to begin work on adding and subtracting with negative numbers.
- Display a large *thermometer* for the class. Move a pointer or counter up and down the scale to demonstrate increases and decreases in temperature.
- Let individual children come up and give each child instructions to move the pointer/counter. For example, place the pointer/counter at 12 °C (*degrees Celsius*) and say: *2 degrees warmer, 6 degrees colder, a drop of 5 degrees, a decrease of 3 degrees, an increase of 5 degrees* and so on. Let the children model each change on a blank scale.
- Turn to **Pupil book 5 page 132**. The children can read through the explanation on their own and then complete question 1 and question 2 independently.

Follow-up
Use **Workbook 5 page 76** to consolidate the work in this lesson. Observe the children as they work to make sure that they are able to calculate temperature increases and decreases on a given scale.

Answers for Pupil book 5 page 132
1 a 28 °C, 10 °C, 22 °C, 15 °C, 8 °C
   b 6 °C, −12 °C, 0 °C, −7 °C, −14 °C
2 a −8 °C   b 7 °C   c 0 °C   d 3 °C

Answers for Workbook 5 page 76
1 a Provided as an example.
   b 3 °C, 0 °C   c 24 °C, −3 °C
   d 36 °C, 3 °C   e −10 °C, 4 °C
   f −11 °C, −14 °C

Add and subtract with negative numbers

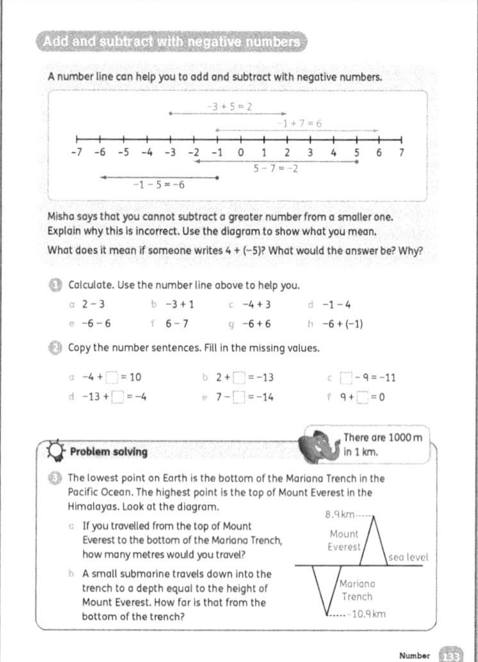

Materials
Number lines; calculators; world map or atlas; photographs of Mount Everest and the Mariana Trench (optional)

Warm-up
Select any 'Mental problem solving' activity from the Activity bank (pages 24–25) as a starter for this lesson.

Focus
This lesson is a good opportunity to have a number talk (pages 17–18)) about what it means to add or subtract a number. Let the children share their ideas.

- Then ask: *What does adding a negative number mean?*
- Let the children talk about this with a partner, then ask them to explain their thoughts and to give examples using number lines if they want to.
- Then ask: *What does subtracting a negative number mean?*
- Again, ask the children to explain their thoughts. Encourage them to use real-life contexts. (For example: 'I owed Miri 3 marbles, so that is −3. She said I didn't need to give them back so I could take away the −3. What that means is that I actually got 3 marbles, so taking away a negative is like adding.') They can also use number lines to show the jumps.
- Work through the example on **Pupil book 5 page 133** to show the children how to use a number line to add and subtract with negative numbers. The question asks what it means if someone writes 4 + (−5). (If you subtract a greater number from a smaller one the answer is negative. 4 + (−5) means 4 − 5 and the answer is −1.)
- Ask questions to remind them that they have already done calculations like this using temperature changes. For example, ask: *How is this like working with temperature?*
- Discuss the questions with the class.
- Let the children use their own number lines or a large display number line as they work through question 1 and question 2.

- Problem solving: Before the children do question 3, use a world map or atlas to locate Mount Everest and the Mariana Trench and show photographs of them.

Challenge
There are lots of interesting graphics and images online showing Mount Everest superimposed on the Mariana Trench. Many of these also show the depth to which different scientists have managed to send submarines. The children could investigate these and prepare a set of problems based on an image for others to answer.

Support
Using inverse operations to find the missing values in question 2 will be confusing for many children. Instead, allow them to make the moves on a number line.

Interesting mistakes
- The children may give the answer −2 km to question 3a because they calculate 8.9 + −10.9 instead of 8.9 − −10.9. Address this mistake using the diagram to show that you are going down 8.9 km and then another 10.9 km.
- Ask questions about the building and elevators from **Pupil book 5 page 128** to help the children understand this concept. For example, demonstrate that to go down from the 2nd floor to level 3 below ground (−3), you go down 5 floors, not −1 floor.

Answers for Pupil book 5 page 133

1 a −1   b −2   c −1
d −5   e −12   f −1
g 0   h −7

2 a 14   b −15   c −2
d 9   e 21   f −9

3 a 19 800 m   b 2000 m

End-of-unit check

Ask some or all of these questions to check that the children have understood the concepts in this unit.
- *What is a negative number?* (a number that is less than zero)
- *What is this temperature?* (Show a temperature on a thermometer scale or number line.)
- *Where is −3 (or any number) on this number line?*
- *I am on the 2nd floor. I get in the elevator and go down 4 floors. What level am I on now?* (−2)
- Give a number sequence and ask questions such as:
  - *What is the rule for this sequence?*
  - *Is there any other way of continuing this sequence? How?*
  - *What are the next three numbers? Why?*
- *The temperature at midday in a cold place was −3 °C. By 4 p.m., the temperature had increased by 4 degrees. Between 4 p.m. and 8 p.m., the temperature dropped 6 degrees. What was the temperature at 8 p.m.?* (−5 °C)
- Give the children a starting number and a rule and ask them to complete the sequence. For example: *Start at 45 and count back in 10s. What is the first number below 0 that you reach?* (−5)

Calculate with decimals

Learning objectives

- Estimate, add and subtract numbers with the same number of decimal places

- Apply place-value knowledge to known additive and multiplicative number facts (scaling facts by one tenth or one hundredth)

- Estimate and multiply decimals by a 1-digit whole number

- Solve problems involving decimals with up to three decimal places

Key words

decimal   whole   tenths   hundredths
place value   square unit

Unit introduction

Materials

Strips divided into tenths; scissors; 1-metre length of rope or string; washer or peg; metre rules or tape measures

Teaching guidance

Do some practical activities with the children to reinforce *decimal* measures and to get them to think about adding decimals in a non-pressurised way.

- Give each child a strip divided into *tenths*. Ask them to cut the strip into two pieces (of any size). They then try to find other children whose pieces will combine with theirs to make whole strips.
- Cut several strips into different numbers of tenths and hand them out to groups. Let them work to arrange the parts into *whole* strips. They do not need to use just two pieces. For example, they could combine 2, 5, and 3 tenths to make one whole.
- Display a piece of rope or string 1 m long. Put a washer or peg on the rope. Move this along and state a measurement, for example, 0.62 m. Let the children take turns to work out the measurement of the remaining section of rope. The children can also play their own game like this using a metre rule or a tape measure marked in centimetres and millimetres.

Pairs that make 1

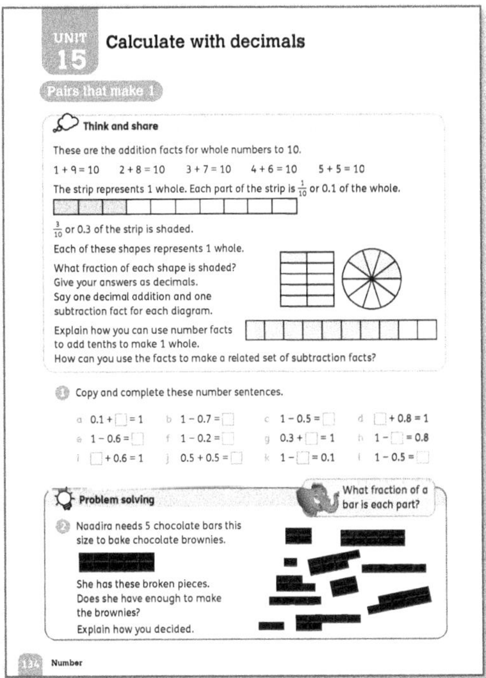

Materials

Strips divided into tenths; squared paper or counters; sets of decimal cards with values from 0.1 to 0.9

Warm-up

Put the children in groups, and give each group several sets of decimal cards. Ask each group to match up pairs that add to 1. Then ask for sets of 3 cards that add to 1. Compare their results. Any set of 3 that no other group has found gets a point. You can also extend to 4 cards.

Focus

- <u>Think and share:</u> Read through the addition facts to 10 on **Pupil book 5 page 134**.
- Give the children time to think about the questions and then let them share how they can use known facts to add and subtract tenths. (for example, 2 + 8 = 10, so 0.2 + 0.8 = 1.0). (For the 5 by 2, rectangle, the fraction shaded is $\frac{1}{10}$ or 0.3; 0.1 + 0.9 = 1.0; 1.0 − 0.9 = 0.1. For the circle, the fraction shaded is $\frac{4}{10}$ or 0.4; 0.4 + 0.6 = 1.0; 1.0 − 0.6 = 0.4. For the 1 by 10 rectangle, the fraction shaded is $\frac{8}{10}$ or 0.8; 0.8 + 0.2 = 1.0; 1.0 − 0.2 = 0.8. The numerators of the fractions are number bonds for 10, because $\frac{10}{10}$ is equal to one whole. Then you can use fact families to find related sets of subtraction facts.)
- Complete the shape activity orally as a class. Allow different children to answer the questions.
- The children can work independently to complete the calculations in question 1. They can use strips divided into tenths if they need to.
- <u>Problem solving:</u> If the children need additional support with question 2, let them model the pieces of chocolate using squared paper or counters and combine them to make whole bars of chocolate.

Support

To reinforce addition to 1, let the children use a set of decimal cards to play a game in pairs. They put the cards face down and take turns to turn over two cards. If the two cards make 1, the player keeps the cards. If not, the next player turns over two cards to see whether they make 1. The player who makes most pairs to 1 wins.

Answers for Pupil book 5 page 134

1   **a** 0.9   **b** 0.3   **c** 0.5   **d** 0.2
   e 0.4   **f** 0.8   **g** 0.7   **h** 0.2
   i 0.4   **j** 1   **k** 0.9   **l** 0.5

2   Each bar has 10 pieces so she needs 50 pieces in total.
She has only 48 pieces, so does not have enough.

Use facts to add hundredths

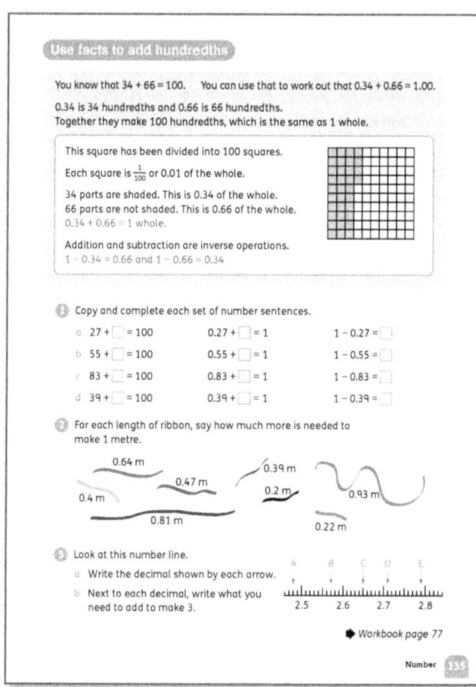

Materials

Real or play coins that add up to one whole currency unit (£1, $1 or 1 of the local currency); 100 chart; counters

Warm-up

Do the 'Fractions, decimals and percentages' activity from the Activity bank (pages 23–24).

Focus

- Let the children work with real or play coins to make amounts less than 1 dollar (or other whole currency unit). For example, give them the amount 27 cents. The children must write this as a decimal ($0.27) and then say how much they need to make a whole dollar ($0.73).
- You can do similar practical activities involving length, mass and capacity.
- Then turn to **Pupil book 5 page 135**. Ask questions as you work through the examples to make sure the children realise that 100 *hundredths* make 1 whole.

- For question 1, let the children use a 100 chart if they do not know the whole number facts to 100. At the end of the activity, let the children check their own work.
- The children can draw number lines or use 100 charts to help them to complete question 2 if they need to. Make sure they remember that there are 100 cm in 1 m.
- Let the children try to work out question 3 on their own. If they need support, spend some time reviewing how decimals with two places are shown on number lines.

Follow-up

Use **Workbook 5 page 77** to consolidate adding pairs of 1-digit decimals (tenths) to make 1 and adding pairs of 2-digit decimals (hundredths) to make 1.

Support

Give each pair a 100 chart and counters. One child covers some of the squares with counters. The other child works out what fraction of the 100 chart is covered and what fraction is uncovered. This will help to consolidate adding hundredths to make 1 whole and subtracting hundredths from 1 whole.

Answers for Pupil book 5 page 135

1   Missing numbers are:
   a 73, 0.73, 0.73   **b** 45, 0.45, 0.45
   c 17, 0.17, 0.17   **d** 61, 0.61, 0.61

2

Ribbon length (m)	Length needed to make 1 m (m)
0.64	0.36
0.47	0.53
0.39	0.61
0.81	0.19
0.2	0.8
0.4	0.6
0.22	0.78
0.93	0.07

3

	Decimal shown	Number to be added to make 3
A	2.48	0.52
B	2.57	0.43
C	2.65	0.35
D	2.71	0.29
E	2.79	0.21

Answers for Workbook 5 page 77

1   0.8 + 0.2, 0.45 + 0.55, 0.9 + 0.1, 0.65 + 0.35, 0.19 + 0.81, 0.76 + 0.24, 0.05 + 0.95, 0.77 + 0.23, 0.15 + 0.85, 0.75 + 0.25, 0.68 + 0.32, 0.5 + 0.5, 0.87 + 0.13, 0.62 + 0.38, 0.49 + 0.51, 0.3 + 0.7, 0.4 + 0.6, 0.58 + 0.42, 0.14 + 0.86, 0.99 + 0.01, 0.71 + 0.29, 0.59 + 0.41, 0.17 + 0.83, 0.84 + 0.16, 0.2 1 + 0.79, 0.41 + 0.59, 0.97 + 0.03

2   0.98; need 0.02
   0.4; need 0.6
   0.65; need 0.35
   0.16; need 0.84

Add and subtract across a whole number

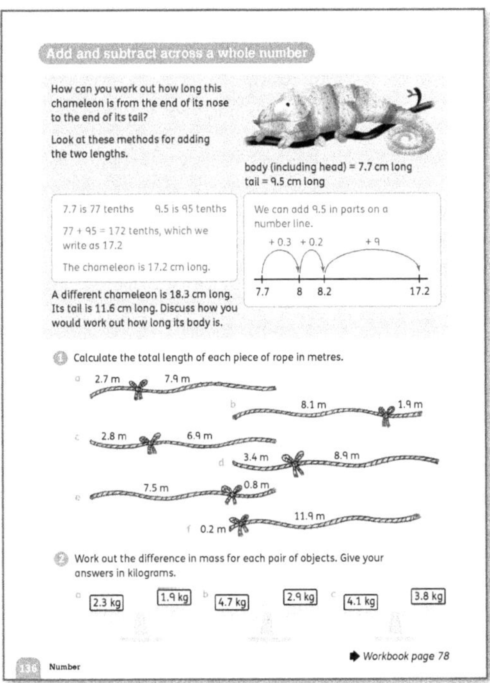

Materials

Cards with money amounts (with two decimal places) that are less than £10 or $10 written on them; blank cards for the children to write on

Warm-up

- Hand out the money cards to the children at random. Tell them this is the price of an item and ask them to make a card showing the amount they would get as change from £10.
- Discuss how they worked this out and ask the children to show any working they did.

Focus

- Do some more work with money amounts and/or measurements to make amounts equivalent to 10, then turn to **Pupil book 5 page 136**.
- Discuss the chameleon problem with the class. The children are asked to find the body length of a chameleon that is 18.3 cm long with tail length 11.6 cm. (Body length = 18.3 cm – 11.6 cm. You could work this out using column subtraction or by counting backwards or forwards on a number line. 11.6 + 0.4 = 12.0; 12.0 + 6.0 = 18.0; 18.0 + 0.3 = 18.3; 0.4 + 6.0 + 0.3 = 6.7. So, 18.3 cm – 11.6 cm = 6.7 cm.)
- Either let the children work on their own through the examples or do this with the class. Make sure that the children can read and understand the number line example.
- Let the children work on their own to complete question 1 and question 2. They can then check their answers using a calculator.

Follow-up

Use **Workbook 5 page 78** to consolidate the concepts from this lesson and to check that the children can derive facts to 10 using decimals.

Challenge

Challenge the children to use the money cards to design a game to teach other children how to add and subtract numbers with two decimal places. They do not have to make the cards add up to £10 in their game unless they think it is helpful to do so.

Interesting mistakes

The most common errors involve putting the decimal point in the wrong position or omitting the decimal point altogether. Remind the children to estimate before they calculate and to think about what the quantities mean. For example, say: *If you are working to find two amounts that add up to £10, your answer cannot be more than £10.*

Answers for Pupil book 5 page 136

1 a 10.6 m   b 10 m   c 9.7 m
  d 12.3 m   e 8.3 m   f 12.1 m
2 a 0.4 kg   b 1.8 kg   c 0.3 kg

Answers for Workbook 5 page 78

1 3.6 cm, 3.5 cm, 7.7 cm, 2.9 cm, 1.8 cm, 0.9 cm
2 7.55 ℓ, 5.1 ℓ, 1.93 ℓ, 8.72 ℓ, 5.01 ℓ
3 $4.53, $8.01, $3.60, $0.79

Add and subtract any decimals

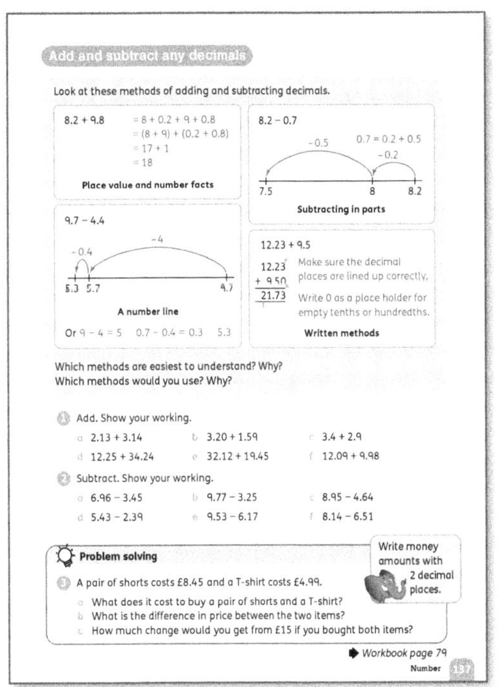

Materials

Tape measures; calculators; squared paper

Warm-up

Select any suitable magic square activity using whole numbers as a starter for this lesson.

Focus

- Use a tape measure to demonstrate adding and subtracting with any decimal amounts. For example, show a measurement of 0.8 m and say: *This is 0.8 metres. I want a length that is 0.4 metres longer than this, what will that be?* (1.2 m) Then show a 1.3 m length and say: *This is 1.3 metres. What is 0.5 metres less than this?* (0.8 m)
- Work through the different methods of adding and subtracting on **Pupil book 5 page 137** with the class. Encourage the children to try all the methods and then explain which they prefer. Their choice will depend on the numbers in the calculation.
- Discuss each method and encourage the children to suggest any other methods they think will work. Let the children use whichever methods they feel most comfortable with to complete question 1 and question 2.
- <u>Problem solving</u>: Let the children use the method they are most comfortable with to answer question 3.
- Give the children some problems involving numbers with 3 decimal places. For example, *I have 0.975 kg of flour. How much more do I need to make 1 kg?* (0.025 kg) *I have 1 litre of juice but I spill 0.225 litres. How much is left?* (0.775 litres)

Follow-up

The children solve the magic square puzzles on **Workbook 5 page 79** by applying what they have learnt about calculating with decimals. They could use calculators to complete these activities, but if they work mentally you will be able to assess how well they have grasped the concepts in this unit so far.

Interesting mistakes

Some children may add or subtract decimals starting from the right without thinking about *place value*. If they do this, encourage them to decompose/partition each number and then say how they could add or subtract the decomposed number. To help with understanding, the children could use place-value tables (page 21) or work on squared paper, using one column for the decimal point.

Answers for Pupil book 5 page 137

1 a 5.27   b 4.79   c 6.3
    d 46.49   e 51.57   f 22.07
2 a 3.51   b 6.52   c 4.31
    d 3.04   e 3.36   f 1.63
3 a £13.44   b £3.46   c £1.56

Answers for Workbook 5 page 79

1 a 223

b Grid B

22.6	21.9	22.4
22.1	22.3	22.5
22.2	22.7	22.0

c Grid C: yes this is still a magic square

17.1	16.4	16.9
16.6	16.8	17.0
16.7	17.2	16.5

2 a Grid D

3.1	12.2	5.7
9.6	7.0	4.4
8.3	1.8	10.9

b Grid E

1.55	6.1	2.85
4.8	3.5	2.2
4.15	0.9	5.45

c Individual answers.

3 Individual answers.

Multiply and divide by 10 and 100

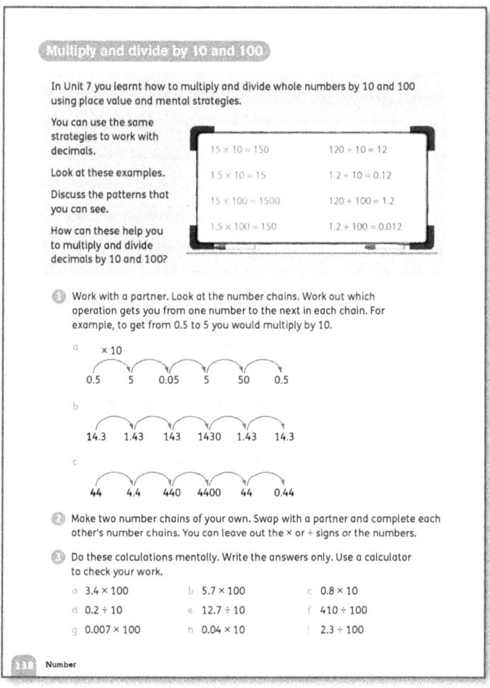

Materials

Place-value tables (page 21): calculators

Warm-up

Use 'Minus the money' from the Activity bank (page 27) as a starter for this lesson.

Focus

- Revise multiplication by 10 using place-value tables and calculators. Record the answers and make sure the children understand that to multiply by 10 we move the digits one place to the left.
- Next, ask: *What happens when we multiply by 10 and we multiply by 10 again?*
- Discuss the children's answers and explain that this is the equivalent of multiplying by 100.
- Do some examples with the class to establish the fact that to multiply by 100 you move the digits two places to the left.
- Repeat these steps for division by 10 and 100.
- Let the children work on their own to read through the examples on **Pupil book 5 page 138** and discuss what the patterns show. The children are asked to explain how the patterns help you to multiply and divide decimals by 10 and 100. (The non-zero digits in the calculation do not change. You can use place value to work out the number of zeros in the answer or the position of the decimal point.) They should realise that working with decimals is no different to the work they have already done with whole numbers.
- The children can work in pairs to do question 1 orally. Let them use calculators to check their answers and to resolve any disagreements as they work.
- The children can work in the same pairs to complete question 2. They should use a calculator to check the chains work and to check their partners' answers.
- For question 3, the children can use place-value tables if they need to.

Challenge

Ask the children to write ten multiplications and/or divisions by 10 and 100 with some mistakes in them. Then they can swap these with a partner and try to find the mistakes in each other's calculations.

Support

- Give the children a set of numbers (for example, 1800, 95 and 210). Ask them to work in pairs to write two multiplications and two divisions by 10 or 100 starting with each number. For example, $210 \times 10 = 2100$, $210 \times 100 = 21\,000$, $210 \div 10 = 21$, $210 \div 100 = 2.1$
- Move on to multiplications and divisions with decimal numbers. For example: $1.8 = 180 \div 100$, $1.8 = 18 \div 10$, $1.8 = 0.18 \times 10$, $1.8 = 0.018 \times 100$
- The children can discuss the problems and use place-value tables if they need to. Let them check their answers using a calculator.

Interesting mistakes

- When children multiply or divide a number such as 4.3 by 10, they sometimes multiply or divide the whole number part and the decimal part of the number separately. For example, they might write $4.3 \times 10 = 40.30$ or $4.3 \div 10 = 0.403$.

- Address such errors by asking the children to use a place-value table and to move the number in its entirety to the left or right (as appropriate).
- You can also have a number talk (pages 17–18) about mistakes like these.

> ### Answers for Pupil book 5 page 138
>
> **1** **a** $\times 10 \div 100$, $\times 100$, $\times 10$, $\div 100$
> **b** $\div 10$, $\times 100$, $\times 10$, $\div 1000$, $\times 10$
> **c** $\div 10$, $\times 100$, $\times 10$, $\div 100$, $\div 100$
> **2** Individual answers.
> **3** **a** 340 **b** 570 **c** 8
> **d** 0.02 **e** 1.27 **f** 4.10
> **g** 0.7 **h** 0.4 **i** 0.023

Multiply decimals by whole numbers

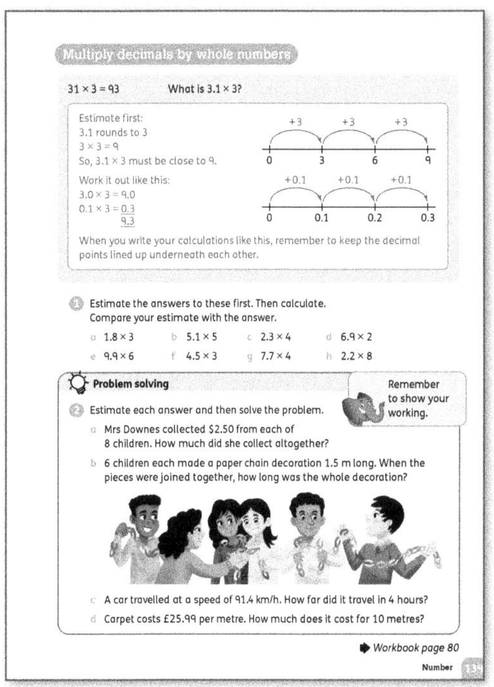

Materials

Calculators

Warm-up

Use 'Round decimals' from the Activity bank (page 24) as a starter for this lesson.

Focus

- Work through the example on **Pupil book page 139** to show the children how to multiply a decimal by a 1-digit number.
- Emphasise the estimating part of the example, as estimating will help the children work out where the decimal point should be in their answers.
- Work through the first few parts of question 1 as a class in the same way to reinforce the ideas. Let the children work independently to complete the question.
- <u>Problem solving:</u> The children can work on their own to solve the problems in question 2.

Follow-up

The children can complete **Workbook 5 page 80** to practise their multiplication skills. In question 1, let them complete one column at a time and then check their own answers and rate their accuracy. Use question 2 to assess whether the children can apply what they have learnt in a problem-solving situation.

Interesting mistakes

The children may make mental arithmetic errors in their calculations. For example, they might make times table errors or place-value errors, such as putting the decimal point in the wrong place. Emphasise the importance of estimating the answer first and then checking their work to decide whether or not the answer is reasonable.

Answers for Pupil book 5 page 139

1   a   5.4   b   25.5
    c   9.2   d   13.8
    e   59.4   f   13.5
    g   30.8   h   17.6

2   a $20   b   9 m
    c   365.6 km   d   £259.90

Answers for Workbook 5 page 80

1

×	2	3	4	5	6	7	8	9
1.1	2.2	3.3	4.4	5.5	6.6	7.7	8.8	9.9
1.8	3.6	5.4	7.2	9	10.8	12.6	14.4	16.2
1.7	3.4	5.1	6.8	8.5	10.2	11.9	13.6	15.3
2.3	4.6	6.9	9.2	11.5	13.8	16.1	18.4	20.7
2.5	5	7.5	10	12.5	15	17.5	20	22.5
2.9	5.8	8.7	11.6	14.5	17.4	20.3	23.2	26.1
3	6	9	12	15	18	21	24	27
3.1	6.2	9.3	12.4	15.5	18.6	21.7	24.8	27.9
3.3	6.6	9.9	13.2	16.5	19.8	23.1	26.4	29.7
3.7	7.4	11.1	14.8	18.5	22.2	25.9	29.6	33.3
4.5	9	13.5	18	22.5	27	31.5	36	40.5
4.8	9.6	14.4	19.2	24	28.8	33.6	38.4	43.2
5.9	11.8	17.7	23.6	29.5	35.4	41.3	47.2	53.1
7.2	14.4	21.6	28.8	36	43.2	50.4	57.6	64.8
9.6	19.2	28.8	38.4	48	57.6	67.2	76.8	86.4
9.9	19.8	29.7	39.6	49.5	59.4	69.3	79.2	89.1

2   Individual answers, for example: B 40 × 0.8; 80 × 0.4; 20 × 1.6; C: 20 × 0.05; 40 × 0.025; 50 × 0.02

More multiplying

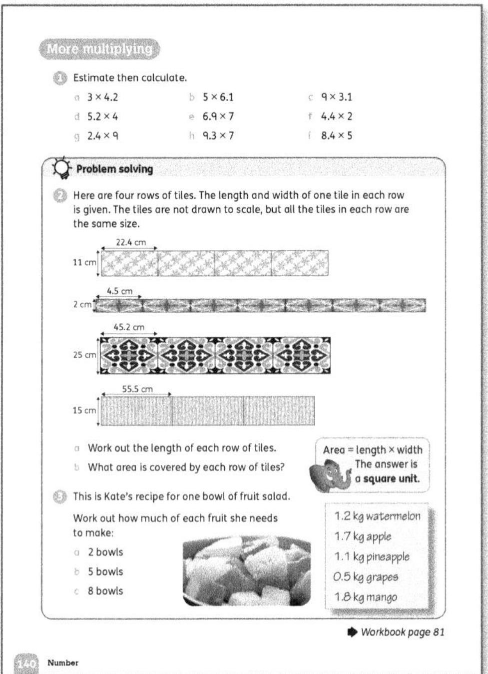

Materials

Calculators

Warm-up

Use 'Decimal times-tables' from the Activity bank (page 26) as a starter for this lesson.

Focus

- No new skills are taught in this lesson. The children need to think about what they have learnt and choose the best methods to answer the parts of question 1 on **Pupil book 5 page 140**.
- <u>Problem solving:</u> Read through the problem in question 2 with the class and check they understand what they are being asked, before they complete the questions on their own. Explain that when two lengths are multiplied, the answer needs to include a *square unit*, for example cm^2.
- The children could draw up a table to record their answers to question 3 so they do not need to write out the recipe three times.

Follow-up

Use **Workbook 5 page 81** to provide additional practice. Let them use calculators to check their own work.

Interesting mistakes

When the children work through multi-step problems like those on **Workbook 5 page 81**, they may make calculation or place-value mistakes if they do not estimate and keep track of their working. Remind them to write down an estimate and encourage them to use bar models to show the steps and help them keep track of what needs to be multiplied and which totals need to be added together.

Answers for Pupil book 5 page 140

1 a 12.6   b 30.5   c 27.9
     d 20.8   e 48.3   f 8.8
     g 21.6   h 65.1   i 42

2 a 89.6 cm; 31.5 cm; 180.8 cm; 166.5 cm
     b 985.6 cm²; 63 cm²; 4520 cm²; 2497.5 cm²

3 a 2.4 kg watermelon, 3.4 kg apple, 2.2 kg
        pineapple, 1.0 kg grapes, 3.6 kg mango
     b 6.0 kg watermelon, 8.5 kg apple, 5.5 kg
        pineapple, 2.5 kg grapes, 9.0 kg mango
     c 9.6 kg watermelon, 13.6 kg apple, 8.8 kg
        pineapple, 4.0 kg grapes, 14.4 kg mango

Answers for Workbook 5 page 81

1 a 16 kg, 25.2 kg, 3.3 kg, 44.8 kg
     b 9.7 kg, 36.4 kg
     c 27 kg, 67.5 kg

Mixed decimal problems

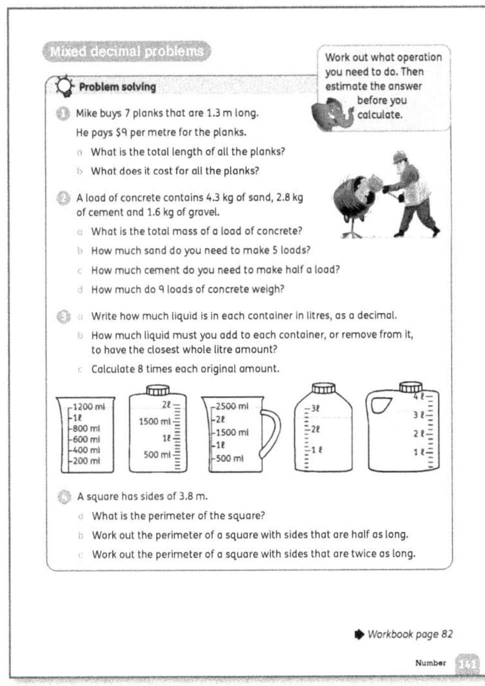

Materials
Poster paper and marker pens

Warm-up
You can use any 'Mental problem solving' activity from the Activity bank (pages 24–25) as a starter for this lesson.

Focus
Problem solving: In this lesson, the children need to combine the calculation skills they have learnt to solve word problems.

- Give the children time to read questions 1–4 on **Pupil book 5 page 141** and take questions from the class before asking them to solve the problems.
- Remind the children to write an estimate and to use bar models or other diagrams to show what they need to find out.

Follow-up
Use **Workbook 5 page 82** to assess how well the children can work with decimals. Let the children complete the work on their own (as a test if you prefer). Then provide the answers. Ask the children to mark their own work and to write a comment about what they understand well and what needs extra attention.

Challenge
Let the children develop a triomino puzzle using decimal calculations.

- Explain that a triomino puzzle is made up of triangles with questions and answers written along the sides. The aim is to match up the edges so that all joined edges show a question next to its answer. For example, here is part of a completed puzzle:

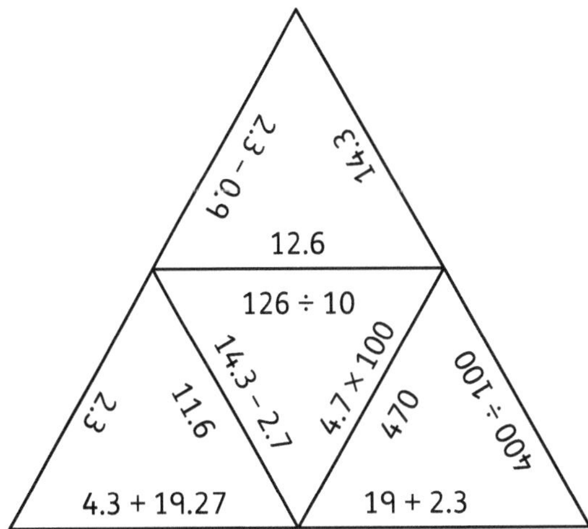

- Provide a template like the one below. The triangles need to be large enough to write calculations on the sides.

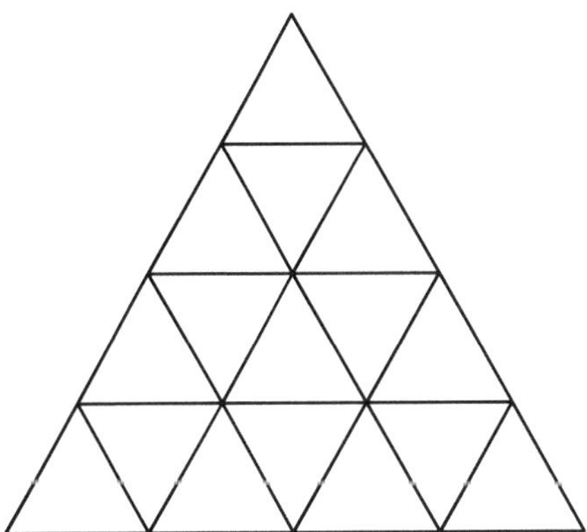

- Ask the children to outline their triangles in a different colour to their partner (so they can tell them apart once they cut the triangles out).
- Next, let the children develop their puzzles. They can be as challenging as they like.
- They can cut out the finished triomino and swap with a partner to solve.
- Store the finished puzzles in envelopes to reuse as a mental starter in later lessons.

Support
Provide poster paper and marker pens. Ask the children to make a poster that clearly shows how to add, subtract and multiply decimals. This will help the

children formalise their learning. They can use diagrams, show different representations and give examples. Display the posters in the classroom.

Answers for Pupil book 5 page 141

1 a 9.1 m   b $81.90
2 a 8.7 kg   b 21.5 kg
    c 1.4 kg   d 78.3 kg
3 a 1.2 ℓ, 1.9 ℓ, 2.5 ℓ, 2.6 ℓ, 3.25 ℓ
    b remove 0.2 ℓ, add 0.1 ℓ, add or remove 0.5 ℓ,
      add 0.4 ℓ, remove 0.25 ℓ
    c 9.6 ℓ, 15.2 ℓ, 20 ℓ, 20.8 ℓ, 26 ℓ
4 a 15.2 cm   b 7.6 cm   c 30.4 cm

Answers for Workbook 5 page 82

1

		29	2.9	0.29	13.5	1.35	1.3
×	10	290	29	2.9	135	13.5	13
÷	10	2.9	0.29	0.029	1.35	0.135	0.13
×	100	2900	290	29	1350	135	130

2 a Mathilde: one pen and one pencil; Jez: two
     pens and one pencil; Nazli: 3 pencils, one pen
     and one ruler

   b 2 pencil cases   £4.50
      5 pens   £9.95
      10 erasers   £2.50
      3 rulers   £3.60
      1 pencil   £0.95
      Total to pay   £21.50

UNIT 16   Volume and capacity

Learning objectives

- Estimate volume and capacity using appropriate standard units

- Use 1 cm³ blocks to estimate and work out the volume of different 3D shapes

- Use all four operations to solve problems involving volume and capacity

Key words

volume   cubic unit

Unit introduction

Materials
A selection of fruits and vegetables; objects and containers available in the classroom (such as pencils, rulers, cups, lunchboxes); sets of 24 cubes of the same size

Teaching guidance
Volume offers a great opportunity to do practical counting and measuring activities, while revising 3D shapes and their properties. Make sure that the children realise that all 3D objects have volume, no matter what shape they are.

- Start by showing children a selection of fruit and vegetables. Try to find some that have a similar volume but different shapes and masses. For example, a green pepper and an orange might have a similar volume, but a different shape.
- Ask the children to order the fruits and vegetables from the one that takes up the most space to the one that takes up the least space. Let them explain how they decided.
- Be clear that we are comparing the amount of space occupied by each object, not its shape, mass or what is inside it.
- Reiterate that every 3D object takes up space. Show some everyday objects and state that pencils, rulers, cups and lunchboxes all have volume because they take up space.
- Some objects can hold other objects. For example, a cup holds a certain amount of liquid. The amount of liquid it can hold is its capacity. A cup may have a capacity of 300 ml, but if it is only half full, the volume of the liquid is 150 ml, which is the same amount of space as 150 cm^3.
- Give each child a set of 24 cubes. The cubes need to be the same size but they do not need to be 1 cm^3 at this stage.
- Place the children in groups of four. Tell the class that each child must use all 24 of their cubes to make a cuboid that is different from the others in their group.
- Once the children have made their cuboids, go around the class and ask individual children to describe the cuboid they made. For example, 'My cuboid is 2 cubes high, 3 cubes wide and 4 cubes long.'
- Ask any other children who have built the same cuboid to raise their hands. Let the children compare to make sure that they agree that the cuboids are the same as the one described.
- You may like to have a class discussion about whether cuboids with the same dimensions but that that look different are actually the same or not. Mathematically, they are the same shape, just in different orientations, for example:

- As different children describe their cuboids, record the dimensions in a table like this one for the class to see:

Height	Width	Length	Total number of cubes
1	1	24	24
1	2	12	24
1	3	8	24
1	4	6	24
2	3	4	24
2	2	6	24

- There are only six possible cuboids using 24 cubes. The table shows all the possible dimensions.

- Discuss the fact that each cuboid takes up space. Point out that all of their cuboids take up the same amount of space: 24 cubes of space. Explain that, in maths, the amount of space that any 3D shape takes up is called the volume.
- You may wish to talk about how to work out the volume of a cuboid without taking it apart and counting all the cubes. Give an example that uses numbers that are easy to work with mentally. For example, show the class a picture of a large cuboid and say: *This cuboid is 24 cubes long, 10 cubes wide and 10 cubes high. How many cubes do you think were used to make the cuboid?* (2400 cubes) *How did you work this out?* ($24 \times 10 \times 10 = 2400$)
- By the end of the activity, most children should realise that to calculate the volume of a cuboid, we multiply the number of cubes in the length, width and height. They will return to this later in the unit.

Volume of 3D objects

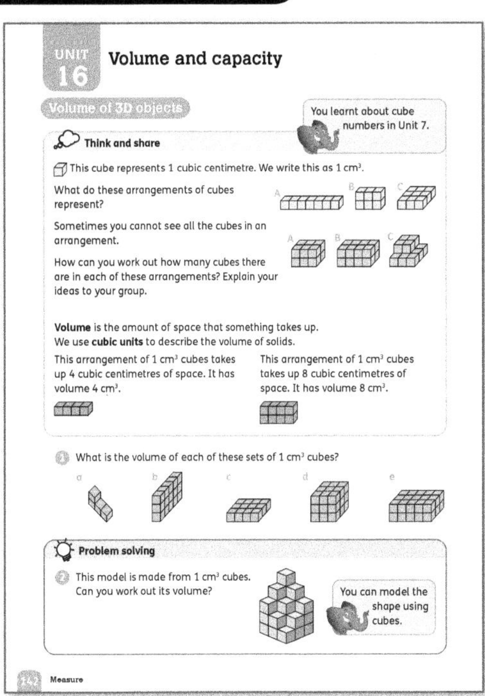

Materials
1 cm^3 cubes; empty box; online video explaining volume (if possible)

Warm-up
- Make up a set of multiplication facts with 3 numbers (for example: $2 \times 4 \times 5$, $3 \times 6 \times 3$, and so on). Ask the children to work out the answers as a starter for this lesson.
- Remind the children to look at the numbers and choose the most efficient strategy.

Focus
- If possible, find a short video online to introduce and explain the concept of *volume*. Check that the video is appropriate and at the correct level before showing it to the class.
- Think and share: Show the class a 1 cm^3 cube. Ask: *What is the length of this cube? The width? The height?* (all 1 cm) Discuss as a class what each arrangement of cubes on **Pupil book 5 page 142** represents. The children should give the

measurements using the unit cm³. (Arrangement A represents 7 cm³, B represents 6 cm³ and C represents 9 cm³. To find the total number of cubes in shapes A and B, you can count how many there are in the top layer and multiply by the number of layers: A has 2 × 6 = 12 cubes and B has 2 × 8 = 16 cubes. For C, you can find how many cubes there are in the bottom layer by working out 3 × 3 and then add on the 4 cubes on the top layer; this gives 13 cubes.)

- Let the children work in groups to discuss how they can work out how many cubes there are in the arrangements where some cubes are hidden. Take feedback when they have finished.
- Remind the class that volume is the amount of space that something takes up. (This is not the same as capacity, which is the amount (usually of liquid) that a container can hold.) Teach the notation for *cubic units*.
- The children can write just the answers to question 1, but if they do a calculation, they should write it down.
- <u>Problem solving:</u> Let the children work in pairs to find the volume of the model in question 2. Allow them to build the model with cubes if they need to.

Challenge

Ask the children to use cubes to make a model with approximately the same volume as one of their shoes. Let them show their models to the class and explain how they worked to make them. The children can use adhesive tack to keep the cubes in place.

Support

Give the children worksheets like the one shown here. Let them label the shaded blocks on each cuboid as length, width or height and complete the rest of the information. They can keep these on their desks for reference as you work through this unit.

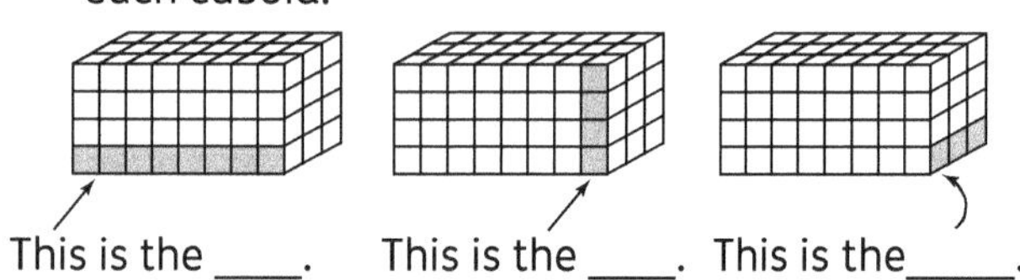

1 Fill in the correct name for the shaded part of each cuboid.

This is the ____.   This is the ____.   This is the ____.

2 Complete this table using the information in the pictures above.

Length	Height	Width	Total number of cubes

3 Complete this sentence.
You can work out the number of cubes in a cuboid if you know the length, width and height. This is how:

____

____

____

Interesting mistakes

- Ask the children to write a definition of *volume* to check whether or not they have grasped the concept. If they say things like 'how loud something is' you

will need to address the mathematical vocabulary. If they say it is how much can fit into something, they have mistaken it for capacity.
- Address mistakes by stating that volume is the amount of space something takes up. Use a closed, empty box and demonstrate that the box takes up space even though it is empty. If you unfold the box to make a flat piece of card, it takes up less space, so its volume is reduced, but it still has a height, a length and a width (but now it cannot hold anything).

Answers for Pupil book 5 page 142

1 a 4 cm³   b 15 cm³   c 12 cm³
     d 27 cm³   e 30 cm³

2 27 cm³

Think about volume

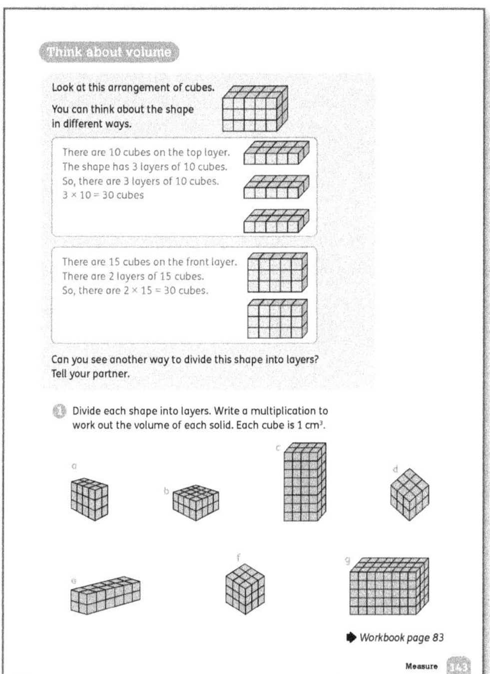

Materials

Cubes; colouring pencils or marker pens

Warm-up

Do any mental multiplication activity as a starter for this lesson.

Focus

- Ask the children to complete **Workbook 5 page 83** to help consolidate their understanding of volume and help them see that solids with the same volume may look quite different.
- Work through the example on **Pupil book 5 page 143** with the class, allowing the children to model the different layers if you think they need to. The children are asked whether they can see another way of dividing the shape into layers to find the volume. (There are 6 cubes in the side layer. There are 5 layers of 6 cubes. So, there are 6 × 5 = 30 cubes.)
- In question 1, the children may divide the shapes into different layers, but they will still get the same volume.

Dividing the shapes into layers provides a first introduction to the concept of volume of prisms as 'area of cross section × length'. The children will use this formula at higher levels, when they calculate the volume of triangular prisms and cylinders.

Interesting mistakes

- Children often omit the units when they work with volume. Remind the children to include the units by asking questions such as: *A volume of 24 what?*
- Some children might count only the cubes they can see and give that total as the volume of a shape. In particular, when they realise that an empty box has volume, they might think that the sides are the same as the volume. Continue to model volume using cubes and real objects to help the children understand the concept.

Answers for Pupil book 5 page 143

1 a $2 \times 3 \times 4 = 24 \text{ cm}^3$
   b $2 \times 4 \times 4 = 32 \text{ cm}^3$
   c $2 \times 4 \times 8 = 64 \text{ cm}^3$
   d $2 \times 3 \times 3 = 18 \text{ cm}^3$
   e $2 \times 2 \times 6 = 24 \text{ cm}^3$
   f $3 \times 3 \times 3 = 27 \text{ cm}^3$
   g $3 \times 5 \times 8 = 120 \text{ cm}^3$

Answers for Workbook 5 page 83

1 a 12 cm^3 b 8 cm^3 c 19 cm^3
   d 24 cm^3 e 31 cm^3 f 26 cm^3
   g 31 cm^3 h 12 cm^3 i 22 cm^3
   j 15 cm^3 k 15 cm^3 l 8 cm^3

Find the volume of containers

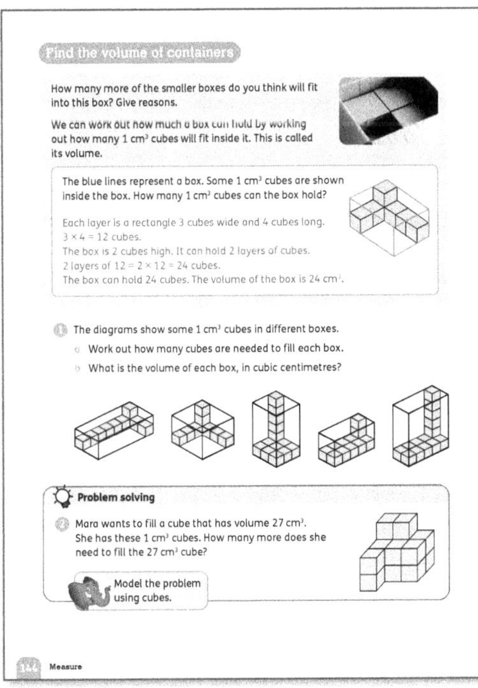

Materials

A small box (for example, a shoe box); cubes, all the same size

Warm-up

As a mental warm-up, give the children a set of three cards with a multiplication on each (no answers). Ask the children to order the cards from the smallest to the greatest product.

Focus

- Show the class a small box. Tell the class that it is full of cubes but that you cannot open it to find out how many are inside.
- Place the box and some cubes on a desk. Ask the children: *How could we use these cubes to work out how many cubes are inside the box?*
- The children may suggest building a cuboid the same size as the box. If so, say: *But what if we do not have enough cubes? How could we work it out then?*
- The children should share their suggestions and you should guide them to the idea that they can use the cubes to measure the length, width and height of the box. They can then use these measurements to work out the volume of the box.
- Ask the children to look at the box in the photograph on **Pupil book 5 page 144** and to share their ideas for working out how many smaller boxes it could hold.
  - They should give reasons for their answers. For example, they might say:
  - 'I can see that the big box is 2 boxes long and 2 boxes wide so there are 4 smaller boxes in a layer.'
  - 'I can see that the smaller boxes are about half the height of the bigger box, so it looks like you can fit in two layers of 4 smaller boxes, which gives a total of 8 smaller boxes.'
  - 'The box is 2 smaller boxes long, 2 wide and 2 high, so it can hold 8 smaller boxes.'
- Let the children read through the example in pairs and then ask them to complete question 1. Remind them to include units when they give the volume of each box.
- Problem solving: Let the children model the solution to question 2 if they need to.

Interesting mistakes

The children may need support to help them interpret diagrams like the ones in the Pupil book activities. For example, they may say that the shape in the first diagram in question 1 is 6 cubes long and 2 cubes wide because they do not realise that there has to be a cube in the corner. Let them use cubes to recreate the shapes to help them see where there are hidden cubes in diagrams.

Answers for Pupil book 5 page 144

1 a 32; 39; 40; 18; 45
   b 42 cm^3; 48 cm^3; 54 cm^3; 30 cm^3; 60 cm^3
2 11

Calculate volume

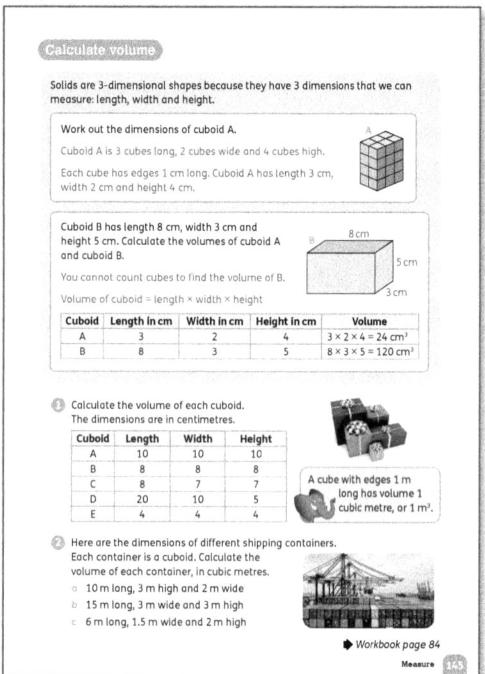

Materials
Selection of cuboid-shaped boxes or other cuboid-shaped solids; rulers; calculators; drawing equipment

Warm-up
Prepare a mental starter for this lesson by thinking of some target numbers (for example, 1000, 48, 320, 160, 212). Put the children in groups and give them one of the numbers. Ask the groups to make up four different three-number multiplication facts with that number as the product. Repeat with other numbers if time allows.

Focus
- Hand out the cuboid shapes (boxes or other solids) and ask the children to measure and record the length, width and height of each in centimetres. They can use decimals for parts of centimetres.
- Ask the class how these measurements can help them work out the volume of each cuboid. Take feedback and suggestions.
- Remind the class that when we are working with real objects it is usually not practical to lay out and count blocks.
- Turn to **Pupil book 5 page 145**. Read through the information and the examples with the class, relating it to their practical work with boxes.
- Let the children work out the volume of each cuboid, using a calculator to check their answers if needed.
- The children can complete question 1 and question 2 on their own.

Follow-up
Use **Workbook 5 page 84** to consolidate work on calculating volume, to revise work on inverse operations and to provide additional practice in drawing 3D shapes.

Challenge
- Ask the children to design a box to hold two cylindrical packets of crackers. The packets have a 6 cm diameter and are 15 cm long.

- The children can either produce a scale drawing of the box or draw a net and make the box actual size. They can then show their boxes to the class, explain how they know that two packets of crackers will fit inside and demonstrate how they worked out the dimensions.

Answers for Pupil book 5 page 145
1 1000 cm³, 512 cm³, 392 cm³, 1000 cm³, 64 cm³
2 **a** 60 m³   **b** 135 m³   **c** 18 m³

Answers for Workbook 5 page 84
1 A = 6, B = 6, C = 360, D = 90, E = 3
2 **a** For example, 3 cm, 3 cm, 4 cm. Other answers are possible.
b For example, 9 cm, 3 cm, 4 cm. Other answers are possible.
3 Individual answers.

Solve volume problems

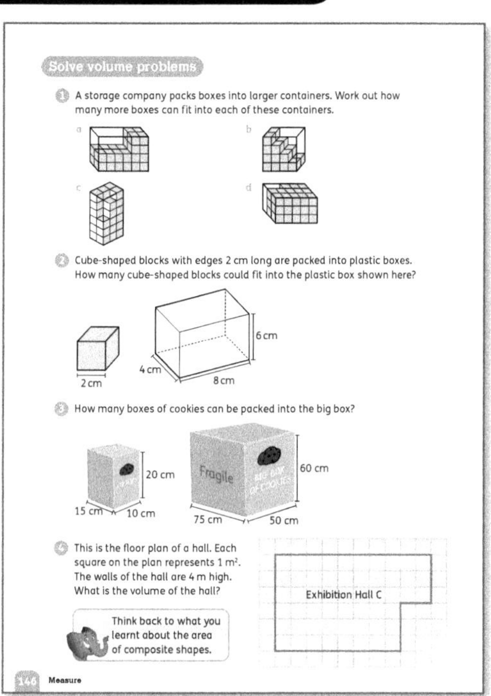

Materials
Cubes

Warm-up
Choose any mental division fact activity as a starter for this lesson.

Focus
- Ask the children to complete the activities on **Pupil book 5 page 146**.
- If the children need support to visualise or count the missing boxes in question 1, let them use cubes to create models of the shapes. They do not actually need to work out the volume of each solid to work out how many boxes are missing. Instead, they have to work out the volume of the missing boxes.
- Encourage the children to draw diagrams to help them to solve the problems in questions 2 and 3.

- To solve the problem in question 4, the children need to divide the floor into two rectangles, calculate the volume for each section of the room and then add the two volumes together. Alternatively, the children could find the volume of the complete cuboid and subtract the volume of the 'missing' section.

Challenge

As a fun activity to finish this unit, the children can work in groups and use squared paper, card, glue, rulers, marker pens and scissors to design and make multi-level model skyscrapers. For example:

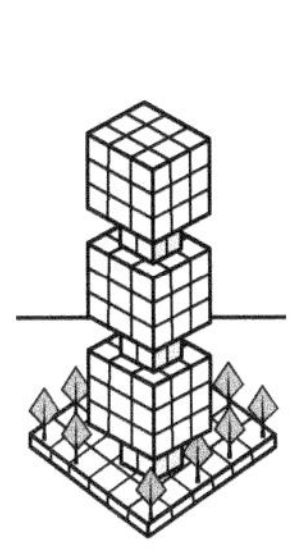

- Encourage the children to look at photos of some interesting buildings to inspire them (for example, the Burj Khalifa in Dubai, UAE, the Petronas Towers in Kuala Lumpur, Malaysia, the CCTV Headquarters in Beijing, China, and the Grande Arche in Paris, France).
- The children should stick the squared paper onto the card before they make the cuboids for their models.
- Once they have done this, they can estimate the volume of each building using the squares on the paper to represent 1 m squares.
- A simpler task would be to build the model using interlocking cubes.

Answers for Pupil book 5 page 146

1 a 16   b 12   c 4   d 6
2 24   **3** 75   **4** 216 m³

Ask some or all of these questions to assess how well the children have understood the concepts in this unit.

- Show the children some 3D shapes and ask: *Which of these 3D shapes takes up the most space? How do you know?*
- *Why are cuboids called 3D shapes?* (They have lengths in three directions.)
- *A cuboid is built using 12 cubes, what could its three dimensions be?* (1 by 1 by 12, 1 by 2 by 6, 1 by 3 by 4, 2 by 2 by 3)
- *A cuboid is made up of 4 layers. Each layer is made of 15 blocks, each with a volume of 1 cm³. What is the volume of the cuboid? How did you work this out?* (4 cm × 15 cm × 1 cm = 60 cm³)
- Show the children some small cubes of different sizes and ask: *Which of these small cubes has a volume of 1 cm³? How can you check this?*
- *If you turn a cuboid so its length becomes its height, how does this affect the volume?* (It stays the same.)
- *Two different cuboids each have a volume of 16 cm³. What could the dimensions of the cuboids be?* (1 cm by 1 cm by 16 cm, 1 cm by 2 cm by 8 cm, 1 cm by 4 cm by 4 cm, 2 cm by 2 cm by 4 cm)
- *A wooden cube has sides 9 cm long. Eight of these cubes fit together to make a cuboid. What is the volume of the cuboid?* (5832 cm³) *How could you arrange the cubes to make the cuboid?* (1 by 1 by 8 cubes, 1 by 2 by 4 cubes, 2 by 2 by 2 cubes)
- *What is the difference between area and volume?* (Area is the space taken up by a 2D shape. Volume is the space taken up by a 3D shape.)
- *What is the volume of a brick that is 12 cm wide, 10 cm high and 20 cm long?* (2400 cm³)
- *What measurements do you need to find the volume of the classroom?* (length, width and height)
- *Estimate the volume of the classroom in cubic metres.*

UNIT 17 Ratio and proportion

Learning objectives

- Understand that a proportion compares a part to a whole and that a ratio compares part to part of two or more quantities

Key words

ratio   proportion

Materials

Print-outs of the golden ratio rectangle in different orientations (see below); compasses and pencils; sheets of thin paper

Teaching guidance

- Ask the class whether they have ever heard of the *golden ratio*. If they have, let them tell the others what they understand by this term.
- Explain to the class that the golden ratio is a maths term for the ratio 1 : 1.618 which is found very often in natural objects.

- Display a large version of the golden ratio rectangle and use this to explain the golden ratio visually.

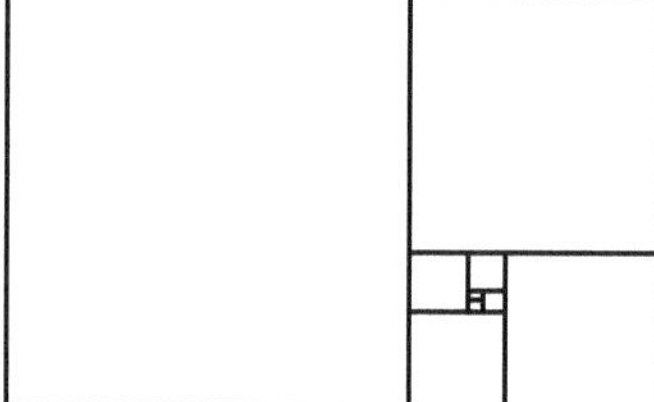

- In the large rectangle, the ratio of width to length is 1 : 1.618.
- If you draw a perfect square inside the rectangle, the leftover part of the rectangle has sides in the same ratio.
- As you continue to draw squares, you get smaller leftover rectangles, which all have sides in the same ratio.
- If you draw a spiral inside the golden rectangle, you get a design with proportions that are visually pleasing.

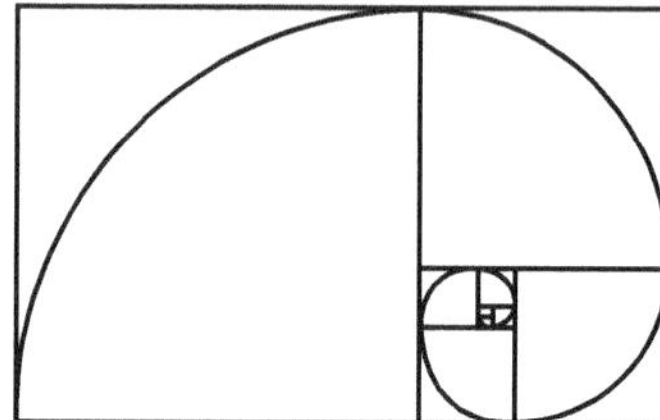

- This spiral design is linked to the Fibonacci sequence of numbers and is found in many different natural objects from fern fronds to the pattern formed by clouds in a hurricane.
- Artists, engineers and designers use the golden ratio to plan their work so that it is visually pleasing.
- The children could search the Internet for examples of the golden ratio in art and design. They will find famous works of art and well-known historical buildings that show this ratio.
- Give the children the golden ratio rectangle in a different orientation (see the one below). Ask the children to use a pair of compasses to draw the golden spiral. You may need to help some children by showing them how to place their compasses and pencil as shown below. They will then need to reposition the compasses and adjust the size to draw a quarter circle in each square.

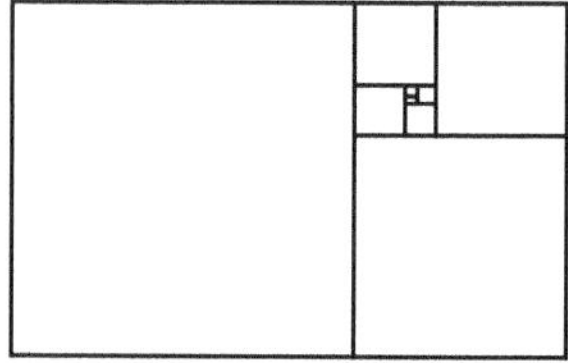

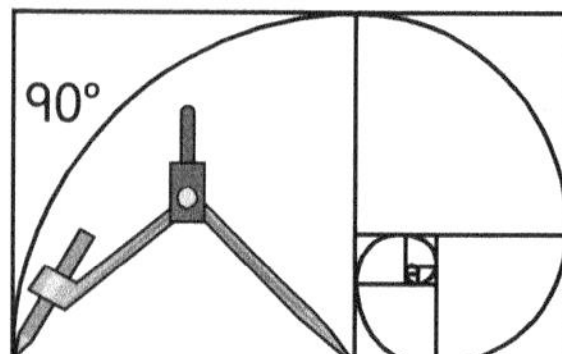

- Once the children have drawn the spiral, ask them to place a thin sheet of paper over the rectangle and spiral and draw a picture or a logo using these proportions. Show an example of a simple landscape silhouette like this one.

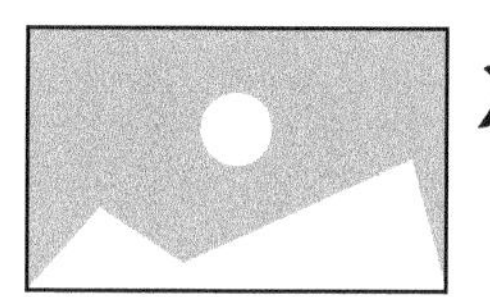

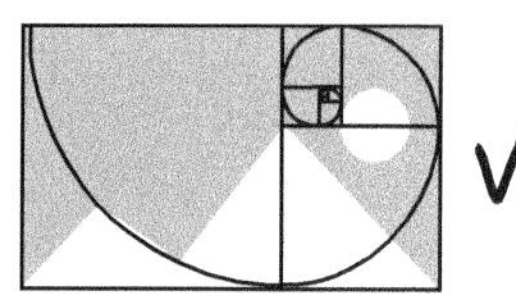

- Display the children's completed designs and pictures in the classroom.

What is a ratio?

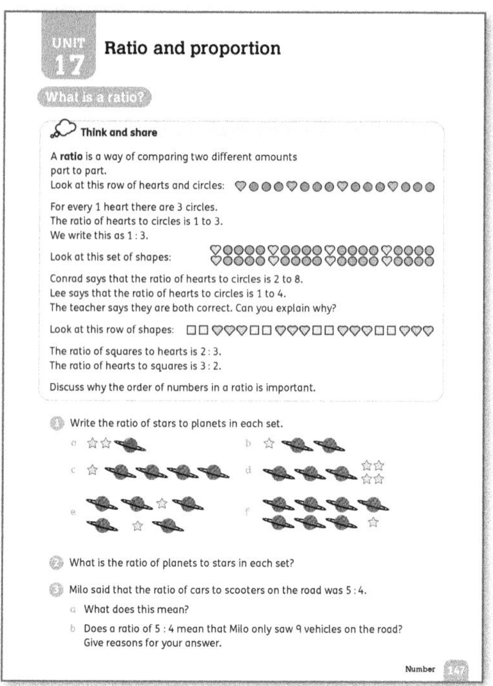

Materials
Sets of beads, counters, cubes or other small objects in different colours; bottle of fruit squash or cordial with mixing instructions; water; measuring jug and glasses; squared paper

Warm-up
Do an equivalent fraction activity as a mental warm-up for this lesson.

Focus
- Tell the class that a *ratio* compares two different amounts part to part. For example, 1 calculator for every 2 children.
- Let the children work in pairs or small groups to use beads or cubes to make patterns with colours in given ratios, such as a ratio of blue to yellow beads of 2 to 3.
- Ask questions to make sure the children understand that the position of the beads in the pattern is irrelevant as long as there are 2 blue beads for every 3 yellow beads. Explore different patterns and ratios.
- Let the children read the mixing instructions on the bottle of cordial and practise measuring out correct amounts of cordial and water in order to dilute the drink in the correct ratio. Ask the class what will happen if you change the order of the ratio. For example, if you mix cordial to water in the ratio 3 : 1 instead of 1 : 3.
- <u>Think and share:</u> Let the children read through the information on **Pupil book 5 page 147** and answer the questions in their groups. Let different children share their answers with the class.
- Make sure that the children realise that a ratio of 1 : 4 is equivalent to a ratio of 2 : 8. Use an example

to demonstrate this: *If we need 1 table for 4 children, we would need 2 tables for 8 children.* Also check that the children can explain why the order of numbers in the ratio matters. (Changing the order of the numbers changes the proportions. A ratio of tables to chairs of 1 : 4 means 1 table for every 4 children, whereas a ratio of tables to chairs of 4 : 1 means 4 chairs for every 1 child.)

- Let the children work through questions 1–3 independently.

Challenge

Ask the children to show that the ratio of the length of a side of a square to its perimeter is fixed and that the ratio $s : p$ will always be 1 : 4.

Support

- Give the children 50 cubes in two different colours and a ratio such as 2 : 3 or 1 : 4. Ask the children to lay out the cubes to show the given ratio. Let the children explain how they worked to do this.
- Alternatively, ask the children to colour squares on a strip of squared paper in a given ratio of blue : yellow.

Interesting mistakes

The children may not make the connection between ratio and the parts of the whole, so they will not understand that a ratio of 3 to 5 means that the whole is divided into two groups, making $\frac{3}{8}$ and $\frac{5}{8}$. Lots of practical activities will help the children understand the concept of ratio.

Answers for Pupil book 5 page 147

1 a 2 : 1 b 1 : 2 c 1 : 4 d 4 : 3 e 2 : 5 f 1 : 7
2 a 1 : 2 b 2 : 1 c 4 : 1 d 3 : 4 e 5 : 2 f 7 : 1
3 a For every 5 cars there are 4 scooters
   b No, he may have seen many more than 9 vehicles. The ratio 'total number of cars : total number of scooters' simplifies to 5 : 4

What is a proportion?

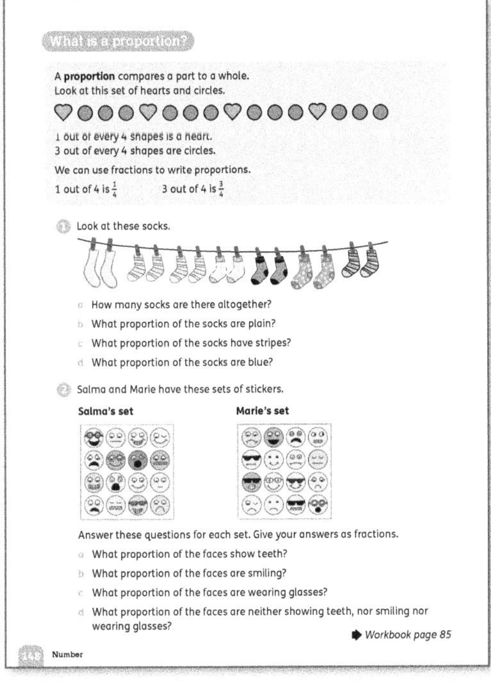

Materials

Colouring pencils or marker pens; materials to make a poster

Warm-up

Use an activity that involves adding fractions with the same denominators to make a total of 1 as a mental starter for this lesson.

Focus

- Introduce the idea of *proportion* by asking the children to colour the shapes on **Workbook 5 page 85** in the given proportions. Discuss how they worked out what this meant. You could also ask the children to give the colours as ratios as well. For example, ask: *You coloured half of this shape blue and half yellow. What is the ratio of blue to yellow squares?*
- Once the children understand the term *proportion*, work through the information and examples on **Pupil book 5 page 148** as a class.
- Do question 1 orally with the class. Let the children think about their answers and then take feedback as a class.
- Let the children complete question 2 on their own. The children can make a table to record the answers for each set of stickers.

Challenge

Ask the children to work in pairs to design a poster that explains ratio and proportion and gives examples of ratio and proportion used in everyday life. Examples could include:

- *The price of posting a parcel is 50 cents for every 100 grams.*
- *One quarter of the sweets in a bag are toffees.*
- *The ratio of toffees to chocolates in the bag is 1 : 3*
- *The teacher-to-child ratio in our school is 1 : 25.*
- *Half of the children go home for lunch.*
- *We can enlarge or reduce a page or picture when we photocopy it. The scale of the enlargement or reduction is a ratio (for example, 1 : 2) and the dimensions of the enlarged or reduced photocopy are in the same ratio as those of the original.*

Answers for Pupil book 5 page 148

1 a 14 b $\frac{2}{14}$ or $\frac{1}{7}$ c $\frac{6}{14}$ or $\frac{3}{7}$ d $\frac{6}{14}$ or $\frac{3}{7}$
2 Salma's set: a $\frac{5}{16}$ b $\frac{6}{16}$ or $\frac{3}{8}$ c $\frac{3}{16}$ d $\frac{7}{16}$
   Marie's set: a $\frac{3}{16}$ b $\frac{4}{16}$ or $\frac{1}{4}$ c $\frac{5}{16}$ d $\frac{7}{16}$

Answers for Workbook 5 page 85

1 Different answers are possible. here are some suggestions.

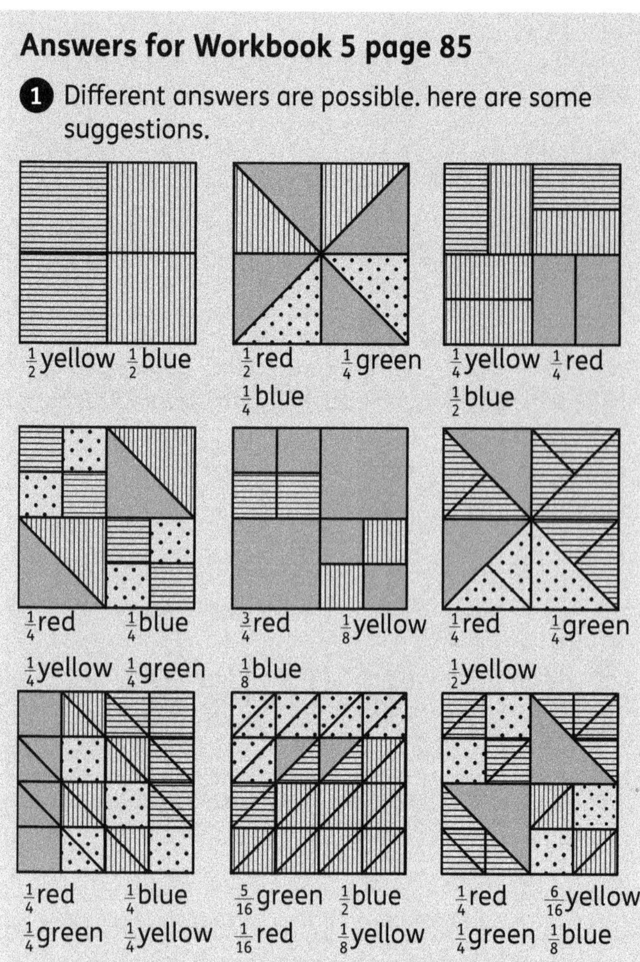

$\frac{1}{2}$ yellow $\frac{1}{2}$ blue	$\frac{1}{4}$ red $\frac{1}{4}$ green $\frac{1}{4}$ blue	$\frac{1}{4}$ yellow $\frac{1}{4}$ red $\frac{1}{2}$ blue
$\frac{1}{4}$ red $\frac{1}{4}$ blue $\frac{1}{4}$ yellow $\frac{1}{4}$ green	$\frac{3}{4}$ red $\frac{1}{8}$ yellow $\frac{1}{8}$ blue	$\frac{1}{4}$ red $\frac{1}{4}$ green $\frac{1}{2}$ yellow
$\frac{1}{4}$ red $\frac{1}{4}$ green $\frac{1}{4}$ blue $\frac{1}{4}$ yellow	$\frac{5}{16}$ green $\frac{1}{16}$ red $\frac{1}{2}$ blue $\frac{1}{8}$ yellow	$\frac{1}{4}$ red $\frac{1}{4}$ green $\frac{6}{16}$ yellow $\frac{1}{8}$ blue

Work with ratio and proportion

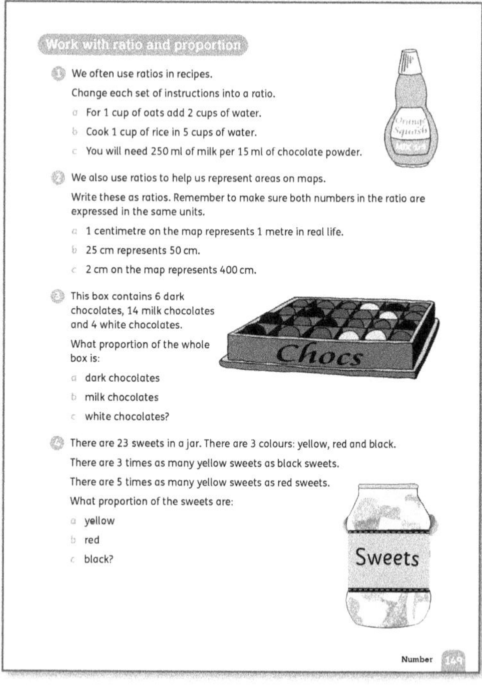

Warm-up
Do some doubling, tripling and halving activities with whole numbers, as a mental starter for this lesson.

Focus
- Work through question 1 and question 2 on **Pupil book 5 page 149** with the class to demonstrate how to express statements given in words as ratios.
- Let the children work independently to complete question 3. Check their answers to make sure that they are able to work with proportion.
- For question 4, remind the children that they can use the 'guess, check and improve' strategy to find the number of sweets in each colour before they try to find the proportions. Drawing a bar model will also help them see that the three parts must add up to 23.

Challenge
Let the children research the amount of sugar in different soft drinks and fruit juices. They will need to decide how to express their findings as proportions (percentages work well). The children also need to present their results. They could do this visually by drawing the appropriate number of small sachets of sugar per bottle or carton.

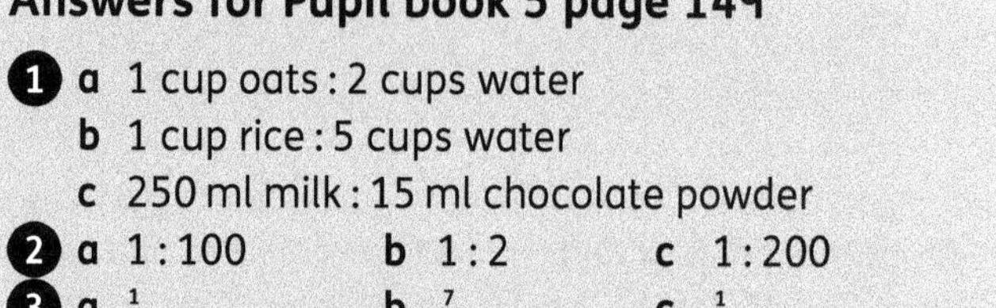

Answers for Pupil book 5 page 149
1 a 1 cup oats : 2 cups water
b 1 cup rice : 5 cups water
c 250 ml milk : 15 ml chocolate powder
2 a 1 : 100   b 1 : 2   c 1 : 200
3 a $\frac{1}{4}$ b $\frac{7}{12}$ c $\frac{1}{6}$
4 a $\frac{15}{23}$ b $\frac{3}{23}$ c $\frac{5}{23}$

Solve ratio and proportion problems

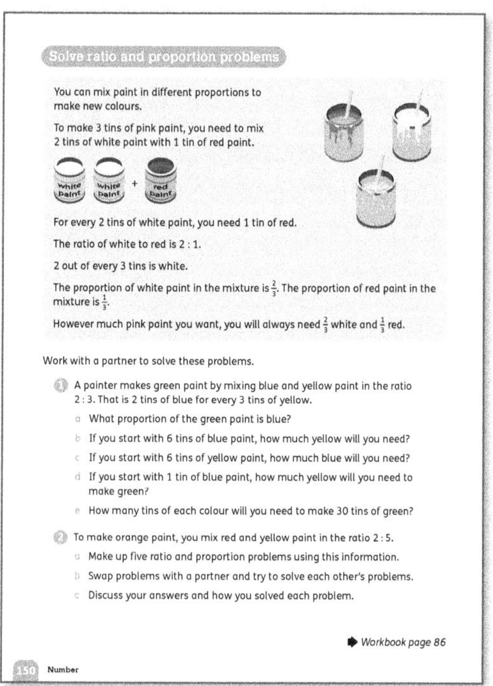

Materials
Paints in different colours (if possible); teaspoons and pots for mixing paint; paint colour swatches (from a hardware or paint store, or online)

Warm-up

Do any mental calculation activity using multiplication and/or division from the 'Calculation skills' section of the Activity bank (pages 25–28).

Focus

- If possible, do an activity in which the children work in small groups to mix paints to make different colours.
- Explain to them that they need to keep track of how much of each colour (for example, how many teaspoons) they use so that they can repeat the experiment and make the same colour. They should record each mixture using ratio and/or proportion.
- The children can test their ratios or proportions by asking another group to mix the colours using the same ratio/proportion. The groups can compare the colours to see if they match.
- Show the class some paint swatches and explain that these colours are mixed using very specific proportions. The amount of each colour you buy (for example: 250 ml, $\frac{1}{2}$ litre or 10 litres) does not affect the proportions.
- Work through the paint-mixing example on **Pupil book 5 page 150** with the class.
- Let the children work through question 1 on their own. Check and discuss the answers.
- For question 2, the children can work in pairs to develop and solve each other's problems.

Follow-up

Use **Workbook 5 page 86** as an informal assessment task (in test conditions if you wish) to check that the children have mastered the important concepts of ratio and proportion.

Interesting mistakes

- The children may not understand how to use multiplicative reasoning to increase or decrease amounts in proportion. Emphasise that doubling, halving, finding three times as much or finding a quarter all involve multiplying or dividing.
- When we compare amounts and express change, we often use multiplicative reasoning. You could give an example such as: *This type of battery lasts 3 times longer than other brands.* The children need to understand that this means it lasts 3 *times* as long as the time for other brands and not the time plus 3.

Answers for Pupil book 5 page 150

1 a $\frac{2}{5}$ b 9 tins
   c 4 tins   d 1.5 tins
   e 12 tins blue, 18 tins yellow
2 a-c Individual answers.

Answers for Workbook 5 page 86

1 1000 g margarine, 1500 g sugar, 1000 g self-raising flour, 1200 g dark chocolate, 20 eggs
2 6 cakes
3 1200 g
4 a $87.50
   b $70
   c $218.75
   d $875
5 29 cakes

End-of-unit check

Ask some or all of these questions to assess how well the children have understood the concepts in this unit.

- *There are 15 sheep and 5 cows in a field. What sentence could you say to describe this situation using the word 'ratio'?* (For example: The ratio of sheep to cows is 15 to 5 or 15 : 5 or three to one or 3 : 1.)
- *What question could have the answer '3 to 2'?* (For example: In this necklace, there are 3 red beads and 2 blue beads. What is the ratio of red beads to blue beads?)
- *When might you use ratio in real life?* (For example: Cooking, mixing juice and water.)
- *This recipe is for 6 people. What must I do to the ingredients if I want to make enough for 3 people?* (Halve the amounts.) *What if I want to make enough for 2 people or 9 people?* (2 people: divide the amounts by 3. 9 people: multiply the amounts by 1.5.)
- *What proportion of this beaded necklace is silver beads?* (Show them a necklace or picture of a necklace.)
- *In which situations do you think we use these ratios?*
  - *Use 1 litre to cover 5 square metres of wall.* (painting a wall)
  - *Mix 1 part flour to 1 part water.* (making glue or dough)
  - *Melt 100 g chocolate for every 200 ml of cream.* (making a chocolate sauce)
  - *You will need 5 ml of butter per slice of bread.* (making sandwiches)
- Draw a set of shapes and shade them using three different colours. *What proportion of the shapes are in each colour?*
- Provide some problems involving ratio and proportion. For example:
  - *The sizes of three angles in a triangle are in the ratio 2 : 1 : 3. What is the size of each angle?* (60°, 30°, 90°)
  - *A triangle with a perimeter or 54 cm has sides with lengths in the ratio 4 : 5 : 3. What are the lengths of the sides?* (18 cm, 22.5 cm, 13.5 cm)
  - *7 bags of dried fruit cost £17.50. The bags are all the same price. What do 5 bags cost?* (£12.50)

Learning objectives

- Describe and compare the likelihood of familiar events using the language of probability

- Position events on a simple probability scale

- Understand that some outcomes are more likely than others and that some outcomes are certain, while others are impossible

- Carry out simple probability experiments and use tables, graphs and other diagrams to represent and interpret the results

Key words

probability   chance   certain   likely
equally likely   unlikely   impossible
probability scale   probability experiment
possible outcomes

Unit introduction

Materials

Flashcards with an everyday event written or printed on each card (Include some events that are certain, some that are impossible, some that are likely and some that are unlikely. For example: I will eat bread today. I will do homework this evening. I will go to bed early tonight. It will rain later today. The sun will rise tomorrow morning. Our school football team will win their next match. I will make a mistake in my calculations.)

Teaching guidance

Do this activity to teach the key words for describing *probability* (*certain, likely, unlikely, impossible*):

- Display the key words, let the children suggest meanings and then explain that in mathematics these terms have specific meanings.

- Give examples to show what each word means. For example: *If it is Monday today, we can be <u>certain</u> that tomorrow will be Tuesday.* Similarly, *If it is Monday today, it is <u>impossible</u> for tomorrow to be Friday. If I usually eat rice every day, it is <u>likely</u> that I will eat it tomorrow.*

- Display one of the flashcards. Ask the children, *What is the chance that this will happen?*

- Discuss any disagreements. For example, one child may say that it is unlikely they will make a mistake in their calculations because they are careful and accurate; another may say it is likely that they will make a mistake because they find it difficult to add in their head.

- After this discussion, display all of the flashcards. Let the class rank the events in order from certain to

impossible. Encourage the children to give reasons for their positioning of the events. They may wish to position some events at the same level.

Do not introduce the probability scale at this stage. The children will learn about that in a later lesson.

How likely is it?

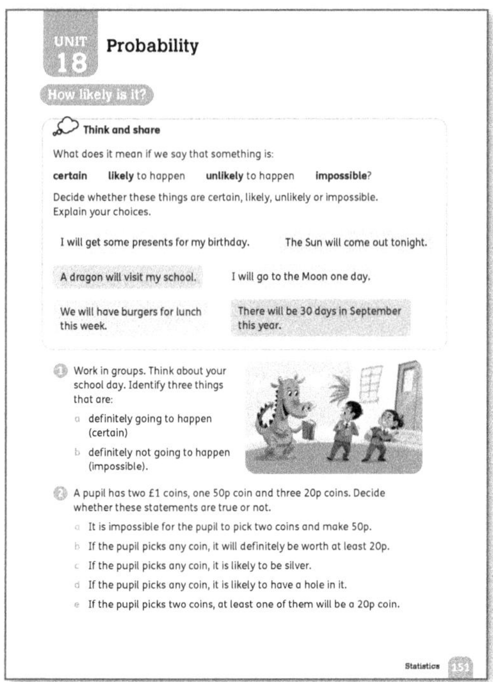

Materials

Coins (or counters marked with the amounts £1, 50p and 20p)

Warm-up

Select suitable activities from the Activity bank (pages 22–29) for each lesson in this unit. Bear in mind that the children will not be doing any number calculation work, so you should try to include some calculation activities.

Focus

- <u>Think and share:</u> Turn to **Pupil book 5 page 151**. Let the children work in groups to revise the meanings of the probability words and use them to describe the given situations. Take feedback from the groups when they have done this.

- The children should do question 1 orally and reach consensus in their groups. Let the groups share their ideas and justify their choices if others disagree with them.

- Give the children a set of coins to help them with question 2 if they cannot think abstractly.

Interesting mistakes

The main focus at this level is on using everyday terms to talk about the chance of something happening. Children whose first language is not English may be unfamiliar with the terms *certain*, *likely*, *equally likely/unlikely* and *impossible*. Make sure that you teach these words and their meanings before you start the activities. Also check with the children as they work through the unit that they remember and can use the terms.

Answers for Pupil book 5 page 151

1 **a** Individual answers.
  b Individual answers.
2 **a** true **b** true **c** true
  d false **e** false

The probability scale

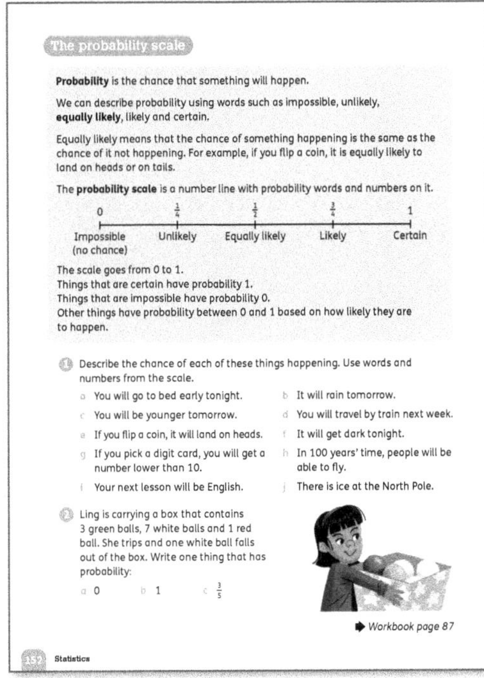

Materials

Large display version of the probability scale, with a slider that can move across it

Focus

- Display the *probability scale*. Explain that we use the words on the scale to talk about how likely something is to happen. These words describe the chance or probability of the event happening. Point out the words on the scale and use the slider to show that events can fit anywhere on the scale.
- Ask the class what it means if there is 'zero chance' that something will happen. Point out that events with zero chance of happening are impossible.
- Then, show them that the scale of probability extends from 0 to 1. Events that are certain have a probability of 1. We can give all other events a value between 0 and 1 to show the chance that they will happen.
- Turn to the probability scale on **Pupil book 5 page 152**.

Point out that an event that is likely has a probability of $\frac{3}{4}$. Ask the class: *What other numbers could you use to describe this probability?* (0.75 and 75%)

- The children can do question 1 on their own and then compare and discuss their answers with a partner.
- The children can discuss question 2 orally in pairs or groups before they write their own answers.

Support

Use **Workbook 5 page 87** to provide extra practice for any children who would benefit from additional support in using the probability scale. Let the children work in pairs to discuss and complete the activities.

Answers for Pupil book 5 page 152

1 **a, b, d, i** Individual answers.
  c impossible, 0 **e** equally likely, $\frac{1}{2}$
  f certain, 1 **g** certain, 1
  h impossible, 0 **j** certain, 1
2 Possible answers:
  a The ball that falls out is blue
  b The ball that falls out is white.
  c The next ball that she takes from the box is white.

Answers for Workbook 5 page 87

1

Impossible	Unlikely	Equally likely	Likely	Certain
0	0.25	0.5	0.75	1

2 Individual answers which should include: if this year is a leap year the probability is 1, and if it is not a leap year the probability is 0; the probability that next year I will be older than I am now is 1.

More probability

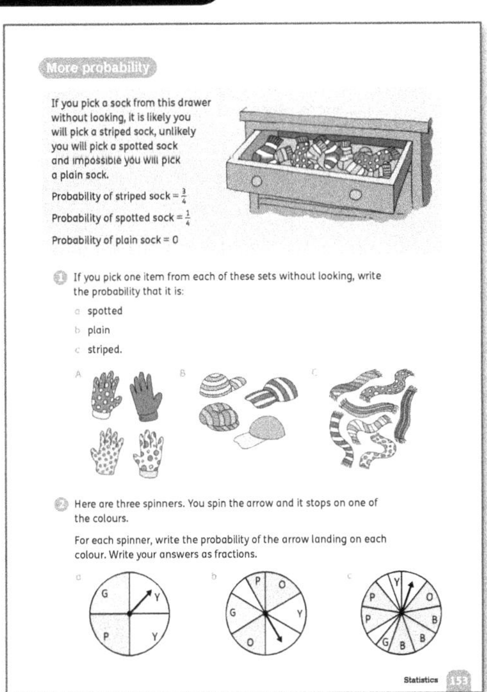

Materials

Coin; 1–6 spinner

Focus

- Read through the example on **Pupil book 5 page 153** with the class.
- Spend some time talking about what 'a good chance' and a 'poor chance' mean in real life and then relate that to the probability scale. Anything above $\frac{3}{4}$ is a good chance. Anything below $\frac{1}{4}$ is a poor chance.
- Remind the children that we can use numbers from the probability scale to describe chance more accurately. For example:
  - Flip the coin a few times. When we flip a coin, we have an equal chance of getting heads or tails. That is a probability of $\frac{1}{2}$.
  - Spin the 1–6 spinner a few times. When we spin a 1–6 spinner, the probability of each number is the same. We have a $\frac{1}{6}$ chance of getting 1 and a $\frac{1}{6}$ chance of getting 6. The probability of getting an even number is $\frac{3}{6}$ or $\frac{1}{2}$.
- It may be interesting to have a class discussion to clarify that a probability of $\frac{1}{2}$ for heads does not mean that you will get heads once out of every two times you flip the coin. It does mean that if you flip the coin a large number of times, you will end up with a more or less equal number of heads and tails.
- The children can work on question 1 in pairs to decide each probability before recording it. They can use fractions, decimals or percentages.
- Before the children complete question 2 on their own, make sure they understand how a spinner works.

Challenge

Give the children some more challenging probability problems. For example:

The numbers 1–20 are written on cards and placed in a box. $\left(\frac{11}{20}\right)$ I take a card at random from the box. What is the probability that the number I take will have the digit 1 in it? Does the probability of taking a number with a 1 in it double if we put the numbers 1–40 in the box? (no) *Explain your answers.* (There are 11 numbers from 1–20 with a 1 and 13 numbers from 1–40. 13 is not double 11.)

Support

You may need to help the children work out what fraction of a circle each sector represents before they work on question 2.

> ### Answers for Pupil book 5 page 153
>
> **1** a A: 0.75; B: 0; C: $\frac{2}{7}$
> b A: 0.25; B: 0.25; C: 0
> c A: 0; B: 0.75; C: $\frac{5}{7}$
> **2** a green $\frac{1}{4}$, purple $\frac{1}{4}$, yellow $\frac{1}{2}$
> b green $\frac{1}{4}$, purple $\frac{1}{12}$, yellow $\frac{1}{6}$, orange $\frac{2}{6}$ or $\frac{1}{3}$, white $\frac{3}{12}$ or $\frac{1}{4}$
> c green $\frac{1}{18}$, purple $\frac{2}{9}$, yellow $\frac{1}{18}$, orange $\frac{1}{9}$, white $\frac{2}{9}$, blue $\frac{3}{9}$

Experiment with probability

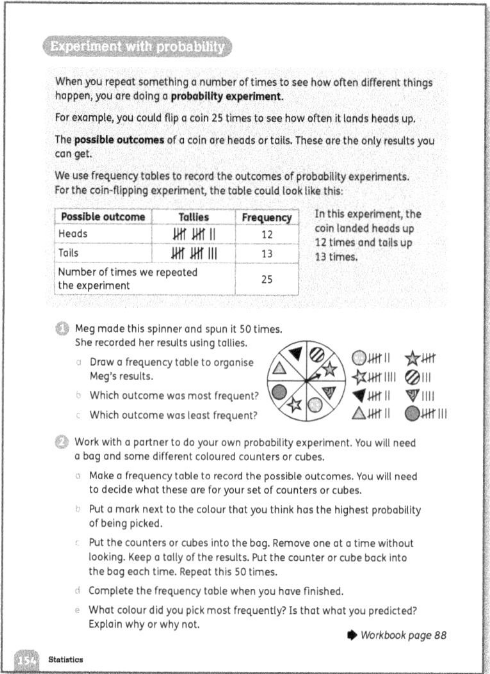

Possible outcome	Tallies	Frequency
Heads	ЖЖ ЖЖ II	12
Tails	ЖЖ ЖЖ III	13
Number of times we repeated the experiment		25

Materials

Small denomination coins; 1-6 spinner; bags, counters, cubes or other equipment for the experiments; circles cut from paper or card

Focus

- If necessary, demonstrate how to flip a coin and let the children practise using a coin of their own. In most currencies, heads is the side with the value of the coin shown on it.
- They should say whether the coin lands heads up or tails up. They could also tally the results for a number of flips to revise tallying.
- Ask the class what results are possible (possible outcomes) when you flip a coin. (only two; heads or tails)
- Ask the children to take turns to come up and spin the 1-6 spinner. Record the outcomes using tallies. After a few children have had a turn, discuss what they did.
- Explain that the children have just done a *probability experiment*. Each time a child spun the spinner, it was one trial in the experiment. The scores were the *outcomes* of the experiment. The number of times they got each number was the *frequency* of that outcome.
- Remind the class that we can use a table to record frequencies.
- Turn to **Pupil book 5 page 154**. Let the children read through the explanation on their own or with a partner.
- The children can complete question 1 on their own.
- Tell the class that they are going to do a probability experiment in pairs. Hand out the equipment for question 2 and let the children discuss what the outcomes of their experiment could be (based on the equipment you have given them). Then, let them work through the steps to do the experiment and record and summarise their results.

Follow-up

Use question 1 on **Workbook 5 page 88** to revise drawing and interpreting a dot plot (in this case, showing the results of a probability experiment). The children can work on their own or in pairs to design and complete the spinner in question 2. If you have time, the children could make spinners and design and carry out their own experiments to find out how closely their results match the given probabilities.

Interesting mistakes

Some of the mistakes that the children might make when working with probability include:

- They might think that later events are affected by earlier ones. For example:
  - 'I have flipped a coin and got 3 heads in a row, so it is more likely to be tails on the next flip.' (It is not more likely; the probability of tails remains $\frac{1}{2}$.)
  - 'I have not spun a 6 for a long time, so I am due to spin one soon.' (Again, this is incorrect; the chance remains $\frac{1}{6}$ every time they spin.)
- When working with spinners with sectors of different sizes, the children might think that all the events are equally likely. They might calculate the probability according to the number of sectors rather than the size of the angles at the centre.

To address any of these mistakes, you could:
- Make a set of incorrect assumption cards and ask the children to explain why these are not correct, forcing them to work out for themselves what is wrong.
- Have a number talk (pages 17–18) around these issues to clarify them.

To address the spinners mistake:
- Use circles with divisions round the outside and ask the children to colour different fractions to make sectors. This will help the children to understand that the size of a sector represents a different fraction of the circle.
  - For example: give the children a circle marked in twelfths. Ask them to colour $\frac{1}{4}$, then $\frac{1}{3}$ and then $\frac{1}{6}$ and then three separate pieces of $\frac{1}{12}$. Discuss the results, making sure that the children understand that the fraction they have coloured is the probability of the spinner landing on that colour.

Answers for Pupil book 5 page 154

1 a

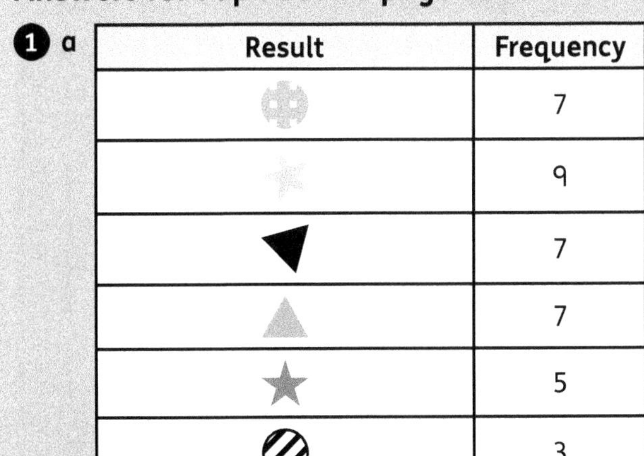

Result	Frequency
	7
	9
	7
	7
	5
	3
	4
	8

b yellow star   **c** striped ball

2 Individual answers.

Answers for Workbook 5 page 88

1 a

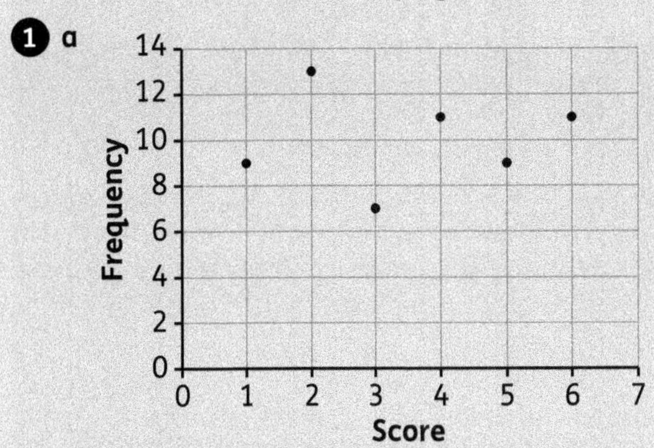

b The spinner landed on each score roughly the same number of times.

2 Many different combinations are possible. Here is an example.

Ask some or all of these questions to check how well the children have understood the concepts in this unit.

- *Explain in your own words what 'probability' means.* (For example: How likely something is to happen.)
- *Tell me something that is certain/impossible.* (For example: Certain: New Year's Day will be on 1 January. Impossible: I will go on holiday to Mars.)
- *What is the probability that these events will happen?* (Give some different events.)
- *What does a probability of $\frac{3}{4}$ mean?* (The event is likely to happen.)
- *How does the probability scale work?* (It is a number line from 0 to 1, going from impossible (probability of 0) to certain to happen (probability of 1). Other things have a probability between 0 and 1, where closer to 0 means that it is unlikely to happen and closer to 1 means that it is likely to happen. Halfway along the scale, things are equally likely to happen as not happen.)
- *Look at this spinner. (Show a spinner.) What is the probability of it landing on* (name a colour or shape)?
- *What is a probability experiment?* (This means repeating something a number of times to see how often different things happen.)
- *I have 3 red beads and 5 blue beads in a bag and I pick one without looking. Which colour am I most likely to pick? Why?* (Blue, because there are more blue beads than red beads.)
- *What are the possible outcomes when you spin a spinner with a square, a red dot and a yellow sun on it?* (square, red dot, yellow sun) *What outcome has a probability of 0 on this spinner?* (For example: Green triangle, purple line.)

Mixed practice 3

Answers to Mixed practice 3 Pupil book 5 pages 155–156

1 735 trees

2 285 paving stones

3 a 1660 + 83
   b 1660 – 83
   c 83, since dividing is the opposite of multiplying
   d 20, since dividing is the opposite of multiplying

4 1050

5 a Yes
   b Possible answer:
     It is easier to divide by 100 than 25. But dividing 800 by 100 gives an answer that is 4 times too small, so she has to multiply by 4 to get the correct answer.

6 a To show 0 °C
   b 3 °C
   c –3 °C
   d Between 06:00 and 08:00; this is when the dashed line slopes upwards most steeply.
   e 2 °C
   f 06:00 and approximately 23:00
   g –7 °C

7 a A net; the diagram folds up to make a 3D shape.
   b

   c 12 cm³   d 6

8 a 10; 9.84   b 45; 44.64
   c 210; 214   d 36; 34.4
   e 6; 6.13   f 5.5; 5.46

9 a Pam
   b Sam has not included the decimal point in his/her calculations. They have not aligned their calculation properly so that numbers with the same place value are in a column.

10 a No; 1 litre is 1000 ml so the ratio is 50 : 1000 or 1 : 20
   b 250 ml   c 400 ml

11 a 15 ℓ/min   b 450 ℓ